RIDING ROCKETS

The Outrageous Tales of a Space Shuttle Astronaut

Astronaut Mike Mullane

SCRIBNER
New York London Toronto Sydney

SCRIBNER
1230 Avenue of the Americas
New York, NY 10020

For information about special discounts for bulk purchases, please contact Simon & Schuster Special Sales:
1-800-456-6798 or business@simonandschuster.com

DESIGNED BY ERICH HOBBING

Text set in Sabon

Manufactured in the United States of America

10 9 8 7 6 5 4 3 2 1

Library of Congress Cataloging-in-Publication Data

Mullane, R. Mike.
Riding rockets: the outrageous tales of a space shuttle astronaut/Mike Mullane.
p. cm.
1. Mullane, R. Mike. 2. Astronauts—United States—Biography.
3. Space Shuttle Program (U.S.) I. Title.

TL789.85.M86A3 2006
629.450092—dc22
[B]

2005056123

ISBN-13: 978-0-7432-7682-5
ISBN-10: 0-7432-7682-5

To my mother and father, who lifted my eyes to space.
To the thousands of men and women of the space shuttle team,
who put me in space.
To Donna, who was at my side every step of the way.

Contents

Acknowledgments

My first and greatest thanks go to my wife, Donna, for her patient and loving support during my writing of this book. My children, Patrick, Amy, and Laura, were also enthusiastic cheerleaders. Thanks!

I am deeply grateful to my agent, Faith Hamlin, of Sanford J. Greenburger Associates, who convinced me to write the story of my life. Thanks, Faith, for holding my hand and so passionately representing *Riding Rockets*.

My Scribner editor, Brant Rumble, poured his exceptional talents into my story and I am indebted to him. Not only did he make me a better writer, he was my number-one fan throughout the publishing process. Also, a heartfelt thanks to the rest of the outstanding Scribner team who had a part in bringing my story to print.

Johnson Space Center Flight Director Jay Greene was the first to read my manuscript and I am thankful for his insight. Thanks also to astronauts Robert "Hoot" Gibson, Rhea Seddon, Mike Coats, Pierre Thuot, and Dale Gardner, who took time from their busy schedules to proof my work. In making these acknowledgments, I am not implying these reviewers agreed with everything I wrote. One thought that I was too hard on the politician astronauts. Another felt that I wasn't severe enough in my criticism of some key NASA managers. While I appreciate all of their opinions, I did not modify my story to accommodate them. *Riding Rockets* is my story created from my memories.

Many of the conversations I relate in the book are decades old. For that reason, the quotation marks should not be construed as enclosing verbatim dialogue. Rather, they contain my recollection of those discussions.

RIDING ROCKETS

CHAPTER 1

Bowels and Brains

I was naked, lying on my side on a table in the NASA Flight Medicine Clinic bathroom, probing at my rear end with the nozzle of an enema. *Welcome to the astronaut selection process,* I thought. It was October 25, 1977. I was one of about twenty men and women undergoing the three-day physical examination and personal interview process that were part of astronaut candidate screening. Almost a year earlier NASA had announced they would begin accepting applications for the first group of space shuttle astronauts. Eight thousand had been submitted. NASA had whittled that pile to about two hundred, a cut I had miraculously made. Over successive weeks all two hundred of us would eventually find ourselves on this same gurney probing at our nether regions as we prepared for our bowel exam.

It was rumored NASA would select about thirty from this group to fly the shuttle. The odds were long that I would be among that blessed few. Not that I wasn't qualified. I had all the squares checked. I was a West Pointer who had taken a commission in the air force. I wasn't a pilot. My eyesight had been too bad for that job. But I had nearly 1,500 flying hours in the backseat of RF-4C aircraft, the reconnaissance version of the F-4 Phantom. Like Goose from *Top Gun,* I was the guy in back. In my ten-year career I was a veteran of 134 combat missions in Vietnam, had completed a master's degree in aeronautical engineering, and was a graduate of the Flight Test Engineer course taught by the USAF Test Pilot School. I was most certainly qualified, but so were a few hundred other applicants. I had been around too many super-achieving military aviators to fool myself. While I might have had a little of the Right Stuff, there were legions of others who had it in abundance, pilots who would make the likes of Alan Shepard and John Glenn look like candy-asses.

Yes, the odds were long, but I was going to give it my best shot. At the moment that best shot was aimed squarely at where the sun didn't shine. I was in the process of preparing for my first proctosigmoidoscopy.

Just before I entered the bathroom I overheard one of the civilian candidates lamenting that he had failed his procto-prep. At the word *failed* my ears perked up. He had skimped in his bowel-cleansing efforts and would have to repeat his test tomorrow.

FAILED PROCTO-PREP. I could imagine those words in big red letters on the man's physical report. Who would see them? Would they count in the selection process? When a selection committee was picking one person in seven and each was a superman or a wonder woman, you didn't want to have the word *failed* anywhere, even in reference to something as innocuous as a procto-prep. My paranoia in this regard was fueled by a military aviator's deathly fear of the flight surgeon. When his stethoscope came to your chest or that blood pressure needle was bouncing, it was your career on the line. A little blip and you could leave your wings on the table. Military aviators looked forward to a physical exam about as much as they looked forward to an in-flight engine fire. We didn't want *failed* on any document concerning anything that came out of a flight surgeon's office. I had known pilots who would secretly visit a civilian off-base doctor for some malady rather than bring it to the attention of a flight surgeon. This had been strictly forbidden, but the only crime was to get caught. I employed the same logic on the flight to Houston for this medical test. In an act of incredible naïveté, the docs at NASA had asked us to hand-carry our medical records from our home bases. This was akin to trusting a politician with a ballot box. As the miles passed, I pulled out pages I felt could generate questions I didn't want to answer. In particular I pulled out references to the severe whiplash I had suffered during an ejection from an F-111 fighter-bomber a year earlier. During that incident my helmet-weighted head had snapped around as if it had been on the tip of a cracking bullwhip. My neck had been badly hyperextended. After a week of wearing a neck brace, the Eglin Air Force Base (AFB) doctors returned me to flight, but I wondered how the neck injury would be viewed by NASA's docs. Certainly it couldn't help me. It was the type of questionable injury that could draw a *disqualified* stamp on my application. In all likelihood there would be 199 other applicants without a history of neck injuries. I wasn't going to take a chance. I liberated the offending pages from my files, planning to reinsert them on the return flight. I had one very slim chance of getting selected as an astronaut. I wasn't going to let a little thing like a felony get in the way. I would alter official government records and, like countless other aviators before, hope I didn't get caught.

Yes, I was going to give this astronaut selection my best shot. I inserted the enema and squeezed the bulb. I was determined when the

NASA proctologist looked up my ass, he would see pipes so dazzling he would ask the nurse to get his sunglasses.

Hold for Five Minutes, read the instruction on the dispenser. *Screw that,* was my thought. That milquetoast civilian who had failed his clean out had probably blown his load at the first contraction. I would hold my enema for fifteen minutes. I would hold it until it migrated into my esophagus. I clamped my sphincter closed, gritted my teeth, and endured bowel contraction after bowel contraction until I thought I would black out. Finally, I blasted the colonic into the toilet. I repeated the process a second time.

Do not repeat more than twice, the label warned. Yeah, right. With the title of astronaut on the line, the warning could have read, *Do not repeat more than twice, death may result,* and I would still have ignored it. I grabbed a third enema and then a fourth. The waste of my last purge was as clear as gin.

I walked from the proctologist like a first grader carrying a gold star on a homework assignment. He had commented several times he had never seen a colon so well prepared. That I didn't shit for the next two weeks was a price I was willing to pay. (And, no, the civilian who had failed his prep wasn't selected.)

Next up was the interview by a NASA psychiatrist. I wondered about this. I had never spoken to a psych in my life. Was there a pass-fail criterion? I considered myself mentally well balanced. (A strange self-assessment given I had just set a world record for holding an enema in a paranoid quest to secure a job.) But how did a psych measure mental stability? Would he be watching my body language? Would a twitch of my eye, a pulsating neck vein, or a bead of sweat mean something? Something bad? In desperation I searched my memory for what *The Right Stuff* had revealed about the *Mercury* 7 astronaut psych evaluations. All I could recall was that they had been given a completely blank piece of paper to "interpret" and that one astronaut had answered, "It's upside down." Was such humor good? I didn't have a clue. I was flying blind.

My first surprise, which added to my fear, was to find out there were to be two psych evaluations by different doctors, each about an hour long. I walked into my first meeting. The doctor rose from behind a desk and introduced himself, shaking hands with a very weak, moist grip. I hadn't been in the room for fifteen seconds and already I was in a panic. Was the grip some type of test? If I echoed it in its limpness, was I indicating I had some latent sexual identity problems? I decided to be firm . . . not crushing, but firm. I watched his face but it was an enigma. I couldn't read anything. I could have been shaking hands with Yoda. His voice was low; low enough I wondered if this was some sneaky hearing test. He motioned me to a chair. Thank God there was no couch. That novelty would have further rattled me.

He held a clipboard and pencil at the ready. I swallowed hard and waited for what I was certain would be a question like, "How many times a week do you masturbate?" But instead he ordered, "Please count backward from 100 by 7s as fast as you can." I heard the click of a stopwatch and the gathering seconds . . . tick . . . tick . . . tick. My chances of becoming an astronaut were racing away with those seconds!

Only because of my plebe training at West Point, where I had learned to instantly obey any order, was I able to respond with lightning reflexes. If he wanted me to count backward by 7s from 100, then I'd do it. At least I didn't have to answer the masturbation question. I began the litany, 100, 93, 86, 79, 72 . . . , then I got off by a digit or two, tried to restart at my last known correct number, stumbled again, and ended in the 60s in a blurring babble of digits. I finally stopped and said, "I think I'm off track." My comment was answered with the *click* of the stopwatch. In the silence it sounded like a gunshot. It might as well have been, I thought. I was dead. At least my astronaut chances had been shot dead. I had failed what was obviously a mental agility test.

The psych said nothing. There was only a prolonged quiet in which all I heard was the scratching of pencil on paper. I had the cleanest colon in the world but my brain was constipated. It had FAILED me. I was certain that was the word being written by the psych. FAILED. Surely the other 199 candidates would breeze through this test. They would probably get to those final numbers . . . 23 . . . 16 . . . 9 . . . 2 . . . in a few blinks of the eye and then ask the psych if he wanted them to repeat the test doing square roots of the numbers. I was certain the man was thinking, *Who let this guy in the door?*

With nothing to lose, and in a desperate attempt to end the maddening silence, I quipped, "I'm pretty good at counting backward by ones."

He didn't laugh. "That won't be necessary." His icy tone confirmed it. I had failed.

After the 7s test, the doc once again positioned his pencil and asked, "If you died and could come back as anything, what would that be?"

More panic seized me. Where was this question leading? Into what minefield of the psyche would I now stumble? I was beginning to wish for the masturbation question.

There was no clock running this time, so I gave the question a little thought. Anything? Should I come back as a person? Alan Shepard? That sounded like a safe answer. But then it dawned on me, Shepard, like all test pilots, hated shrinks. Was it Shepard who had dismissively suggested the blank page "was upside down"? I couldn't recall, but I didn't want to take a chance. I'd better stay away from a wish for reincarnation as an astronaut icon who might be infamous among psychs for trivializing their profession.

I asked for clarification. "When you say anything? Do you mean as another person or object or animal?"

He merely shrugged with body language that said, "I'm not giving you any leads." Clearly he wanted me to step on one of those psyche mines by myself.

I toyed with the idea of saying I would like to come back as Wilbur Wright or Robert Goddard or Chuck Yeager or some other aviation/rocketry pioneer. Perhaps this would send a signal that being an astronaut was my destiny. But again, my mind's voice whispered caution. Maybe such a reincarnation wish would identify me as a megalomaniac in search of glory.

Then, in a burst of inspiration I had it. "I would like to come back as . . . an eagle." It was a brilliant answer. Clearly it conveyed my desire to fly yet didn't give the doctor a door to crawl farther into my synapses. (I later heard one interviewee say he was tempted to answer the question with "I'd like to come back as Cheryl Tiegs's bicycle seat." It would have been interesting to see how the psych would have responded to that.)

My eagle answer was acknowledged by more pencil scratching.

His next question was an obvious attempt to have me judge myself. "Tell me, Mike, if you died right now, what epitaph would your family put on your headstone?"

Boy, was this going to be easy, I thought. After faking some serious deliberation I replied, "I think it would read, 'A loving husband and devoted father.'" I was sure I had scored some points. Could there have been a better answer to convey the message that my family came first, that I had my priorities right? In reality I would have sold my wife and children into slavery for a ride into space. I thought it best not to mention that fact.

"What is it that you feel is your unique strength?"

I wanted to reply, "I can hold an enema for fifteen minutes," but instead said, "I always do my best at whatever I do." For once, I told the truth.

Psych One's interview continued. I was asked whether I was right- or left-handed (right) and what religion I professed (Catholic). He also inquired about my birth order (number two of six). He seemed to write a long time after hearing these answers. I would later learn a disproportionate number of astronauts (and other super-achieving people) are firstborn, left-handed Protestants. Maybe my exclusion from all of those groups was the reason I couldn't count backward by 7s.

Finally, I was excused to Psych Two. I walked slump-shouldered to another door, certain my astronaut dream was stillborn on the report that Yoda was finishing . . . *Candidate Mullane unable to count backward by 7s.*

Psych Two was the good cop in the good-cop/bad-cop routine. Dr. Terry McGuire welcomed me with a robust handshake and an expansive smile. I've seen that same smile on the faces of used car dealers. I looked for the diamond ring on McGuire's pinky but it was absent.

Dr. McGuire was outgoing and talkative. He didn't have a pencil and pad in hand. "Come in. Take a load off. Have a seat." Another chair, thank God. Everything about his voice and mannerisms said, "I apologize for that other bozo you had to contend with. He's got the skills of a chiropractor. I'm different. I'm here to help you." Just as it is on the car dealer's lot, I was certain it was all an act. He wasn't after my wallet. He sought my essence. He wanted to know what made me tick, and, like Captain Kirk facing a Klingon battle cruiser, I was ordering, "Shields up!" My astronaut chances might already be headed down in flames but I was going to continue to give it my best shot until the rejection letter arrived.

After some small talk about the weather and how my visit was going (fine, I lied), the good doctor finally began his assault on those shields. He asked just one question. "Mike, why do you want to be an astronaut?"

I had always assumed I would be asked this question somewhere in the selection process, so I was prepared. "I love flying. Flying in space would be the ultimate flight experience." Then, I added some bullshit to make it sound like love of country was a motivator. "I also think I could best serve the United States Air Force and the United States of America as an astronaut."

Boy, did I slam-dunk that question, was my thought. The only way I could have done better was if I'd brought in Dionne Warwick to sing the national anthem in the background.

But I was wrong. My slam dunk was rejected. I couldn't blow off Dr. McGuire with that rehearsed dribble. He looked at me with an all-knowing smirk and replied, "Mike, at the most fundamental level, we're all motivated by things that occurred in our youth. Tell me about your childhood, your family."

God, how I hated essay questions.

CHAPTER 2

Adventure

I was born a week after the end of World War II, September 10, 1945, in Wichita Falls, Texas. "He looks like a monkey" was my grandfather's first impression. I had a mop of shaggy black hair and, just like a chimp, outward-deployed adult-size ears. Through my early childhood, my mom fought to correct this defect. At bedtime she would adhesive-tape the billboards to the sides of my head, hoping they would grow backward. But it was a lost cause. Somewhere in the night, nature would overcome adhesive and my ears would sprong outward like speed brakes on a fighter jet.

As I had told Psych One, I was the second child in a Catholic family that would ultimately include six children—five boys and a girl. When I was born my dad was serving as a flight engineer aboard B-17s in the Pacific, so it was left to my mom to name me. She picked Richard. I was only a few hours out of the womb and already burdened for life with wing-nut ears and the handle Dick. It was no wonder when my dad returned home he began to call me by my middle name, Mike. *Christ, give the kid a break,* I imagine him thinking.

Though the war had ended, my dad remained in the service. My earliest memories are of weekend visits to air-base flight lines and sitting in the cockpits of C-124s and C-97s and C-47s and other cargo planes, where Dad allowed me to grab the control yokes and "steer" the parked monsters. He also took me to base operations, where aircrews were arriving from all corners of the globe. They would give me silver wings right off their chests, brightly colored patches, and strange coins from faraway lands. In my eyes they were heroic beyond anything Hollywood could conjure.

My dad was a New Yorker, an Irishman born and raised in Manhattan. He was full of amazing, colorful, exaggerated, and frequently untrue stories. No doubt his blarney was a source of inspiration for me. He made flight out to be a thing of grand adventure, particularly his flying experiences in the Pacific theater of WWII.

He described being attacked by Washing Machine Charlie, a Japanese pilot who kept the Americans from getting any rest by flying over their Philippine base at night and dropping pop bottles from an antique bi-plane. The air whistling over the openings would produce the scream of a bomb, sending everybody out of their bunks and into shelters.

"Boys, we named him Washing Machine Charlie because that damn Jap [with my dad, Japs were always "damn"] had the worst-running engine we ever heard. I know he tuned the engine wrong just to make it sputter and backfire and keep us awake. It sounded like a dying washing machine." Then Dad would put on a goofy Red Skelton–like face, purse his lips, and produce a litany of fart sounds to describe the offending machine. My brothers and I would laugh and laugh and beg him to "pretend to be Washing Machine Charlie."

A boom of thunder would put us on another flight. "Boys, one time our damn navigator [like Japs, navigators were always "damn"] got us lost in a thunderstorm. Lightning hit our plane. I could feel it crawling all over my body. My hair exploded off my head, which is why I don't have any today. It heated the fillings in my teeth and I burned my tongue when I touched it to the silver."

On other occasions he would swoop through our room with arms outstretched describing how gooney birds (albatrosses) would perch on the wings of his B-17 and hitch a ride during takeoff. The birds would spread their own giant wings and use the rush of air to achieve flight.

I suspect my dad, flying late in the war, never really saw a Japanese fighter aircraft but you would have never known it from his stories. He told of being shot down and parachuting into an island jungle. He and his crew teamed up with native freedom fighters and fought their way to the coast, where an American submarine rescued them. I now know this never happened, but his colorful fiction planted a seed in my soul. I wanted to live this same adventure. I wanted to fly.

Every year or two my dad would be transferred to another base and like Bedouin tribesmen we would pull up stakes and head for a new horizon. Locales in Kansas, Georgia, Florida, Texas, Mississippi, and Hawaii would ultimately boast a Hugh J. Mullane mailbox. For me every move was eagerly anticipated. I couldn't wait for the moving van to drive away and a new adventure to start. Curled in a blanket in the back of a car, like puppies in a basket, my brothers and I would fall asleep to the rhythmic *thump-thump-thump* of the pavement. It was the heartbeat of anticipation, of the unknown. Sometimes I would awake in the middle of the night and savor the smells of a new climate or watch lightning flash in the distance. During the day we'd stop at weathered signs advertising fresh fruit and buy buckets of ice-cold sweet cherries. We'd stop at gas stations with signs reading, "Last gas for 100 miles." I would watch my dad fill a canvas bag with water and hang it over the Indian-head hood ornament of our Pontiac station wagon. I was giddy with the thought of a highway that would be empty for one hundred miles. Later I learned there was a gas station every twenty miles with the same sign. But at that age, twenty miles was as good as a hundred. I would lean forward in my seat and stare over my dad's shoulder at horizons so crystalline they looked as if they were painted with a single-haired brush. I'd watch watery mirages sheen the blacktop in front of us and spinning dust devils, and blue-black thunderstorms pregnant with rain walking on stilts of lightning. And there was that unending song leading me into the emptiness, *thump-thump-thump*.

Another source of adventure was family camping trips into the wilds of the west. Why my New Yorker dad had such a love for the outdoors remains a mystery to me. Perhaps it was the very fact he had lived for so long in an asphalt jungle. My mom was even more afflicted. She was born a hundred years too late. I could easily see her striking out from Independence, Missouri, in a wagon train bound for Oregon. As a seventy-five-year-old widow she drove with another septuagenarian girlfriend from Albuquerque to Alaska. I was only surprised that she didn't walk. She craved a tent and rock-ringed fireplace more than most women pined for a remodeled kitchen. Happiness for her was standing over a smoky campfire and cooking pancakes and bacon while dancing from foot to foot trying to shake off the morning chill.

In preparation for these trips we would pile the accouterments on the roof of our car: a couple of Coleman coolers, a white gas stove, lanterns, tents, fishing poles, aluminum lawn chairs, and bags of charcoal. There were axes, shovels, thermos jugs, cooking utensils, and sleeping bags all cinched into place and covered with a tarpaulin. We were carrying a canvas iceberg. The car interior, containing a brood of kids and two dogs, was no less cluttered. If they could have seen us, Okies right out of *The Grapes of Wrath* would have felt sorry for us.

We set sail on the roads of the American west. And when I say roads, I don't mean interstate highways. My parents avoided those like watered-down gas. There was no adventure in traveling an interstate. Those were for wimps. Instead, they would search for the most obscure byways, take forgotten trails through sleepy towns and gravel-covered mountain passes. The sight of a sign reading, "Danger: Unimproved Road," might as well have said, "Gates of Heaven Beyond." My dad steered for those passages like an ancient Greek heeding the call of a Siren. I recall one occasion when a locked chain between two wooden posts held just such a warning sign. My dad took it as a challenge and dispatched his army of boys to rock one of the posts back and forth until it was loosed from the earth. We pulled it up, drove the car through, and replanted the post. Not only was it an unimproved road, now it was *our* road.

My parents navigated trails in a Pontiac station wagon that a modern army tracked vehicle would not attempt. A fallen tree or boulder in the way? Not a problem. Like Chinese coolies, the Mullane boys would saw, hack, lever, or sheer-muscle any obstacle out of the way.

Not that some of these excursions didn't put us in peril, like the time we were deep into the mountains of southern New Mexico when the radiator boiled over. It was obvious from the virgin dust there had been no traffic for many days, possibly weeks, maybe never. This was long before the days of cell phones. There would be no call to a tow truck. We were facing Donner Party extinction.

My dad, an expert at repairing planes, always carried an extensive set of tools in the car. Unfortunately, it seemed every time we broke down we were missing that one tool we needed. Apparently, our station wagon didn't have the engine of a C-124.

On this occasion, though, no tool was going to help us. We needed water and there was none around. But my dad was nothing if not resourceful. He directed us to tear apart the car looking for anything wet. A couple cans of Coke and beer went into the radiator. A jug of cherry cider that my older brother had purchased from a roadside fruit stand was quickly enlisted. Several oranges were squeezed into this Prestone.

Then, my dad noticed my younger brother wandering away. "Where are you going?"

"I have to pee."

Soon we were all standing on the grillwork of the car peeing into the radiator. "Steady, boys. That's a damn Jap Zero you're aiming at. Don't waste a round."

It was enough. Smelling like an overripe Porta-Potty, our hissing wagon limped into a gas station. The gagging mechanic asked if something had died under the hood.

On another occasion, overheated brakes threatened our descent from a mountain pass. No doubt recalling his earlier success with the radiator, my dad had each of us boys pee on a wheel to cool the brakes. Nobody has ever used urine as skillfully as my father.

These were wonderful and formative times in my life. I became my parents. I had to know what was *beyond* . . . beyond the next mountain, beyond the next bend, beyond the next canyon. There wasn't a national park, monument, snake farm, meteor crater, volcano, or rock shop we didn't visit. The side-rear windows of our car were covered with the colorful decals of our journeys: gift-shop stickers of Yellowstone's Old Faithful, the jagged spires of the Grand Tetons, the Grand Canyon, Glacier National Park, Canyon de Chelly, White Sands, Death Valley, Monument Valley, Glen Canyon, and innumerable other sights on the byways of the west. We had July snowball fights at mountain passes named Engineer, Independence, and Imogene. We tumbled in sand dunes and fished mountain streams and hiked cloud-scraping peaks just to see over them. We collected treasures of feldspar and fool's gold and quartz and petrified wood. My mom's scrapbooks are full of photographs of the family under state welcome signs—Arizona, Colorado, Utah, Wyoming, Montana, Nevada, California. In these states and more I would crawl into a sleeping bag wrapped in the smells of adventure—wood smoke and tent canvas—and watch the star shine outline our forest cradle. And my dreams would be of the greatest adventure of all . . . flight.

CHAPTER 3

Polio

The seminal moment in Mullane family history occurred on June 17, 1955, while we were stationed at Hickam Field in Hawaii. I was nine years old. Dad was now a flight engineer aboard C-97 and C-124 cargo aircraft of MATS, the Military Air Transport Service. He returned from a mission with a high fever and was admitted to Tripler Hospital. The diagnosis was polio. Dad, thirty-three years old, a vibrant 6-foot, 200-pound athletic man, would never walk again.

We remained in Hawaii for six months while he recovered. On New Year's Day 1956, the family was flown by Air Force hospital plane to

Shepard AFB near my mom's parents in Wichita Falls, Texas. There, Dad's convalescence continued.

During this time my parents tried to shield us from the trauma they were experiencing and, for the most part, they were successful. I recall only two occasions when my dad's private hell was revealed to me. On one, he had taken my brothers and me on a car ride. A new Pontiac had been modified with hand controls so he could drive. He stopped at a store window, a man passed him a bottle, then he drove into the Texas prairie, parked, and drank. To this day whenever I smell bourbon I am taken back to this moment. He told us of Washing Machine Charlie and of paddling a native canoe to a submarine. But this time it was different. He was crying as he told the stories. I had never seen my dad cry and I couldn't understand why these great stories were making him sad now.

Finally, he tossed the bottle from the window and steered the car toward home, turning the drive into a carnival ride. He would race the car and then jam on the brakes so my brothers and I would tumble over the seats and come up giggling. Again and again he would accelerate and then skid to a stop. By some miracle we made it to my grandmother's house uninjured. Dad rose onto his braces and crutches and was singing an Irish ballad with a drunken slur as he slowly made his way up the sidewalk. He threw his crutches forward and dragged his useless legs behind. Yard by yard the rhythm took him toward the front door. Then my grandmother burst from the house and began to beat him with a broom, screaming that he was a drunk and should be damned for it. He tried to grab her weapon but missed and toppled onto the cement. I had never seen adults behave this way. My brothers and I began to cry. Neighbors rushed out to watch the spectacle. My mom was screaming. My grandmother, a teetotalist, strict German woman, was a demon from hell. In her mind there was no excuse for drunkenness, even if the drunk in question was struggling to come to grips with polio. Dad cursed and grabbed at her but his useless legs were an anchor. She easily kept out of his reach. The broom came down on his head and his glasses spun off. She circled to his back where he couldn't defend himself and beat him some more. My mom herded my brothers and me inside so we wouldn't witness any more of the mayhem. We left Dad facedown on the sidewalk bawling like a child while my grandmother continued to punish him with the broom.

Several months later I caught another glimpse of how polio was torturing my dad. My mom was grocery shopping and as I wandered the aisles of the store I encountered a man showing a friend his artificial legs.

I overheard him speaking: "Lost both my legs in the war." He punctuated the comment by rapping on each shin with a cane. I was fascinated at the freakishness of the injury and stared. The men separated, the veteran walking away with the aid of his cane and a hip-swinging gait. For a moment I was struck as still as Lot's wife. The man had no real legs but he was walking. I broke from my shock and ran after him. "Mister!" I called. He stopped. "Mister, how can you walk? You don't have any legs." He smiled at my innocence. "Son, a German mortar blew off my legs. The doctors strapped on these pretend legs." Again he slapped his cane against one of the limbs. It made a hollow sound.

I smiled. "Thanks, Mister."

I sprinted past my mom. "Where are you going, Mike?"

"I'm going to see Dad."

I ran to the parking lot and climbed into the front seat of the car and breathlessly explained to my father what I had seen. "Dad, there's a man in the store who doesn't have real legs. They got cut off in the war. The doctor gave him pretend legs. But he's walking!" I blurted out the news while wearing a huge smile.

My dad stared at me, total confusion written on his face. He didn't get it. He didn't understand the significance of what I had just discovered. But I would be the hero and explain it to him. "Dad, all you have to do is ask the doctor to cut off your legs and then he can give you pretend legs and you can walk again, just like that man in the store!"

Immediately tears welled in his eyes. He tried to smile. "Mike, thank you for that idea, but it's not my legs that are the problem. It's my nerves. The polio germs ate them up. Cutting off my legs wouldn't help me."

My face fell in crushing disappointment. I was certain I had stumbled on the secret to putting my father back on his feet. Dad hugged me and cried into my neck. I was so confused. I didn't understand germs or nerves. I only understood what my eyes had told me, that it was possible to walk without real legs.

"You go on and help your mom with the groceries."

I climbed from the car and walked backward a couple steps, staring at my dad. His head was resting on the steering wheel and he was sobbing.

This was the last time he ever revealed to me or my siblings what must have been a titanic struggle to adapt to life without legs. But he did. Within a year he had returned to being the man I remembered before polio. His Washing Machine Charlie imitation got even better.

After several months of recovery in Texas my parents were faced with the decision of where to settle the family. Wichita Falls was not an option. It was an oil and cattle town with limited employment opportunities for the handicapped. And there was no thought given to returning to my dad's hometown, New York City. The vertical landscape there would make a wheelchair life difficult. So they picked Albuquerque, New Mexico, as their post-polio home. During our travels we had driven through the city a couple times and Mom and Dad had always liked it. There was a VA hospital, work opportunities, and a climate that made wheelchair life a little more tolerable.

Albuquerque was my last childhood move and I thank God for it. We were permanently in the west. No longer did we have to drive for days to reach the deserts and mountains we had all come to love. Now we could satisfy our collective need to see over the next horizon on a weekly basis. The fact my dad could not walk was hardly an impediment to our adventures. He had five boys and we could carry him and his wheelchair anywhere. And we did. We hefted him over steep and rugged terrain to an isolated lake and put him in my brother's canoe. We carried him across streams and along trails. We sat him in meadows to enjoy a sunset or the majesty of a distant thunderstorm.

In addition to the wheelchair-bearer duties, my dad's polio also turned us into bartenders and urine-dumpers. My dad was a 7-and-7 man and at the end of the day we would mix him a highball of Seagram's Seven whiskey and 7-Up. He trained each of us to pour two fingers of the liquor, add ice and then the soft drink. He also trained us to empty containers of his urine on road trips.

On one drink-making occasion, my little brother found an already opened bottle of 7-Up. He brought the drink to my dad, who tossed back a swallow. "Christ, Mark, how much whiskey did you put in this? It's really strong." My brother explained he had used a two-finger measurement, just like always. Only after Dad was down to slurping on the ice did the lightbulb come on. Earlier in the day, when nobody had been around, he had peed in an empty 7-Up bottle and left it on a table. He jerked the highball from his mouth and sniffed it like a dog at a hydrant. "Mark! Where did you find the 7-Up for my drink?"

"There was a bottle of it on the table."

"Jesus Christ! I've been drinking a 7-and-piss." Moments later, with a fresh drink in his hand, Dad philosophically mused, "I guess if you have to drink urine, it's best you drink your own."

There was one item from my dad's walking life that he adamantly refused to let polio take from him: his beloved toolbox. It came with us on every trip. In spite of past experience, Dad was convinced he would someday save us all from a serious car malfunction with his tools. He got his chance on a blistering day in June 1963. We were on a trip through a remote area of desert in northwest New Mexico. (The term *remote* will always be redundant in describing the venues of Mullane adventures.) The car began to surge. For a moment there would be acceleration, then coast, then acceleration. My dad frantically oscillated his accelerator hand control up and down, hoping to clear the problem. At this point most drivers would probably have exclaimed, "There's something wrong with the car!" But it was impossible for my aviator father to avoid the lexicon of his prior life. "We're losing power!" was his cry. It would not have surprised me in the least to have heard, "The goddamn Japs have got us. Feather number two!"

We coasted into the dirt parking area of a small adobe Indian trading post. Faded signs advertised turquoise jewelry, rugs, and other curios. A wooden porch provided the only shade within a hundred miles. A half dozen Indians sitting on chairs and several scruffy dogs lying on the planking had staked their claim to that shade. I recall how fascinated my brothers and I were by the fact the Indians were dressed like cowboys. They wore denim and boots and large western hats. My mom admonished us not to stare but we did nevertheless. The Indians were as still as carvings. Our clattering, sputtering arrival seemed invisible to them. They didn't move, scratch, wave away a fly, or speak. They just sat on their chairs staring into the shimmering heat in utterly silent, regal repose.

My dad instantly diagnosed the car problem. "Balls! It's the goddamn fuel pump." For some strange reason, the testicular epithet *balls* was my dad's favorite obscenity.

"Hugh, not in front of the children!" was my mother's favorite response to his swearing. Hers was a cry I heard many, many times in my youth, but it never altered my dad's vocabulary.

Dad locked his leg braces and rose onto his crutches. In spite of this strange sight, the six Indians remained as still as a Remington bronze. The station wagon, the jabbering kids, the man on crutches appeared no more remarkable to them than a dust devil.

Three of us boys joined my dad at the fender and popped the hood. "I'll show you kids how to test the fuel pump." For once we had the right tools and my dad leaned into the engine and began to work. He disconnected the output of the fuel pump. Then he pulled the distributor wire. He explained his actions: "I'll have Mom turn the key. The car won't start because I've pulled the distributor wire. But the starter will turn the cam shaft, which will cause the fuel pump to operate. If it's working, we should see gas from this hose."

He then called for my mom to start the car. I was so proud of my dad. How many other fathers could do roadside maintenance like this in the middle of the desert? And my dad was doing it on crutches and braces.

As pride swelled my soul, gas began to squirt from the output line of the pump and spray across the hot engine. A vapor of fuel immediately enveloped us. With a gratifying grunt my dad proclaimed, "The fuel pump's okay." As he was stating the obvious, I noted a blue spark coming from the distributor wire and flashing to the steel of the engine block. It made a *tick, tick, tick* sound. I was just about to comment on this spark when an explosion flashed under the hood. KA-BOOM! The fuel vapor had combined with the surrounding air to form an explosive pocket. The spark from the disconnected distributor wire had provided an ignition source. We had just become all too intimate participants in the first test of a fuel-air weapon. Four decades later, the air force would develop just such a weapon for use against terrorists hiding in caves. The press ballyhooed the device. *Nothing like it has ever been developed,* they crowed. Not exactly. Let history now show my dad was the first to test such a weapon. He did it under the hood of a 1956 Pontiac station wagon at the Teec Nos Pos trading post in northwest New Mexico on June 14, 1963. It worked.

My brothers and I stumbled backward with blasted eardrums and flash-blinded eyes. The smell of burned hair wafted in the breeze. My dad, unable to retreat because of his braces, had fallen straight backward like a cleanly felled tree. He was now a cartoon character with a blackened face and the few remaining hairs on his bald pate burned to cinders. The engine block was on fire. My mom worked furiously to get my youngest brothers and sister and the dogs away from the burning car.

On the trading post porch, the previously lethargic dogs were on their feet barking madly. And the response of the Indians? They roared with laughter. I mean fall-on-the-floor, breath-stealing, tear-welling laughter. Jabbering in their native tongue they pointed at the white man's folly as if it were a free fireworks show (which it was) and laughed and laughed and laughed.

My dad finally recovered enough to hear the laughter and understand its source. He rolled onto his stomach and made a valiant attempt to crawl to the nearest Indian, no doubt to kill him with his bare hands. John Wayne, leading a silver screen cavalry charge, had never looked as fierce. Thank God, at that moment, Dad couldn't walk for he certainly would have ended up in prison for manslaughter. Finally, he braced himself on one forearm, shot out an upraised middle finger, and roared, "Shove it up your ass, you bastards!" Even the great cry of "balls" didn't fit this affront.

Over the barking dogs I again heard my mom's plaintive cry, "Hugh, not in front of the children."

Even near-death experiences like this did nothing to dampen my enthusiasm for family trips into the great emptiness of New Mexico. It was an emptiness that pulled the sky closer. I could climb *through* the clouds and sit on a mountain peak and look down and imagine I was in a jet skimming over the white. I would lie in a meadow and watch thunderstorms build over the Sangre de Cristo Mountains and dream of someday soaring between the canyons of vapor.

But it was the New Mexican night skies that most captured my imagination. I needed only to step into my backyard to be in space. Unlike the light-polluted, haze-shrouded heavens that had domed our previous homes, Albuquerque's sky was dry and stygian in its blackness. On one occasion, my grandmother visiting from New York City stood with me looking at the night sky and commented, "You won't be able to see any stars tonight, Mike. It's cloudy."

I was baffled by her comment since to my eye the sky was spotlessly clear. "Grandma," I had said, "there are no clouds. It's clear."

"Oh, Mike, sure there are . . . a long one that goes clear across the sky." And her hand swept from horizon to horizon. Then I realized she was referring to the Milky Way. The edge view of our galaxy was a fog of stars, a "cloud" to the unpracticed observer.

My dad taught me how to take time exposures of the night sky. I would set one of his cameras in the desert and open its aperture and let the turn of the Earth trace the starlight onto film. Later, with the developed photos in hand, I would marvel at the circles of varying brightness and color the stars had inscribed about Polaris. Occasionally I would catch the streak of a shooting star, a recording that would excite me as much as if I had found buried treasure.

Above Albuquerque the stars and planets blazed with a clarity I had never seen before. They seemed so *accessible*. With the help of a small Sears telescope and my imagination I traveled this sky on a nightly basis. I would stare at the crescent of Venus and the red circle of Mars and count the brightest moons of Jupiter. The thrill of those observations was no less than what Galileo must certainly have experienced. When a wire-thin sketch of moon was part of the fresco, I would train my telescope on it and imagine I was walking its deeply shadowed craters and mountains. When meteor showers were predicted I would drag a sleeping bag into the desert and lie awake to watch their flashes of fire and pray that one would miraculously land nearby.

But another night sight was soon to thrill me even more.

CHAPTER 4

Sputnik

On the morning of October 4, 1957, I entered my dad's bedroom to say good-bye before departing for school. As usual he was drinking coffee, smoking his pipe, and reading the paper. This morning, however, he was purple with rage. "The goddamn Reds have put some type of moon around the earth. Balls! What the hell is Eisenhower doing? What if there's an H-bomb on the damn thing?"

I picked up the paper and read about the orbiting Sputnik and how the Russians were saying it was just the beginning of their space program. They were working to put men into space. There were interviews with American scientists who predicted our nation would do the same. A smaller sidebar explained that Sputnik would be visible as a moving dot of light over Albuquerque just after sunset.

That evening I stood with the rest of the city population in the cold October twilight to watch the new Russian moon twinkle overhead. My dad watched from his wheelchair, cursing Eisenhower for being asleep at the switch. I was struck dumb by the spectacle. The paper had said the object would be moving at 17,000 miles per hour, 150 miles in the sky. I was mesmerized by the thought of traveling at such speeds and altitudes. The newspaper had said men would someday do it. The science fiction movies of my youth had long depicted manned spaceships flying to distant planets. Now Sputnik proved that was really going to happen. There would be spaceships! I couldn't imagine a more exciting adventure and I wanted to be part of it. I wanted to fly in space.

Within weeks I was launching my own rockets in the New Mexico desert. These were not the cardboard and balsa-wood rockets seen in today's hobby stores. Those didn't exist in my youth. My rockets were multistage steel-tubed devices, five feet in length with welded steel fins, and filled with wicked home-brewed propellants. Basically, my rockets were pipe bombs. How I survived this period of my life, I have no idea. I was a preteen boy mixing rocket fuel in glass jars and tamping it into steel tubes. You couldn't get closer to mayhem and death than that.

I searched for any source of steel tubing. An early find was the extension piece of my mom's vacuum cleaner. Its stainless-steel gleam and lightweight construction had *rocket* written all over it. "It's perfect, Mom!" I cried. Without hesitation she handed it over. She and my dad were oblivious to the dangers of my experiments. In response to the new celestial Red menace, school-sponsored rocket clubs were organized to get kids interested in science and engineering and the formula for rocket fuel was passed out like raffle tickets. If the school was involved, then it must be safe, was my parents' erroneous thinking.

Not only did my mom give up her vacuum extension (and henceforth had to vacuum like a scoliosis victim), she also let me use her iron to heat plastic to form parachutes for my capsule payloads of ants and lizards. That work destroyed her iron. She let me use her oven to bake recipes of foul-smelling, fertilizer-based rocket fuel. My dad's extensive tools were at my disposal, as was his time. He would drive me to machine shops to have nozzles turned on lathes, and to chemical supply companies to purchase rocket fuel ingredients. He would drive me and my bombs into the desert. There he would rise onto his braces and crutches and hold a Super-8 movie camera on the action. I would set up the rocket and lay wire to the car battery. Dad would offer an irreverent prayer that the rocket would land on Khrushchev's head, then he would recite a short countdown. At zero I would touch the wires to the battery and hope for the best. Sometimes that best was a perfect streak of smoke leading a thousand feet into the azure blue. A white puff would mark the firing of the parachute extraction rocket and we would be treated to the glorious sight of my Maxwell House coffee can capsule descending on a parachute of dry-cleaning plastic and kite string.

More frequently, however, my launches closely mirrored NASA's. At the count of zero an explosion would rend the air and a sulfurous cloud would be all that remained of my creation. My dad, ever the optimist, would opine that the rocket had gone into orbit or possibly hit the moon. We would remain silent and listen for any sound of a whine or thud of an impact, but there would be none. That was proof enough for my father that the rocket was on its way to the Kremlin.

During one launch my dad was nearly a casualty. He was whooping and hollering at what appeared to be a perfect ascent until the missile arced into a dive straight for us. "Jesus Christ, Mike! It's coming right at us!" He was ten feet from the car and did his best to sprint for its cover. With the clatter of steel braces and aluminum crutches he sounded like a machine gone wild. I was powerless to help and abandoned him like a dropped Popsicle. I dove under the car and turned to look up, certain I was going to witness my dad getting shish-kebabed by a five-foot smoking skewer. I made one final, not so helpful cry, "Watch out, Dad!"

"Balls!" he roared. As a whistling sound gained in decibels, Dad suddenly stopped and jerked himself into a ramrod-straight pose, holding his crutches as close to his body as possible, trying to minimize his target footprint. His face was scrunched up, eyes squeezed to slits, teeth bared. There was a brief whooshing sound followed by a loud *whoomph*. The rocket embedded itself in the sandy soil no more than ten feet away. A thin curl of smoke corkscrewed from its nozzle.

"Christ on a crutch, that was close, Mike."

Dad christened that rocket the "Kamikaze," which set him off on another story about how a damn Jap kamikaze had almost rammed his plane. "I emptied my twin 50s at the bastard and never landed a hit. But he missed us anyway, just like that rocket."

As 1957 drew to a close, I was filled with anticipation. Not because of any party to celebrate the New Year, but rather because 1958 was to be the IGY, the International Geophysical Year. If there was ever a measure of how consumed I was by space, this was it . . . that I was as impatient for the IGY to arrive as most kids were for the end of school. I had read about it in several science magazines. A host of countries were to cooperatively investigate space with sounding rockets and instrumented balloons, and the United States was going to launch its own satellites. I couldn't wait.

My greatest treasure of this epoch was the book *Conquest of Space* by Willy Ley. Forget Homer and Shakespeare and Hemingway. They were hacks. For me Willy Ley was the greatest writer of all time. His descriptions of spaceflight supplemented by Chesley Bonestell's magnificent space paintings launched me into orbit decades before a NASA rocket.

" . . . there will be zero hour, zero minute, and zero second, and then the roaring bellow from the exhaust nozzles of the ship . . . the ship will ride up on the roaring flames, disappearing in the sky in less than a minute . . .

"The earth will be a monstrous ball somewhere behind the ship, and the pilot will find himself surrounded by space. Black space, strewn all over with the countless jewels of distant suns, the stars. Stretching across the great blackness the pilot will see the Milky Way."

To my twelve-year-old brain there was no more wonderful prose written in any language anywhere. I could *see* that great ball of Earth. I could *see* those stars. I could *see* the deep blackness. I read the book again and again. I read it until the pages came unhinged. I consumed Bonestell's paintings like other boys ogled the breasts of African natives in *National Geographic*. There were illustrations of astronauts watching a "canalled" Mars from one of its moons, Deimos. Other paintings showed spacesuited explorers walking among the mountains of our moon and on the gravel desert of Saturn's moon, Mimas. The subtitle on Ley's book said it all, "A preview of the greatest adventure awaiting mankind."

As soon as there was a NASA, I became its number-one fan. I watched every launch on TV. I wrote for photos and stuck them on the walls of my bedroom. I knew at sight every missile in the U.S. arsenal . . . Redstone, Vanguard, Jupiter, Thor, Atlas, Titan. I could recite their height, thrust, and payload weight. I learned NASA's vocabulary: apogee, perigee, payload, LOX, A-okay. I sent NASA drawings of my own rockets and gave them helpful suggestions on how they might build better missiles. I followed the trials and tribulations of NASA's program with the same passion other kids followed their favorite ball team. When the *Mercury 7* astronauts were announced I memorized their biographies and pored over *Life* magazine's photo-essays on them and their machines. I couldn't wait to be one of them and constructed a fantasy in which I would actually *replace* them. One of the points continually raised in the news was the anemic thrust of NASA's rockets. The United States launched grapefruit-size satellites while the Russian payloads were measured in tons. I was convinced, when all was said and done, Alan Shepard and John Glenn and the other astronauts would be too heavy to be blasted into orbit. In my dream NASA would be unable to find any adult test pilot light enough for one of their rockets to lift. They would then search among the skinny kids of America to staff the astronaut corps. I wrote to NASA with that very suggestion, making sure my name and address were prominent.

The rockets and posters and sky watching weren't enough. Astronauts were pilots. I had to fly. At age sixteen I began flying lessons. After a dozen hours my instructor deemed me safe enough to solo. There are some memories so seared into our synapses we carry them to our graves—our first sexual experience, the birth of our children, combat, the death of a loved one. I can include my first solo flight in those memories that will play in full Technicolor in my age-addled brain. Four decades later I can still feel the adrenaline-boosted flutter of my heart as

I taxied onto the runway and glanced at that empty right seat. My left hand gripped the yoke so tightly I'm surprised I didn't liquefy the plastic. My right hand was welded to the ball of the throttle. I lifted my feet from the brakes, slid the throttle to the firewall, and the machine slipped down the runway. I had never experienced a sound more sweet than the roar of that 100-horsepower engine. I eased the yoke back and watched the Earth fall away. Even in my later space shuttle launches I doubt my heart was pounding as it was at this moment in my life. I was flying! Later I walked to the car with the three most wonderful words in the English language written in my logbook: *Cleared for solo*.

Unfortunately, there was one major impediment to continuing my flying lessons. Money. I had a little saved from summer jobs but flying lessons were expensive. It has been said that necessity is the mother of invention. For teenage boys, necessity is the mother of idiocy. At the time I had a teen friend who had also achieved solo status in his flying lessons and, like me, was struggling to find the funds to continue. We put our heads together and came up with a way to make our money go further. He would rent a plane from one of the Albuquerque airports and fly it to another city field. There, I would meet him and we would fly together, sharing expenses and flying time. There was just one small problem with our plan: It was felonious. As student pilots we could only fly solo or with our instructors. On every flight we would be violating FAA rules. But we quickly concluded that getting caught would be the only crime, and with him flying to a secondary airport to pick me up, we had reduced the chances of discovery.

We decided to see how high we could climb and nursed the Cessna to 13,000 feet, violating yet another FAA requirement that supplemental oxygen be used above 12,000 feet. On another day we wanted to feel the thrill of speed and skimmed just yards above the cholla cactus of the Albuquerque deserts. The spires of the nearby Sandia Mountains were a lure and we weaved through those, all the while buffeted by severe downdrafts.

And on every flight I dreamed of someday flying higher and faster, of doing what Willy Ley had described. I dreamed of feeling the crush of a rocket's G-forces on my body and of seeing the great globe of Earth behind my ship. I dreamed of the day I would fly a rocket as part of the "Conquest of Space."

"Mike, at the most fundamental level we're all motivated by things that occurred in our youth. Tell me about your childhood, your family." A smiling Dr. McGuire awaited my answer. But I kept the shields up. I said nothing about Washing Machine Charlie or polio or near-death experiences in the wilds of the west or exploding rockets or violating FAA rules. What would those stories have said about Mike Mullane? That I had been emotionally scarred by my dad's struggle with polio? That I was an out-of-control risk taker? That I scorned rules? There was no way I was going to reveal that history. So I lied.

"I was raised in a Beaver Cleaver family," I said. "No divorces. No anxieties. No emotional baggage. My dad was an air force flyer and his influence excited me about flight. I was a child of the space race and that exposure excited me about spaceflight. As soon as there were astronauts, I wanted to be one." End of story.

It was probably the same story he heard from every military flyer. No doubt some of the civilians, unaccustomed to the reality that doctors of any stripe can only hurt your flying career, broke down in tears as they revealed they were breast-fed by their mothers until they were six or were abandoned or beaten or molested or sucked their thumbs or wet their beds. But military flyers knew better. We would have lied about a wooden leg or a glass eye. *You find it* would have been our attitude. I had a one-in-seven chance of making the astronaut cut. I didn't want anything to stand out in any report coming out of my medical exams. I wanted to be so normal that when somebody looked up that word in the dictionary, they would see my picture. So I lied. I didn't mention pissing in radiators or exploding car engines or dodging mountains in a Cessna 150. I lied even when the truth might have helped my cause.

CHAPTER 5

Selection

On February 1, 1978, the first space shuttle–era astronauts, thirty-five in number, stood on the stage in the auditorium of Building 2 at Johnson Space Center (JSC) to be formally introduced to the world. I was one of them.

The actual press announcement had come two weeks earlier. At that time I had been on temporary duty from my Florida base to Mt. Home AFB, Idaho, testing the new EF-111 aircraft. Like the other

208 people who had gone through the astronaut selection process, I had had an ear tuned to the telephone for several months. Not that I expected to be picked. Far from it. I felt it had been a fluke I had made the interview cut in the first place. In their more studied deliberations the NASA committee would finally realize what they had in Mike Mullane: an above-average type of guy; nothing spectacular; a twelve-hundred-and-something SAT guy; a 181st-in-his-West Point-class type of guy; a guy incapable of counting backward by 7s. There was no way I was going to twice fool an organization that had put men on the moon. But, like a lottery player who knows he is going to lose, I was still going to check the numbers.

On Monday morning, January 16, 1978, the numbers came up and I was a loser. I was certain of it. While dressing for work, I turned on the TV. I wasn't on it. But Sally Ride and five other women were. NASA had announced the newest group of astronauts, including the first women astronauts. There was video of newshounds jostling for position in front of their homes. Vans with brightly colored TV call letters crowded the streets. Curious neighbors circled the houses. And these smiling, radiant, joyful young women answered questions shouted by the press, "What's it like to know you are one of the first women astronauts? When did you want to be an astronaut? Did you cry when you heard the news? Will you be scared when you ride the shuttle?"

I went to my living room and drew back the curtain to see if there was a squadron of news vans parked in my driveway. Nope. No vans. No frothing press. No neighbors. Nothing. I was alone to dwell on my rejection. I tried to rationalize my loss:

It wasn't meant to be.

I had given it my best shot.

Maybe I'll get selected next time.

You never know unless you try.

I sought comfort in these and a hundred other motivational platitudes. But none helped. The winners were on TV. The losers were watching them. I thought of calling my wife, Donna, in Florida and telling her the bad news, but decided it could wait. I just didn't feel like talking about it.

I drove to my Mt. Home AFB office to find Donna had tried to reach me there. She had left a message, "Mr. Abbey at NASA called this morning and wants you to call him back." George Abbey was chief of Flight Crew Operations Directorate (FCOD), the NASA JSC division that included the astronaut office. He had chaired the astronaut candidate interview boards. I was certain this would be his "thanks for the effort" call.

I dialed the number and got Abbey's secretary. After a moment of holding (more proof of rejection) he came on the line. "Mike, are you still interested in coming to JSC as an astronaut?"

In the moments that followed I proved it is possible to live with a stopped heart. Over the previous hour I had built a precise rejection scenario. The women on TV were proof NASA had notified the lucky winners. (Actually, the women had been notified first so there would be thorough news coverage of their novelty.) Now, NASA was just following up with courtesy calls to the rest of us. But George's lead-in question certainly didn't sound like a prelude to a rejection.

There wasn't enough spit in my mouth to wet a stamp but somehow I managed to croak a reply, "Yes, sir. I would definitely be interested in coming to JSC."

Interested?! What the hell was I saying?! I was *interested* in having Hugh Hefner's job. I would *kill* to be an astronaut.

Abbey continued, "Well, we'd like you to report here in July as a new astronaut candidate."

I don't recall anything else from that conversation. I was blind, deaf, and dumb with joy. NASA had selected Mike Mullane as an astronaut.

I immediately called Donna with the news. "I told you! I told you, Mike! Didn't I always say everything would work out for the best? I told you!" And she had. Again and again she had. She had never lost faith. I wanted so much to be with her to share in the thrill, but it wasn't to be. I wouldn't be home for another couple days.

I called my mom and dad and they were as stunned as I. My dad and I laughed as we reminisced about launching my homemade rockets. I could sense that my mom, ever the pragmatic parent, was already anticipating the danger this new job would bring. No doubt her rosary was going to get a workout over the next couple years.

I called my commander in Florida. After offering his congratulations, he said Brewster Shaw and Dick Covey, both test pilots in the squadron, had also gotten Abbey calls. They were in. But other pilots were receiving rejection calls. I hurt for them. But not for long. My boundless, intoxicating joy roared back.

That night I bought beer for the rest of my Mt. Home AFB office and included them in my celebration. At that particular moment I was glad I was away from my home squadron. Most of the Idaho EF-111 flyers were from the USAF Tactical Air Command and none had applied for the astronaut program. My celebration with them was unalloyed. That was not going to be the case at my Eglin AFB flight-test squadron, which was filled with test pilots and test engineers. Virtually everyone had applied. The losers' disappointment was going to be as crushing as my joy was over-the-top. Shaw and Covey would have their celebration tempered by the presence of people who were dying inside.

When I was sufficiently sober, I left for my apartment. The base was far out in the desert and the road was deserted. I honked the horn and screamed like a teenage girl at a rock concert. I rolled down the window and screamed into the icy wind. I detoured into the desert, got out of the car, and screamed some more. I couldn't calm down. I punched the air with my fists. I jumped and sprinted and kicked the sand and laughed out loud. Finally, I hopped onto the warm hood, lay back, and watched the stars turn over my head, just as I had done on countless occasions as a child. When a satellite twinkled over, my heart gave a small lurch. God willing, in a few years, I would be riding rockets. I would be in a satellite . . . the space shuttle.

Now, two weeks later, I was standing with the other thirty-four astronauts of my group. Though our official report date wasn't until July, NASA had gathered us all together for an early, formal introduction to the world.

The Astronaut Class of 1978
(towns and cities are birthplaces)

Pilot Astronauts

Daniel Brandenstein, Watertown, WI, Lieutenant Commander, USN, age 34

Michael Coats, Sacramento, CA, Lieutenant Commander, USN, age 32

Richard Covey, Fayetteville, AR, Major, USAF, age 31

John "J. O." Creighton, Orange, TX, Lieutenant Commander, USN, age 34

Robert "Hoot" Gibson, Cooperstown, NY, Lieutenant, USN, age 31

Frederick Gregory, Washington, D.C., Major, USAF, age 37

David Griggs, Portland, OR, Civilian, age 38

Frederick Hauck, Long Beach, CA, Commander, USN, age 36

Jon McBride, Charleston, WV, Lieutenant Commander, USN, age 34
Steven Nagel, Canton, IL, Captain, USAF, age 31
Francis "Dick" Scobee, Cle Elum, WA, Major, USAF, age 38
Brewster Shaw, Cass City, MI, Captain, USAF, age 32
Loren Shriver, Jefferson, IA, Captain, USAF, age 33
David Walker, Columbus, GA, Lieutenant Commander, USN, age 33
Donald Williams, Lafayette, IN, Lieutenant Commander, USN, age 35

Military Mission Specialist Astronauts
Guion "Guy" Bluford, Philadelphia, PA, Major, USAF, age 35
James Buchli, New Rockford, ND, Captain, USMC, age 32
John Fabian, Goosecreek, TX, Major, USAF, age 38
Dale Gardner, Fairmont, MN, Lieutenant, USN, age 29
R. Michael Mullane, Wichita Falls, TX, Captain, USAF, age 32
Ellison Onizuka, Kealakekua, Kona, HI, Captain, USAF, age 31
Robert Stewart, Washington, D.C., Major, U.S. Army, age 35

Civilian Mission Specialist Astronauts
Anna Fisher, New York City, NY, age 28
Terry Hart, Pittsburgh, PA, age 31
Steven Hawley, Ottawa, KS, age 26
Jeffrey Hoffman, Brooklyn, NY, age 33
Shannon Lucid, Shanghai, China, age 35
Ronald McNair, Lake City, SC, age 27
George "Pinky" Nelson, Charles City, IA, age 27
Judith Resnik, Akron, OH, age 28
Sally Ride, Los Angeles, CA, age 26
Margaret "Rhea" Seddon, Murfreesboro, TN, age 30
Kathryn Sullivan, Paterson, NJ, age 26
Norman Thagard, Marianna, FL, age 34
James "Ox" van Hoften, Fresno, CA, age 33

Actually, I was standing with thirty-four other astronaut *candidates*. Our group, ultimately to be known as the TFNGs or Thirty-Five New Guys, became the first to have the suffix *candidate* added to our astronaut titles. Until the TFNG handle stuck, we would be known as

Ascans. (A later class would call themselves Ashos for *Astronaut Hopefuls.*) NASA had learned the hard way that the title *astronaut* by itself had some significant cachet. In one of the Apollo-era astronaut groups, a disillusioned scientist had quit the program before ever flying into space and had written a book critical of the agency. Since his official title had been astronaut, his publisher had been able to legitimately promote the book with the impressive astronaut byline. Now NASA was hedging its bets with our group. For two years we would be candidates on probation with the agency. If one of us decided to quit and go public with some grievance, NASA would be able to dismiss us as nothing more than a candidate, not a real astronaut. Personally, I felt the titling was an exercise in semantics. In my mind you weren't an astronaut until you rode a rocket, regardless of what a NASA press release might say.

Dr. Chris Kraft, the JSC director, welcomed us. As a teenager I had seen his picture in *Life* magazine articles about the Apollo program. Now, he was welcoming me into the NASA family. *Pinch me,* I ordered my guardian angel.

A NASA public relations officer began to read each of our names and an audience of NASA employees applauded. There were fifteen pilot astronauts. I was one of twenty mission specialist (MS) astronauts. MSes would not be at the stick and throttle controls of the shuttle. In fact, most of us were not pilots. Our responsibilities would include operating the robot arm, performing experiments, and doing spacewalks. As the name implied, we would be the specialists for the orbit activities of the mission.

As the role call neared the "Ms," my heart was trying to make like an alien and explode out of my chest. I still couldn't believe this was for real. When he got to it, I expected the announcer to pause on my name, look bewildered, consult with Chris Kraft, and then say, "Ladies and gentlemen, there's a mistake on this list. You can scratch R. Michael Mullane. He's a typo. He couldn't count backward by 7s." Then, two burly security guards would grab me by the elbows and escort me to the main gate.

But the announcer read my name without hesitation. He didn't stumble. He didn't consult Dr. Kraft. He read it like I was *supposed* to be on the list. *It's truly official now,* I thought. I had to believe it. I was a new astronaut . . . candidate.

The diversity of America was represented on that stage. There was a mother of three (Shannon Lucid), two astronauts of the Jewish faith (Jeff Hoffman and Judy Resnik), and one Buddhist (El Onizuka). There were Catholics and Protestants, atheists and fundamentalists. Truth be known, there were probably gay astronauts among us. The group included three African Americans, one Asian American, and six females. Every press camera was focused on this rainbow coalition, particularly the females. I could have mooned the press corps and I would not have been noticed. The white TFNG males were invisible.

Another first was the political diversity of the group. Military pilots, the mainstay of prior astronaut selections, were almost always politically conservative. They were highly educated, self-reliant, critical thinkers who scorned the "everybody's a victim" ethos of liberalism. But the reign of the right ended with the large number of civilian astronauts standing on that stage. Among their ranks were people who had probably protested the Vietnam War, who thought Ted Kennedy's likeness should be on Mount Rushmore, who had marched for gay rights, abortion rights, civil rights, and animal rights. For the first time in history, the astronaut title was being bestowed on tree-huggers, dolphin-friendly fish eaters, vegetarians, and subscribers to the *New York Times*.

There was another uniqueness about the civilians . . . their aura of youthful naïveté. While the average age difference between the military and civilian astronauts wasn't extreme (approximately five years), the life-experience difference was enormous. Some of the civilians were "post-docs," a title I had first heard that inauguration day. Literally, they had been perpetual students, continuing their studies at universities after earning their PhDs. These were men and women who, until a few weeks ago, had been star gazing in mountaintop observatories and whose greatest fear had been an A- on a research paper. Their lives were light-years apart from those of the military men of the group. We were Vietnam combat veterans. One helicopter pilot, told of making low-level rocket attacks and having exploded body parts hit the windshield of his gunship and smear it with blood. We were test pilots and test engineers. In our work a mistake wasn't noted by a professor in the margin of a thesis, but instead brought instant death. Rick Hauck, a navy pilot, had barely escaped death in an ejection from a crashing fighter. I had my own fighter-jet ejection experience.

It wasn't just this proximity to war and death that differentiated the military flyers from the post-docs, it was also the civilians' lack of exposure to life . . . at least their exposure to the rawer side of life. On a stopover in the Philippines on my way to Vietnam I checked into a hotel and was handed a San Miguel beer and a loose-leaf binder with photos of the available prostitutes. It was room service. Place your order now. As they say, one of the first casualties of war is innocence. To sit at a Vietnamese bar was to have a woman immediately at your side stroking your crotch, trying to make a sale. Everybody had a favorite number at the Happy Ending Massage Parlor (to simplify identification, available girls wore numbered placards around their necks) and I knew many aviators in Vietnam who had their PCOD circled on their calendar. This was their Pussy Cutoff Date, the date at which they would have to stop their whoring to allow the incubation period for STDs to pass (and for a cure to be achieved) before going home. One navy TFNG told of sitting in a dirt-floored Southeast Asia bar while a naked GI got it on with a prostitute on an adjacent table. It was just a wild guess on my part, but I doubted any of the post-docs had similar experiences in the Berkeley SUB. There was a softness, an innocence in their demeanor that suggested they had lived cloistered lives. It was hard for me to look at some of them and not think they were kids. Some might still have been virgins. Steve Hawley, George (Pinky) Nelson, and Anna Fisher were exceptionally young in the face. There was no way they were going to get inside a bar without being carded. Jeff Hoffman was the picture of academia. He had arrived at NASA with a beard and a collapsible bicycle suitable for the Boston subway. He didn't even own a car. He rode to work on his bike and carried a lunch pail. All that was missing were suede elbow patches on his suit coat and a pipe in his mouth to make the "professor" picture complete.

I felt a subtle hostility toward the civilian candidates. I know many of the military astronauts shared my feelings. In our minds the post-docs hadn't paid their dues to be standing on that stage. We had. For us, it had been a life quest. If someone had told us our chances of being selected as an astronaut would improve if we sacrificed our left testicle, we would have grabbed a rusty razor and begun cutting. I couldn't see that passion in the eyes of the civilians. Instead, I had this image of Sally Ride and the other post-docs, just a few months earlier, bebopping through the student union building in a save-the-whales T-shirt and *accidentally* seeing the NASA astronaut selection announcement on the bulletin board and throwing in an application on a lark. Now they were here. It wasn't right.

As the photographers continued to flash-blind the females and minorities, I watched Judy Resnik and Rhea Seddon (pronounced "Ray"). Between them, there would be one more first represented in our group: the first "hotty" in space. Judy was a raven-haired beauty, Rhea a striking Tennessee blonde. No TFNG male was looking at them and fantasizing about their PhDs.

CHAPTER 6

The Space Shuttle

As we TFNGs gloried in our introduction, we were woefully ignorant of the machine we were going to fly. We knew the space shuttle would be different from NASA's previous manned rockets, but we had no clue just how different or how the differences would affect the risks to our lives.

Before the space shuttle, every astronaut who had ever launched into space had ridden in capsules on throwaway rockets. The only thing that had ever come back to Earth was the capsule bearing the astronauts. Even these capsules had been tossed aside, placed in museums across America. While the capsules had grown in size to accommodate three men, and the rockets to carry them had grown bigger and more powerful, the basic Spam-in-a-can design, launched with expendable rockets, had been unchanged since Alan Shepard said, "Light this candle," on the first Mercury-Redstone flight.

We would fly a winged vehicle, half spacecraft and half airplane. It would be vertically launched into space, just as the rockets of yesteryear, but the winged craft would be capable of reentering the atmosphere at twenty-five times the speed of sound and gliding to a landing like a conventional airplane. Thousands of silica tiles glued to the belly of the craft and sheets of carbon bolted to the leading edge of the wings and nose would protect it from the 3,000-degree heat of reentry. After a week or two of maintenance and the installation of another 65,000-pound payload in the cargo bay, it would be ready to launch on another mission.

The space shuttle orbiter (the winged vehicle) would have three liquid-fueled engines at its tail, producing a total thrust of nearly 1.5 million pounds. These would burn liquid hydrogen and liquid oxygen from a massive belly-mounted gas tank or External Tank (ET). Eight and a half minutes after liftoff the empty ET would be jettisoned to burn up in the atmosphere, making it the only part of the "stack" that was not reusable.

As powerful as they were, the three Space Shuttle Main Engines (SSMEs) did not have the muscle to lift the machine into orbit by themselves. The extra thrust of booster rockets would be needed. NASA wanted a reusable liquid-fueled booster system but parachuting a liquid-fueled rocket into salt water posed major reusability issues. It would be akin to driving an automobile into the ocean, pulling it out, and then hoping it started again when you turned the key. Good luck. So the engineers had been faced with designing a system whereby the liquid-fueled boosters could be recovered on land. It quickly became apparent that it would be impossible to parachute such massive pieces of complex machinery to Earth without damaging them and posing a safety hazard to civilian population centers. So the engineers looked at gliding them to a runway landing. One of the earliest space shuttle designs incorporated just such a concept. Like mating dolphins, two winged craft, each manned, would lift off together, belly to belly. One would be a giant liquid-fueled booster/gas tank combination, the other, the orbiter. After lifting the smaller orbiter part of the way to space, the booster would separate and two astronauts would glide it to a landing at the Kennedy Space Center (KSC). The astronauts aboard the orbiter would continue to fly it into space using internal fuel for the final acceleration to orbit velocity.

However, designing and building this manned liquid-fueled booster was going to be very expensive at a time when NASA's budget was being slashed. The agency had won the race to the moon and Congress was ready to do other things with the billions of dollars NASA had been consuming. In this new budget reality NASA looked for cheaper booster designs and settled on twin reusable Solid-fueled Rocket Boosters (SRBs). These were just steel tubes filled with a propellant of ammonium perclorate and aluminum powder. These ingredients were combined with a chemical "binder," mixed as a slurry in a large Mixmaster, then poured into the rocket tubes like dough into a bread pan. After curing in an oven, the propellant would solidify to the consistency of hard rubber, thus the name *solid* rocket booster.

Because they were the essence of simplicity, SRBs were therefore cheap. Also, because after burnout they were just empty tubes, they could be parachuted into salt water and reused. There was just one huge downside to SRBs: They were significantly more dangerous than liquid-fueled engines. The latter can be controlled during operation. Sensors can monitor temperatures and pressures, and if a problem is detected computers can command valves to close, the propellant flow will stop, and the engine will quit, just like turning off the valve to a gas barbecue. Fuel can then be diverted to the remaining engines and the mission can continue. This exact scenario has occurred on two manned space missions. On the launch of *Apollo 13* the center engine of the second stage experienced a problem and was commanded off. The remaining four engines burned longer and the mission continued. On a pre-*Challenger* shuttle mission the center SSME shut down three minutes early. The mission continued on the remaining two SSMEs, burning the fuel that would have been used by the failed engine.

Solid-fueled rocket boosters lack this significant safety advantage. Once ignited, they cannot be turned off and solid propellant cannot flow, so it cannot be diverted to another engine. At the most fundamental level, modern solid rocket boosters are no different from the first rockets launched by the Chinese thousands of years ago—after ignition they have to work because nothing can be done if they don't. And, typically, when they do not work, the failure mode is catastrophic. The military has a long history of using solid rocket boosters on their unmanned missiles, and whenever they fail, it is almost always without warning and explosively destructive.

The SRB design for the space shuttle was even more dangerous than other solid-fueled rockets because their huge size (150 feet in length, 12 feet in diameter, 1.2 million pounds) required them to be constructed and transported in four propellant-filled segments. At Kennedy Space Center these segments would be bolted together to form the complete rocket. Each segment joint held the potential for a hot gas leak; there were four joints on each booster. Redundant rubber O-rings had to seal the SRB joints or astronauts would die.

Yet another aspect of the design of the space shuttle made the craft significantly more dangerous to fly than anything that had preceded it. It lacked an in-flight escape system. Had the *Atlas* rocket, which launched John Glenn, or the *Saturn V* rocket, which lifted Neil Armstrong and his crew, blown up in flight, those astronauts would have likely been saved by their escape systems. On top of the *Mercury* and *Apollo* capsules were emergency tractor escape rockets that would fire and pull the capsule away from a failing booster rocket. Parachutes would then automatically deploy to lower the capsule into the water. The astronauts riding in *Gemini* capsules had the protection of ejection seats at low altitude and a capsule separation/parachute system for protection at higher altitudes.

The shuttle design did accommodate two ejection seats for the commander and pilot positions, but this was a temporary feature intended to protect only the two-man crews that would fly the first four shakedown missions. After these experimental flights validated the shuttle design, NASA would declare the machine *operational,* remove the two ejection seats, and manifest up to ten astronauts per flight. Such large crews would be necessary to perform the planned satellite deployments and retrievals, spacewalks, and space laboratory research of the shuttle era. *These crews would have no in-flight escape system whatsoever.* These were the missions TFNGs were destined to fly. We would have no hope of surviving a catastrophic rocket failure, a dubious first in the history of manned spaceflight.

The lack of an escape system aboard operational space shuttles—indeed, the very idea that NASA could even apply the term *operational* to a spacecraft as complex as the shuttle—was a manifestation of NASA's post-Apollo hubris. The NASA team responsible for the design of the space shuttle was the same team that had put twelve Americans on the moon and returned them safely to Earth across a quarter million miles of space. The Apollo program represented the greatest engineering achievement in the history of humanity. Nothing else, from the Pyramids to the Manhattan Project, comes remotely close. The men and women who were responsible for the glory of Apollo had to have been affected by their success. While no member of the shuttle design team would have ever made the blasphemous claim, "We're gods. We can do anything," the reality was this: The space shuttle itself *was* such a statement. Mere mortals might not be able to design and safely operate a reusable spacecraft boosted by the world's largest, segmented, uncontrollable solid-fueled rockets, but gods certainly could.

It would be more than just the unknowns of a new spacecraft that TFNGs would face. NASA's post-Apollo mission was also uncharted territory. Having vanquished the godless commies in a race to the moon, the new NASA mission was basically a space freight service.

NASA sold Congress on the premise the space shuttle would make flying into space cheap and they had good reason to make such a claim. The most expensive pieces of the system, the boosters and manned orbiter, were reusable. On paper the shuttle looked very good to congressional bean counters. NASA convinced Congress to designate the space shuttle as the national Space Transportation System (STS). The legislation that followed virtually guaranteed that every satellite the country manufactured would be launched into space on the shuttle: every science satellite, every military satellite, and every communication satellite. The expendable rockets that NASA, the Department of Defense (DOD), and the telecommunications industry had been using to launch these satellites—the Deltas, Atlases, and Titans—were headed the way of the dinosaur. They would never be able to compete with the shuttle on a cost basis. NASA would be space's United Parcel Service.

But this meant that, of all the planned shuttle missions, only a handful of science laboratory missions and satellite repair missions would actually require humans. The majority of missions would be to carry satellites into orbit, something unmanned rockets had been doing just fine for decades. Succinctly put, NASA's new "launch everything" mission would unnecessarily expose astronauts to death to do the job of unmanned expendable rockets.

As we TFNGs were being introduced, NASA had to have been feeling good. They had a monopoly on the U.S. satellite launch market. They also intended to gain a significant share of the foreign satellite launch market. The four shuttles were going to be cash cows for the agency. But the business model depended on the rapid turnaround of the orbiters. Just as a terrestrial trucking company can't be making money with vehicles in maintenance, the shuttles wouldn't be profitable sitting in their hangars. The shuttle fleet had to fly and fly often. NASA intended to rapidly expand the STS flight rate to twenty-plus missions per year. And, even in the wake of post-Apollo cutbacks, rosy predictions said they had the manpower to do it.

The shift from the Apollo program to the shuttle program represented a sea-change for NASA. *Everything* was different. The agency's new mission was largely to haul freight. The vehicle doing the trucking would be reusable, something NASA had no prior experience with. The flight rate would require the NASA team to plan dozens of missions simultaneously: building and validating software, training crews, checking out vehicles and payloads. And NASA would have to do this with far less manpower and fewer resources than had been available during Apollo.

I doubt any of the TFNGs standing on that stage fully comprehended the dangers the space shuttle and NASA's new mission would include. But it wouldn't have mattered if we had known. If Dr. Kraft had explained exactly what we had just signed up to do—to be some of the first humans to ride uncontrollable solid-fueled rocket boosters, and to do so without the protection of an in-flight escape system, to launch satellites that didn't really require a manned rocket, on a launch schedule that would stretch manpower and resources to their limits—it wouldn't have diminished our enthusiasm one iota. For many of us, our life's quest had been to hear our names read into history as astronauts. We wanted to fly into space. The sooner and the more often (and who gave a shit what was in the cargo bay), the better.

CHAPTER 7

Arrested Development

On my first official day as an astronaut candidate I faced two things I had never faced before: picking out clothes to wear for work and working with women. In my thirty-two years of life, beginning with diapers, there had always been a *system* to dress me. I had gone to Catholic schools for twelve years and worn the uniforms of that system. In four years at West Point I never had a piece of civilian clothing in my closet. The air force also told me what to wear. Not once on a school or work morning had I ever stood in front of my closet and pondered what I should wear that day. As a result, I was a fashion illiterate. And I wasn't alone. I had already seen a handful of the veteran astronauts wearing plaid pants. Even I, completely clueless on the subject of style, sensed this might be a little too retro. When my children saw one of the plaid-panted victims, they hid their faces and giggled. To this day, whenever my adult children see a golfer wearing plaid, they'll comment, "Hey, Dad, check it out . . . an astronaut." A military astronaut might show up for a party dressed in a leisure suit or Sansabelt slacks and, most telling, no other military astronaut present would know there was anything even slightly amiss.

Fortunately for my children, my limited wardrobe did not include plaid. Rather, my first attempt at workday attire required me to satisfactorily combine one of four solid-colored slacks with one of four solid-colored shirts. I failed. At the breakfast table my wife recoiled as if I had walked in with a nose ring. "You're not going to work dressed like *that,* are you?" It was a question I would hear many times in the first few weeks of my NASA life. Donna even threatened to put Garanimals hang tags on my slacks and shirts. She would mock me: "The lions go with the lions and the giraffes go with the giraffes." I had no problem with the suggestion. I thought it was an excellent idea.

If I had no clue about how to dress myself, I was in another galaxy when it came to working with women. I saw women only as sex objects, an unintended consequence of twelve years of Catholic school education. The priests and nuns had pounded into me that females were equated with sex, and sex brought eternal damnation. Girls were never discussed in any other context. They were never discussed as real people who might harbor dreams. They were never discussed as doctors or scientists or astronauts. They were only discussed as "occasions of sin." The shortcut to hell was through a woman's crotch was all I learned about the female gender as a teenager. Their breasts would earn you an introduction to Beelzebub, too. In fact, just *fantasizing* about their breasts and their other parts (the soul-killing mortal sin of *impure thoughts*) would also send you straight to hell. Only in marriage did the rules change. Then, sex was fine—productive sex. In marriage a woman achieved her highest state in life—getting on her back and producing children. "The primary purpose of marriage is procreation of children" was dogma in my wife's 1963 "Marriage Course" curriculum guide from St. Mary's High School.

The same guide also includes a lesson on "Masculine and Feminine Psychology" with a table of "characteristics." *Males are more realistic, females more idealistic. Men are more emotionally stable, women are more emotionally liable. Man loves his work, woman loves her man.* And, my favorite, *Men are more likely to be right, women more likely to be wrong.*

I accepted these twisted sexist messages of Catholicism so completely that in my senior year of high school I wrote a term paper on why women should not be allowed to attend college. After all, I eloquently reasoned, they never did anything with an education. They only attended college to find a husband. They were needlessly filling classes and taking seats from males who would require the education to get jobs . . . real jobs. I received an A on the paper. I had learned my lesson well.

The Hollywood movies of my childhood did nothing to dispel what I was learning in school. The men were always depicted as the *action* gender, be they cowboys in action against the Indians, a soldier in action against the Japanese, or an astronaut saving humanity. The women were always the passive gender, waiting at home, cooking, and caring for the children. They were only active when the letters and telegrams came that their heroic men had taken an arrow, bullet, or meteor. Then they cried. After all, they were more *emotionally liable.*

To further guarantee my ignorance of females, I had lived and worked in environments awash in testosterone my entire life. I was raised in a family of one girl, five boys. My sister was nine years younger than me. This was no Brady Bunch house filled with teen girls to give me a clue about how to act and what to say around females. I recall one of my wildlife-enthusiast brothers had a basketball on which he had written, *I hate girls. Bighorn sheep, I like.* That about summarized the Mullane boys' attitude toward females. We were more comfortable around four-legged animals than a human carrying an X chromosome.

I didn't go to either my junior or senior prom. At the few dances I did attend I stood against the gym wall with the other nerds, dweebs, and losers attempting to fight off impure thoughts. When there would be a girl-ask-boy dance, I would be left against that wall. The girls at St. Pius X High School weren't stupid. In my entire four years of high school I doubt I accumulated more than a couple hours talking to girls. My senior yearbook doesn't have a single entry from a girl. In fact, it bears only one dedication. It's from a fellow dweeb that reads, *You missed Korea, but here's hoping you make Vietnam*. It probably comes as no surprise that when I graduated I was as virginal as Mary the Mother of God.

The West Point of my era was another all-male bastion. Cadets joked that the mortar in the granite walls was actually semen. My Company K-1 marching song included the lyrics "We make the cum fly." West Pointers of my era looked at women with the same leering eye as a convicted felon doing thirty to life. Flirtation Walk, so romantically depicted in the movies, was littered with more used rubber than a Firestone test track. My West Point experience reinforced what Catholicism had taught me—women were nothing more than sex objects.

The USAF officer corps that I entered in 1967 was also a male organization. I never encountered a female flyer. Strippers entertained us at the O'club on Wednesday and Friday nights. Military flyers saw women only as receptacles. If anyone from my era says otherwise, they must be running for Congress. Women may be from Venus and normal men may be from Mars, but military flyers are from the planet *ARRESTED DEVELOPMENT* (AD).

When I walked into the astronaut office on day one, I didn't have the slightest sense of how to incorporate the six TFNG females into my work life. My behavior and vocabulary around them was exactly as it was around men. I recall an early incident of telling a joke to a TFNG audience including Sally Ride that had the word *tits* in it. Sally hardly said another word to me for the next ten years. But, at the time, I didn't have a clue. I had no other experiences to draw upon. Professional women were as unknown and unknowable to me as sea life at the bottom of the Arctic Ocean.

I wasn't the only one afflicted with an atrophied brain when it came to interacting with the women. On one occasion a few of the men found a live grass snake near the gym. They entered the ladies' locker room and shoved the slithering creature into Judy Resnik's purse. After she returned from her run, every man present gathered at the locker room door giggling like middle-schoolers. They heard the shower run, stop, then the shower door open. A few minutes later a scream right out of the movie *Psycho* echoed through the gym. The guilty parties evaporated faster than you could say "Tailhook."

A navy TFNG probably best summarized the military male attitude about women. We were standing outside Sally Ride's office. She was absent and I took the opportunity to point out the bumper sticker on the front of her desk. It read, "A woman's place is in the cockpit." My Top Gun companion looked at the sticker and chuckled. "A woman *is* a COCK pit." That was exactly how most of the military astronauts saw women in general and good-looking women in particular. We were flying blind when it came to working with professional women. And now, we were thrown into a group of women who weren't just professional. They were pioneers. They would be carrying the banner of feminism into the final frontier.

Who were these alien creatures we confronted?

Judy Resnik, twenty-eight, hometown Akron, Ohio, had a doctorate in electrical engineering from the University of Maryland and was a classical pianist. She had been married briefly and was divorced. Prior to selection she had been working with the Xerox Corporation. Judy and I would fly our rookie mission together, an experience that would make us close friends. She would die on *Challenger,* her second mission.

Rhea Seddon was a thirty-year-old unmarried surgeon from Murfreesboro, Tennessee. Besides being easy on the eye, she had an alluring Tennessee accent as smooth as Wild Turkey. She was a Berkeley graduate and held a doctorate in medicine from the University of Tennessee.

Anna Fisher, twenty-eight, was another very attractive medical doctor. She was born in New York City but called San Pedro, California, her hometown. She held a doctorate in medicine and a master's in chemistry from the University of California, Los Angeles. When Anna entered NASA she was married to another doctor, Bill Fisher, who would later be selected as an astronaut in the class of 1980. Anna would have her first baby a few years after her TFNG selection and would become the first mother to fly in space.

Sally Ride was a twenty-six-year-old physicist PhD from Stanford University. She was also a superb tennis player, having achieved national ranking on the junior circuit. At the time of her TFNG selection she was unmarried but would later marry and divorce Steve Hawley, another TFNG. Her hometown was Los Angeles, California.

Kathy Sullivan, twenty-six, was unmarried and held a doctorate in geology from Dalhousie University (Halifax, Nova Scotia). Kathy would later become the first American woman to do a spacewalk. Her hometown was Woodland Hills, California.

Shannon Lucid was from Bethany, Oklahoma, and had a cement-thick Sooner accent. I once heard her speaking with Apollo astronaut and fellow Okie, Tom Stafford, and wondered if they were speaking Klingon. At thirty-five, Shannon was the oldest TFNG female and the only mother (three children) at the time of her selection. She held a doctorate in biochemistry from the University of Oklahoma.

All of these women were feminists in the sense they were out to prove they were as good as any male astronaut. Only Sally Ride struck me as an activist, a woman bent on making a political statement as opposed to a personal one. She seemed to view the world through NOW prescription lenses. Every action had to be gender sanitized. Before her first space mission I heard her say there could be no live TV downlink of her during orbit food preparation because it would show her in a traditional female role, even though food preparation, like toilet cleaning, was a shared crew responsibility. After the mission, at a JSC welcome for the crew, a NASA PR spokesperson brought out a bouquet of roses for Sally. She refused to accept them, as if to do so would be an affront to women. After all, the males weren't being given roses. Every military TFNG quickly learned to be careful in word and deed around Sally. She had about as much tolerance for our arrested development as Billy Graham did for a Wicca.

The other five females cut us varying degrees of slack. Rhea Seddon was a model of tolerance. She had to be. A couple years into our TFNG lives she married Robert "Hoot" Gibson, an F-14 Tomcat fighter pilot. Forget the James Carville and Mary Matalin marriage as one of polar opposites. Compared with Rhea and Hoot, James and Mary are paragons of blissful compatibility.

It was easy to see the mutual attraction between Rhea and Hoot. Rhea was a petite, confident surgeon. She was blonde, beautiful, outgoing, and a classy dresser. Hoot was the Chuck Yeager of the TFNGs, capable of flying anything with a stick and throttle, and flying it as if it were a natural extension of his body. He didn't so much strap into a cockpit as meld with it. He was a natural-born leader and would ultimately rise to the position of chief of the astronauts a few years after *Challenger.* On appearance alone, women were drawn to him. He had a Tom Selleck look about him, with a large blond mustache and an easy smile. In his California childhood he had learned to surf and play guitar. He rode motorcycles. He built his own formula-one racing plane in his garage, was a frequent winner in air races around the country, and held several world speed and altitude records. He also flew super–high power WWII fighter planes and Russian MIG jets in air shows as part of a flying museum. He was the consummate fun-lover. When I returned from my third shuttle mission, my wife told me how Hoot, a family escort for that mission, had driven the crew wives into the Edwards AFB desert and spun doughnuts in the sand in a government van. That was Hoot, always ready to thrill the women. Occasionally at parties, he and a few other navy pilots would grab a microphone and serenade a skirt with "You've Lost That Lovin' Feeling." Next to Hoot, Tom Cruise's *Top Gun* character, Maverick, looked like a *Show Boat* chorus member. Oh, did I mention he was the lead singer and guitarist in the astronaut band, Max Q? I hated him. I'm kidding of course. I had great respect for Hoot and thoroughly enjoyed being part of his crew on my second shuttle mission, STS-27. But in his company another man couldn't help but ask himself, *What would any woman see in my sorry ass?*

But there was another characteristic that came with the guitar, surfboard, and airplanes. Hoot was, like all military aviators, a male chauvinist pig. If NOW had a ten-most-wanted pig list, he would have been at the top. If they had ever caught him, though, it would have only taken a few minutes before the NOW politburo fell into a hair-pulling cat fight screaming, "I want to have his baby!" Hoot was that charming. Lest you think I exaggerate, even Sally Ride went out with Hoot when both were single, which says a lot about his charm factor.

Hoot's hallmark was his "snorting." Whenever he saw a young, attractive woman, he would discreetly make a sound like a pig snort. This was a physical manifestation of one of his favorite expressions, "I'd like to snort her flanks." He did this snorting so often that when he was assigned as the commander of STS-27, our mission was nicknamed Swine Flight by the office secretaries. I'm sure Gloria Steinem and Sally Ride would have thoroughly agreed.

So, Rhea and Hoot's marriage was one of the world's great mysteries, like the rise of life on earth. If the pope were ever to beatify a woman as the patron saint of wifely patience, it would have to be Rhea. Indeed, we called her Saint Seddon for putting up with Hoot.

If Hoot were number one on NOW's most-wanted pig list, I would have been number two. As test pilots would say, I operated at "the edge of the envelope." It was as if I had sexist Tourette's syndrome. The joker in me would leap from my mouth. Only around Sally did I keep myself somewhat throttled. I had a sixth sense about the danger there, like a dog knows not to paw at a snake. But Sally really wasn't an issue. After my tits joke, she avoided me like I was criminally insane.

I definitely tested Shannon Lucid's feminist tolerance a few times. I liked Shannon. She always struck me as indifferent to office politics, whereas her five peers were clearly vying for that most coveted of titles: FIRST AMERICAN WOMAN IN SPACE. Shannon was just there to do a good job. Whatever came of that, so be it, was the attitude she projected. I admired her for that. Her philosophy would serve her well. Ultimately she would fly five times in space, including a six-month stay aboard the Russian Mir space station. (Sally Ride only flew twice and departed NASA after *Challenger.* But it's her name in Billy Joel's song, not Shannon's. Life isn't fair.)

Shannon's first flight had Saudi Arabian Prince Sultan Salman Al-Saud aboard and after the mission he invited the crew and their spouses to visit Saudi Arabia. Shannon's husband could not make the trip. Shannon wasn't concerned. She didn't need a man to hold her hand. Wrong. Saudi Arabia did not allow women to enter the country alone. She had to have a male escort. When Shannon heard this she told headquarters she wasn't going. NASA HQ and the State Department were concerned about the potential press photo featuring only the men from the mission being greeted in Riyadh by King Fahd, so they asked the Saudis to look into their laws for a loophole. I was in my office when a TFNG came in with word they had found one. The Saudis would allow Shannon to enter as Dan Brandenstein's honorary daughter (Dan was the mission commander). Or, she could enter as John Fabian's honorary sister (John was another crewmember). Or, they might make a special exception, as they had when the queen of England had visited the country, and designate Shannon an *honorary man.*

When the men in my office heard this, we exploded in laughter. What greater insult could a feminist hear than to be told she must take on the label "man" to get some respect. When I heard this, I couldn't contain the joker in me. I immediately went to Shannon's office and congratulated her on having achieved the highest honor a woman could ever hope to achieve . . . to be designated an honorary man. Shannon had a lively sense of humor and laughed at my antics, but I made certain not to walk down the stairs in front of her for the next few weeks.

Shannon later came into my sights at a Bible study meeting. The astronaut office was filled with devout persons of several faiths. Some of the most religiously committed astronauts were marines, a fact that shocked and awed me. Marines were known for eating their young, not for their "praise Jesuses." But several had organized a weekly Bible study. Shannon was a member of the group, as were Donna and I. The topic of one meeting was how people who had never "known Jesus Christ" might be treated by God in the afterlife. One group member posed these thought-provoking questions: Could a native from the jungles of Indonesia, who had never heard of Jesus Christ, enter God's Kingdom? Or how about a mentally ill person or someone born with half a brain?

The last part of this question was a setup the joker in me could not let pass. I jumped on it. "Yeah, Shannon, what about women?"

Suddenly Shannon had the mark of the beast. I was a dead man. She bore into me with a look that said, "I'm going to stake you to an anthill!"

After the meeting it was suggested I find God somewhere else . . . somewhere far, far away from this Bible study group. Donna and I were excommunicated.

When I die, I'm going to two hells. On Mondays, Wednesdays, and Fridays, I will be in Bible hell, the one with all the fire and brimstone. There, demons will torture me with their fiery pitchforks. But during the rest of the week I'll be damned to feminist hell, where some high-value parts of my body will be placed in a red-hot vise and Shannon Lucid, Sally Ride, and Judy Resnik will take turns cranking the vise tighter and tighter while I plead, "Mercy! Mercy! I was raised on the planet Arrested Development. I couldn't help myself!"

CHAPTER 8

Welcome

On a hellish July day in 1978, properly dressed by my wife and handicapped with a brain from Planet AD, I drove through the gates of the Johnson Space Center to begin my TFNG life. If NASA ever needs to test a space probe designed to survive on the surface of Venus, a Houston parking lot in summertime would suffice. Air-conditioning isn't a luxury in Houston. It's a life support system. Until I arrived in Houston I would laugh at those supermarket tabloid reports of people walking down a sidewalk and spontaneously combusting. But after one day of a Houston summer, I no longer laughed. It could happen.

Besides a small rocket park featuring a *Saturn V* moon rocket horizontally displayed near the entrance, there was nothing to suggest JSC had anything to do with space. There were no towers or gantries or blockhouses. A passerby could easily think it was a university campus or a corporate headquarters. The architecture screamed "low bid." Except for size, every building was identical, each featuring a façade of exposed aggregate concrete. The major buildings were positioned around a duck pond landscaped with pine and oak trees to relieve the otherwise flat, boring terrain of southeast coastal Texas.

Johnson Space Center was located in the far south of Houston's urban sprawl. It was nearly as close to Galveston as it was to Houston's city center. The community in which many NASA employees lived was the suburb of Clear Lake City—implying a lake nearby, and a clear one at that. Wrong. Clear Lake was neither clear nor a lake, but rather a chocolate-tinted, humidity-shrouded inlet from the nearby Gulf of Mexico that served as a time-share for a couple billion vacationing mosquitoes. Obviously Clear Lake City had been named by a real estate developer. If there was truth in advertising, Clear Lake City would have been named Fire Ant City. In its abundant grasses were legions of these insects, which should be on the UN's list of weapons of mass destruction. Fire ants have been known to kill babies, the elderly, and newly born animals (I'm not kidding). To step in one of their mounds was to understand what it feels like to be napalmed.

If only the fire ants preyed on the Olympic-size roaches, which were equally ubiquitous, then at least one pest would have been eliminated. But the ants did not. In some kind of insect pact, the ants stayed outdoors, leaving the roaches free to turn homes into vast roach motels. Every morning brightly colored exterminator trucks poured into suburbia like tanks coming ashore at Normandy. Technicians donned moon suits and slung flamethrower-like tanks to their backs to enter the combat zones of kitchens and baths. But theirs was a lost cause. The roaches thrived on their powders and gases and liquids. Even the old standby, the shoe, proved ineffective because these roaches were masters of land *and* air. They flew. I recall an early incident at a TFNG party where the hostess chased a four-incher into a corner and chortled with glee as she aimed her toe at it. "Eat leather, you bastard!" But as her foot came down the monster spread its wings and launched itself straight at her face. She screamed and fled, flailing her arms as if her hair were on fire. Meanwhile, the victorious roach broke off its attack, made a clattering turn, and settled on the mantle, tucking its wings back into its body like a majestically perched eagle. For the rest of the party it remained on that mantle, its antenna waving back and forth like semaphores, daring anybody to attack. There were no takers.

Looking at the southeast Texas topography, weather, flora, and fauna, I doubt a single TFNG thought, *Of all the places in America I would like to live and work, I would choose Houston, Texas*. But nobody was complaining. We had just been blessed with the best job on Earth. If our offices had been in a landfill, we wouldn't have cared.

The astronaut offices were on the top (third) floor of Building 4. They ringed the outer perimeter of that floor, leaving the interior offices for the coffee bar, bathrooms, mail room, photo archives, conference rooms, and other administrative functions. Like the exterior of the buildings, the offices had a low-bid, cookie-cutter sameness about them. Dilbert would have been right at home. The walls were movable steel panels. Magnetic picture hangers were needed for any *Sports Illustrated* swimsuit calendars, of which there were a few. The office décor was straight out of *Designing Bureaucrat*: faux-wood desks and credenzas, cheap swivel chairs, battleship-gray steel filing cabinets, fluorescent lights. Clearly NASA was spending its money on rockets, not astronaut office frills. On the hallway walls hung magnetic nameplates, decorative space photos, and a bulletin board. The latter was intended to disseminate administrative information, but also served as a battlefield for gender wars. In one instance an article on bone loss in weightlessness appeared. The MD author concluded that female astronauts would be more vulnerable to such loss as they aged. One of the women had circled that point and handwritten a note, *This is why women should be first in line to fly the shuttle*. A resident of Planet AD had answered, *This is why NASA should hire younger women*.

On another occasion, someone had pinned up a magazine article on reproduction in zero-G. The author had hypothesized that it would require a threesome to copulate due to the repelling effect of Newton's law, which dictated the first "action" would produce an equal and opposite "reaction." One wit had written, *No! This is why God gave us arms and legs*. Another had tacked up a sign-up list next to the article requesting volunteers to "participate in 1-G simulations." Someone, almost certainly one of the women, had scrawled across it, *Grow up*. I would come to love the B-board wit. It was frequently laugh-out-loud funny.

Pairs of TFNGs shared offices. I had no clue how the pairings were made, or who made them, or what they might imply for future flight assignments. My roommate was Mike Coats, an Annapolis graduate and navy pilot noted (by my teenage daughter) for looking like the Superman character played by Christopher Reeve. Mike quickly acquired the handle Superman. He was also legendary for his ability to continuously flip a pen (and always catch it) while talking or studying or standing at the urinal or just about doing anything. He never watched the pen and he never missed. Up and down the spinning pen would fly, always landing precisely in his fingers, to be immediately flipped upward again. I wondered how that had played with the psychiatrists. I couldn't imagine Mike had stopped his flipping while talking to those doctors. He would have exploded.

The Monday morning all-hands meeting was our introduction to the essence of the astronaut business. Held in the main conference room of the astronaut office and chaired by Chief of Astronauts John Young, these weekly meetings were a venue to air important issues. I entered the room with the same trepidation a student feels on the first day of class. Where to sit was the first issue I had to address.

A large table dominated the room. On it sat some conference phones and an overhead projector. A screen hung on the wall at the front of the room. Chairs ringed the table, but I gave no thought to taking one. This was the sacred table of Apollo. Alan Shepard and Jim Lovell and Neil Armstrong had sat here. At the moment moonwalker John Young was sitting at its head. There was no way one-day-old Ascan Mike Mullane was going to sit at that table. Perhaps the chairs were assigned to the veteran astronauts and I would be embarrassingly evicted like a Cheers' patron being asked to move from Norm's bar stool. I looked elsewhere. Several rows of chairs had been placed at the back of the room and I aimed for these cheap seats. Most of my fellow TFNGs did likewise. Most, but not all. Rick Hauck, the senior ranking TFNG pilot and our class leader, took a seat at *the* table. *You didn't get more alpha male than this,* I thought. Rick was already lifting a leg to mark his territory. We hadn't been on the job for fifteen minutes and the competition was already fierce. He was making a statement: I'm going to be the first TFNG in space. Every one of us glared at him and wondered if we shouldn't have shown some balls (or ovaries) and parked our asses at that moon-dusted table.

Also seated around the table were other spaceflight veterans, the big men on campus, who every freshman longed to be. Besides John Young, there was Alan Bean, the only other moonwalker remaining in the office. There were also some astronauts from the Skylab program: Owen Garriott, Jack Lousma, Ed Gibson, Paul Weitz, and Joe Kerwin. One astronaut remained from the Apollo-Soyuz program, Vance Brand. One of the *Apollo 13* crew was still aboard, Fred Haise. Ken Mattingly, the original *Apollo 13* astronaut who was exposed to German measles and replaced at the last moment, was still with NASA and at the table. He had later earned his wings on *Apollo 16*.

The rest of the office included seventeen astronauts who were still waiting for their first spaceflight. Seven had been dumped on NASA in 1969 by the USAF after their Manned Orbiting Laboratory program was canceled. The others had been selected in the late years of the moon program and had been in line to fly on *Apollo 18* through *20*. But Congress had pulled the plug after *Apollo 17*. Most of these unlucky seventeen had been at NASA for more than seven years and hadn't been any closer to space than I had. And they were still many years away from earning their wings. *Please, God, spare me that fate,* was my prayer.

Though we were brand-new, TFNGs understood the coin of the realm. Spaceflight. Those who had ridden rockets were rich beyond measure. Those who hadn't were paupers. There was no astronaut "middle class." We had assumed a job in which rank, wealth, awards, degrees, and all other measures of success were absolutely meaningless. In that regard the unflown older astronauts in the room were as proletariat as us wet-behind-the-ears Ascans. Forget their near decade of service at NASA. It didn't count. A lifetime of flying a desk, even a NASA astronaut desk, couldn't put a gold astronaut pin on your lapel. Every one of us rookies in that room, regardless of age or title, were classless peons and we would remain so until that glorious day when the hold-down bolts were blown and our ride began. In that split second we would become kings.

As I looked at the crowded table, I knew every TFNG was thinking the same thing: *Why don't these old farts just leave or die or something?* We were the brash teenagers in the company of seniors who were slowing us down. We couldn't fly until they did. How many missions would they consume? How many years would I have to wait before they were up and out? Though we would soon form tight friendships with these vets, no rookie astronaut ever shed a tear when a member of the older generation decided to move on. At their retirement parties we were the happiest ones there, knowing that one more cockpit seat had just opened up. Don't let the door hit you in the ass, was our attitude.

Later, I would learn how these seniors feared us. We had been selected by George Abbey, director of Flight Crew Operations (FCOD), John Young's boss. They hadn't been. They were astronauts long before George assumed his position. If rumor was true, George would be making shuttle flight assignments. The older astronauts wondered if they would ever fly. George might just skip right to us, his protégés, and flush those also-rans onto the street. There was no astronaut contract guaranteeing a spaceflight. So the seniors in that room saw us as threats to their place in line. It wasn't just the TFNGs who were sniffing one another and lifting a leg. Everyone was. We all were in a lather to find our place in line for a ride into space and guard it with fang and claw.

Even though the six females couldn't metaphorically lift a leg, they were certainly looking at their five peers and measuring the competition. It was a no-brainer one of them would be aboard the first shuttle carrying any TFNG crewmember. The NASA PR machine was chomping at the bit to get a woman in space. While I doubted it would come to hair pulling and face scratching, there was bound to be as much competition among the fair sex as there was among the males.

John Young welcomed us with a few forgettable words, all delivered while he looked at his shoes. Dealing with life-and-death situations as a test pilot and astronaut hadn't endowed Young with any public speaking skills. He seemed nervous and hesitant to make eye contact with his audience. It was a personality trait we would learn wasn't just associated with welcoming speeches. (The things *Life* never mentioned.) His stature and voice made him even less compelling. Like all the earliest astronauts, he was short and small framed. He was a Florida boy, and he had the accent and vocabulary of one. He frequently used the expression "them boys" in reference to anybody outside the astronaut office. He wasn't warm or approachable. *Reclusive* wouldn't be far from the mark. But he did have a great understated humor. When Florida named one of Orlando's main thoroughfares the John Young Parkway, John said, "Them boys shouldn't have done that. I ain't dead yet."

He didn't leave his wit behind when he flew in space. On STS-9, when two of *Columbia*'s computers failed in orbit, causing a major and potentially life-threatening problem, John looked at his pilot, Brewster Shaw, and said, "This is what they pay us the big bucks for." He was probably making $70,000 a year at the time.

The meeting proceeded with Young and Crippen, the crew for the first shuttle flight, discussing their mission preparations. We TFNGs were still naïve enough to believe the NASA press releases proclaiming the first shuttle mission would fly in 1979. NASA HQ was loathe to admit to Congress that the machine was well behind schedule, and so they published wildly optimistic timelines about as likely to be achieved as the Chicago Cubs winning the World Series. (STS-1 would ultimately launch on April 12, 1981.) One of the veterans would tell us what the acronym NASA actually stood for: Never a Straight Answer. We learned to add years to any dates provided in a NASA press release about the shuttle schedule.

There was little we newbies could understand in the discussions that swirled around the table. The language of NASA was so laden with acronyms that it took many months to become fluent. The commander of a mission wasn't called the "commander." He was the CDR, pronounced as the individual letters . . . C, D, R. And a pilot wasn't called a "pilot." He was a PLT, again pronounced as the individual letters. A pulse code modulation master unit wasn't called such; it was a "puck-a-moo." I have heard entire conversations between astronauts without a single recognizable noun in them. "I was doing a TAL and Sim Sup dropped the center SSME along with the number-two APU. Then MS2 saw an OMS leak, we got a GPC split . . ."

So we just listened in silence to the technobabble. At the close of the meeting, when Young asked if any TFNGs had anything to say, we all sat on our hands. All except Rick Hauck. There was a stir among us as he raised his hand. *Surely he wasn't going to make a technical contribution?* was our collective thought. Could he be that far ahead of us? Our competitive paranoia roared to life.

But Rick's comment wasn't technical. He just asked if all the TFNGs would remain in the room to cover some administrative items. A secretary entered and passed around copies of our official NASA photos for our review. We had posed for these as part of our in-processing. Now the mail room was filled with thousands of lithographs in which we had been perpetually frozen as smiling, thirtyish, flight-suited youths. Decades later, when there was little resemblance to the actual living person, these first photos were still being sent out. No doubt they have been a source of great confusion when used by American Legionnaires and city officials and others waiting to identify their astronaut luncheon speaker exiting an airport jet way.

Next, the secretary placed a few paper tablets on the table and asked us to give a sample signature for the auto-pen machine. Autographs! Apparently the world would be clamoring for our autographs in such quantities that NASA had a machine to automatically pen them. If there was ever an indication of the new world we had entered, it was this. Except on checks, I had never been asked for an autograph in my life.

I took a page and penned *Mike Mullane.* It didn't look right. Too small, too tight, too anal, I thought. My "Ms" in particular looked like they had been made with a nun standing over me. They were too legible. Each was composed of symmetrical double humps that would have fit perfectly into the capital line guides of my third-grade Red Chief tablet. Such a signature would never do. It seemed to me famous people always had illegible signatures. I took another page and tried a radical swipe and imagined how it would look on a photo on some collector's wall. It appeared as fake as it was. Another page bit the dust. I tried signing faster, slower, with more slant, less slant . . . I wanted an autograph that would dazzle. Then it dawned on me. *Everybody* was doing the same thing. An act that had been as casual as, well, signing your name had suddenly become a quest, a personal challenge. I looked around and saw several TFNGs intensely studying their pages. A few tongues worked around the corners of mouths. To produce the perfect autograph was hard labor. I was witnessing the definition of astronauts . . . competitive to the nth degree. They even beat the shit out of their own muses. *Why can't you come up with a memorable autograph, goddamn you!* I could hear the buzz of pages disappearing from the tablet, ammunition being expended in thirty-five private wars to produce the perfect signature. By the time the secretary had her autographs for the auto-pen machine, a small forest had been wasted.

CHAPTER 9

Babes and Booze

Over the next several months we continued our agency indoctrination by visiting NASA "centers" around the country. Like all government agencies, NASA spreads its operations over multiple states to gain the largess of as many congressional delegations as possible. We flew in private NASA jets to Kennedy Space Center, to NASA Ames Research Center in Mountain View, California, to Marshall Spaceflight Center (MSFC) in Huntsville, Alabama, and to several other NASA and contractor facilities scattered around the country. At each location we were introduced to the workers, took tours, and received briefings on the operations of each facility.

There was a social agenda as well. Many nights would find us at a cocktail reception or dinner hosted by a local community official. Some of these events were more work than fun. Attendees clamored for autographs and photos. Or the press would be invited and they would squeeze us for interviews. For the most part, though, we were good ambassadors for NASA and warmly welcomed the attention. The women probably welcomed it less so. With each passing day it was becoming more evident that the major focus was on them. Even the black TFNGs would become as invisible as us white guys whenever Judy or Rhea or Anna—the triumvirate of TFNG beauty—walked into the room. They were particularly dazzling when they were dressed in their dark blue patch-covered flight coveralls. There wasn't a man or woman in any public setting who didn't stare. I recall one local politician questioning several of us men at a party. He was totally focused on our comments until Judy walked by in her flight suit. Then he interrupted us, said, "Excuse me," and hurried to catch up to Judy. We were abandoned like the out-of-state voters we were.

What was it about the women in their flight suits? It wasn't like the clothing flattered their figures. NASA ordered them off the shelf. The nuns of my high school would have loved them. They were baggy in all the right places, effectively neutering the female form. But in them, Judy, Rhea, and Anna stole the audience. The flight suits seemed to transform them into fantasy creatures like Barbarella or Cat Woman or Bat Girl. If Madonna had walked into a room in a jewel-bedecked Prada special, dripping Tiffany diamonds, and stood next to a coverall-clad Judy, Rhea, or Anna, the Material Girl would have paled to "ordinary." Everybody, men and women alike, wanted to be seen with the flight suit–dressed women and pose for photos with them. Occasionally they would be so bothered and exhausted by the attention, they would use us men as human shields. At one of the parties I was standing with Dale Gardner, Norm Thagard, and a few others when Judy Resnik ducked behind our backs and whispered, "Close it up. I don't want that press guy to find me." A moment later we saw the stalker, pen and pad in hand, searching the room for his quarry. He eventually camped out at the exit to the ladies' room, expecting Judy had fled there.

Eating an uninterrupted meal in public in a flight suit quickly became impossible for the TFNG females. Patrons would approach them and ask for autographs, scrounging for any scrap of paper, including napkins, sugar packets, or bank deposit slips from the back of their checkbooks. At one meal the entire kitchen staff came out to meet Judy. The proud establishment owner, a large Italian woman, fawned over her as if she were royalty while ignoring me and the other men as if we were Judy's foot servants. In jest I interrupted their love fest and said, "Hey, what am I . . . chopped liver?" Moments later the woman brought out a plate of exactly that, raw chopped liver, and dropped it in front of me. Judy laughed. So did I. I like a good joke even when it is on me.

Besides the open bars at our soirées, there were other attractions for the males . . . young, beautiful women. Lots of them. At a Florida event one of the coarser TFNGs observed, "Mullane, look at this party. It's a potpourri of pussy." I had been in enough officers' clubs in my life to know that aviator wings had more babe-attracting power than Donald Trump's twelve-inch wallet. The Navy SEAL insignia had the same effect. One SEAL told me that some of the young women who frequented their officers' club were nicknamed Great White Sharks because they had swallowed so much SEAL meat. The male TFNGs were learning there was an even more powerful pheromone than jet-jockey wings and the SEAL insignia: the title astronaut. The fact that none of us had been any closer to space than an airline flight attendant didn't seem to matter. To the space groupies the title was good enough. We males found ourselves surrounded by quivering cupcakes. Some were blatantly on the make, wearing spray-on clothes revealing high-beam nipples, and smiles that screamed, "Take me!" The few bachelor TFNGs must have experienced some Zen-like ecstasy. In fighter pilot talk, they operated in a "target-rich environment." They should have just donned a full-body latex suit and gotten a "please take a number" dispenser.

Even the gold bands on the fingers of us married TFNGs were no deterrent to many of these women. They were equal opportunity groupies. Of course it was easy to see who was taking advantage of the situation. During the head count on the bus to return to a hotel, some MIAs would be noted. "He said not to wait for him. He got a ride."

"Yeah, I'll bet he got a ride" would be the rebuttal and a wave of snickers would follow.

It was also easy to see who was traumatized by the body swapping . . . the post-docs. I doubt any of them had ever met a married colleague with red-blasted "all-nighter" eyes, trailing the odor of alcohol and sex as he exited a motel room with a smiling young woman. Sensing their shock, Rick Hauck spoke to them on a bus returning from a meet-the-astronauts mingle. "Everybody needs to understand their moral standards aren't necessarily shared by others in the group. If you see something on one of these trips that offends you, keep it to yourself. It's none of your business. You could damage somebody's marriage."

How different was Rick's speech from John Glenn's "keep your peckers stowed" speech of twenty-five years earlier. As documented in *The Right Stuff,* Glenn cautioned his six peers against adulterous activity because of the scandal that would result if they were discovered. Now, a quarter century later, Rick's comments were aimed at the spectators, not the perpetrators. Zip your mouth, not your pants. How the moral compass had swung. Adultery and divorce had lost their stigma. Neither was going to affect a TFNG's career.

Philandering wasn't the only thing shocking the post-docs on these trips. The art of alcohol abuse was another, and some military TFNGs were true Picassos.

"Who wants to try a flaming hooker?" was Hoot Gibson's question at a Cape Canaveral bar one night. The recipe for the drink included a prodigious quantity of high-proof alcohol served in a brandy snifter. The drink was served *on fire.* I stuck around for this. Fire and intoxicated astronauts were material for David Letterman's stupid human tricks.

As always, there had to be competition. Winners were those who could throw back the complete shot in one gulp without burning themselves, then slam down the glass with the residual alcohol still burning. Needless to say, it helped to be at the bulletproof level of intoxication before attempting this trick.

Like a circus barker, Hoot roped in a crowd of unsuspecting post-docs. None thought it was possible. Hoot smiled at the challenge, unstuck a cigar from his mouth, slicked his mustache into order, grabbed the flaming drink, and quaffed it back. He slammed down the glass. A blue flame hovered over it.

The gauntlet had been thrown down and several suckers readied themselves to duplicate the feat. The bartender served up more glasses and torched them. With fear-tightened faces the post-docs picked them up and hesitantly brought them to their lips. Soon a new smell mingled with the miasma of cigar smoke, perfume, and beer . . . burning facial hair. There were cries of pain as flaming alcohol scorched mustaches, lips, and chins. Through it all Hoot smiled and puffed his cigar with an expression saying, "Why do I do this?" Periodically he would down another drink to keep enticing the wounded scientists back to the flame. Each time he remained uninjured and the glass retained the blue flicker of success. Each time it emboldened another post-doc to attempt self-immolation. As the hour drew late, Hoot finally explained the trick. "You have to be fearless. Toss the entire glass. Don't sip. There isn't enough oxygen in your mouth to feed the flame so it'll go out. If you do it fast enough, the flame will stay with the glass."

The formula for success had come far too late. At breakfast the next morning a few embarrassed, miserably hungover post-docs sat at the table nursing multiple blisters on their faces. Some of those victims, no doubt, were dreading having to explain to their spouses the source of their injuries. "Honey . . . you're not going to believe how this happened." Indeed, they wouldn't.

At every opportunity the military TFNGs also introduced the civilians to our lively, sometimes sick, sense of humor. During our tour of NASA's California facilities, Steve Hawley made the mistake of asking Loren Shriver, Brewster Shaw, and me to dinner with a former colleague of his. In the course of the meal Steve's friend, a male astrophysicist, became overawed with the Vietnam aspect of our past lives. Like me, Loren and Brewster were combat veterans of that conflict. The young scientist was relentless in probing for information on our experiences. "Mike, what did you do in Vietnam?"

I couldn't pass up the opportunity to play with his head, so I seamlessly replied, "I flew a candy bomber."

"A candy bomber? What was that?"

I had a fish on the line and began to reel it in. "In the villages the women and children would hide in their spider holes and trenches. You could never get them in the open. So I flew a plane loaded with canisters of candy and would swoop low over the villages and drop them nearby. This would bring the women and children out of their holes to scoop it up." At this point in my story I pointed to Loren and Brewster. "And these guys would be thirty seconds behind me loaded wall to wall with napalm and would lay it down on those villagers. It got them every time."

The scientist's eyes widened in shock and outrage. I could just imagine the scene playing out in his brain: images of women and children dipped in jellied gasoline running around on fire. He snapped his head to Loren and Brewster, anticipating a denial. At this point I expected my twisted joke to come undone but Brewster and Loren picked up my lead. They assumed the steely eyes of professional killers and silently nodded in the affirmative. Every Vietnam atrocity this young scientist had ever heard of was now confirmed.

Hawley tried to calm him. "That's bullshit. They make up these stories all the time. Don't believe them. They didn't kill any women and children."

At that comment, Brewster shrugged. He didn't say a word but his body language did: "You can believe what you want." There was no doubt in any of our minds Steve's friend walked away from dinner believing he had just socialized with war criminals.

On a trip to Los Angeles it was Jeff Hoffman who felt the sting. At breakfast he asked Brewster and me what we had done the night before. While we had actually been at a bar having a few beers, I immediately replied, "We visited that museum."

"What museum?"

I made up an incredible story about a museum of "cultural art." Loren Shriver picked up on my lead and added his own embellishments about famous paintings by Picasso and sculptures by Michelangelo. Dick Scobee joined in with more bullshit. Through it all Jeff expressed his disappointment at missing such a rare and wonderful opportunity. Finally he asked, "Where's the museum?"

I replied, "It's right next to the Christian Science Reading Room. We did some studying there before going to it."

Even this over-the-top BS didn't immediately register in Jeff's brain. He continued to lament he had missed one of America's greatest museums. A minute later he jerked up from his coffee. "You guys made all that up, didn't you?" We laughed.

Jeff would prove to be the most enduring TFNG scientist. Over the years, many of the other civilians would become enamored with the military aviator mystique and would take on varying degrees of its form. But, to the very end, Jeff remained an unpolluted scientist—a fact that presented some great opportunities for us AD retards. I recall a Monday meeting in which he made an impassioned request for better attendance at an astronaut office science lecture series. Attendance was voluntary and few of the military TFNGs were showing up. Jeff begged, "Guys, we're going to have coffee and doughnuts and the visiting professor really has some fascinating stuff to tell us. You really should be there." He then expanded on the science that would be covered. I watched the pilots. Their faces were pictures of disinterest. The only thought running through their brains was *I wonder where happy hour will be?*

Jeff finally finished. "Do you have any questions?" He looked so hopefully at his tuned-out audience, it about broke my heart. He was desperate for any indication that we had paid the slightest attention to his pleas. "Any questions? Any questions at all?" But the room remained as silent as an OMS burn.

I slowly raised my hand and Jeff's face lit up like a sunspot. "Yes, Mike."

"I was just wondering. . . . What type of doughnuts are you going to have?" The walls of the room nearly blew apart with laughter. It was one of Jeff's many lessons that the military aviator brain was a science wasteland.

Like Hoot with the flaming hookers, I wondered, *Why do I do this?* and smiled that I had. But I will ultimately pay the price. Besides Bible hell and feminist hell, I'll also burn in post-doc hell.

CHAPTER 10

Temples of History

In our early TFNG months we were introduced to the Outpost Tavern, a temple of space history. The Outpost was the astronaut after-work hangout located a few blocks from JSC's front gate. It was aptly named. To say the Outpost was "rustic" was like saying King Tut has a few wrinkles.

The building was a shack of weather-beaten boards, its parking lot as cratered as the Ho Chi Minh Trail. Some of these water-filled holes could have swallowed a small sedan. After stepping around a minefield of fire-ant mounds, patrons entered the Outpost through two saloon-style swinging doors cut out in the shape of curvaceous bikini-clad girls. The bar ran around two walls. A griddle and deep-fryer served up burgers and fries certain to deposit a couple millimeters of plaque in every artery of the body. The low ceiling trapped a cloud of atomized grease and cigarette smoke like pollution in a temperature inversion. A dartboard, a shuffleboard table game, and a pool table offered entertainment. The interior décor consisted of space posters and astronaut photographs stapled to the walls and ceiling. The Outpost was the only bar in America where the pinups were smiling flight-suited women astronauts.

Why the Outpost was picked as the astronaut hangout has been lost to antiquity, but it is almost a tradition for flying units to have such a retreat. For Chuck Yeager and the rocket plane pilots of the '50s and '60s there was the Happy Bottom Riding Club near Edwards AFB; for the early astronauts, it was the Mouse Trap Lounge in Cocoa Beach, Florida. Most likely the Outpost became the unofficial watering hole for shuttle-era astronauts because of the sanctuary it offered. I never saw anybody approached for an autograph or interview in the Outpost. Perhaps outsiders were intimidated by the obstacle course of potholes or they assumed the building was condemned.

Every Friday happy hour, many TFNGs would be at the Outpost. The building would ultimately be the scene of our crew-selection parties, our landing parties, and our promotion parties. It would be the place where we traded gossip and bitched about our management. We would meet with our payload contractors and refine checklist procedures on the backs of napkins. And the Outpost would ultimately serve as a refuge, where we would grieve for our lost friends. The Outpost has been a witness to so much of the astronaut experience it should be moved in its entirety to the National Air and Space Museum in Washington, D.C. It is as much a part of space history as the rocket planes hanging from the museum ceiling.

Our TFNG apprenticeship also introduced us to the loftiest temple of space history, the Mission Control Center (MCC). As we stepped into the deserted, silent room, I imagined we experienced the same sense of awe a rookie baseball player experiences when he jogs onto the field for his first Major League game. We were in the "Show," stepping where legends had stepped before. Here was where cigars were smoked in celebration of the *Apollo 11* landing. Here was where the words, "Houston, we have a problem" were first received when an explosion shattered the *Apollo 13* service module.

Like pennant flags hanging in a stadium, large renditions of patches of the missions controlled from the facility decorated the walls. The front of the room was dominated by a floor-to-ceiling rear projection screen. This was where the sinusoidal orbit traces, spacecraft location, and other engineering data would be projected during an actual mission. From the floor in front of this screen to the back of the room were consecutive rows of computer consoles. Each row was terraced to be slightly higher than the one in front. On top of these consoles were signs with acronyms that labeled the function of the particular station. FDO referred to the Flight Dynamics Officer's station, where a handful of men and women would monitor the trajectory of a launching and reentering spacecraft. INCO was the label for the Instrumentation and Communication Officer. PROP referred to the Propulsion systems controller. There were other labels: EVA, PAYLOADS, SURGEON, PAO, DPS, and more.

Our veteran astronaut escort had us take seats at the consoles and instructed us on how to wear the internal earpiece and microphone that were part of the MCC intercom system. He then began to explain the organization and function of each of the MCC stations. Every shuttle system, from the electrical system to the hydraulic system, from the environmental control system to the robot arm, had a controller who was an expert on that system and monitored its performance via the shuttle's data stream. These MCC controllers were supported by their respective "back rooms," which were filled with more specialists who had telephone access to the system contractors. In an emergency each controller had a wealth of brainpower to tap into.

Each MCC controller reported to the flight director, who occupied a console in the back of the room. "Flight" had overall responsibility for the conduct of the mission. They were the ones who faced the possibility of time-critical decisions carrying life-or-death consequences for the astronauts. It had been Flight Director Gene Kranz who had issued the famous edict "Failure is not an option," and had led his team in saving the lives of the *Apollo 13* crew. In my dozen years as an astronaut I would never meet a flight director I didn't think was cut from the same mold as Kranz. There are no superlatives too great to describe the MCC teams.

The escort shifted our focus to the CAPCOM position. This was the only MCC position that astronauts filled. CAPCOM was the "Capsule Communicator," the term *capsule* a carryover from the days in which astronauts flew in capsules. Early in the space program it was correctly determined that only one person should be in voice contact with flying astronauts. To have each of the MCC controllers talking to a crew would be chaos. The logical person to be the astronaut "communicator" was another astronaut. It had been this way since Alan Shepard's first flight when Deke Slayton had served as his CAPCOM. CAPCOMs, our leader explained, would work hand in glove with the flight director to make sure mission crews got the exact information they needed, nothing more and nothing less. As part of our training, we would all shadow a CAPCOM before filling that position ourselves.

The TV cameras mounted on the MCC walls were next brought to our attention. During missions these were always aimed at the CAPCOM and flight director positions. An indiscrete nose pick or crotch scratch might end up as material for one of the late-night comedy shows.

After answering some questions, our escort asked us to remain on the MCC intercom. He then called for a technician to "roll the audio." What we heard were the voices of Gus Grissom, Ed White, and Roger Chaffee. The tape was from January 27, 1967. The three astronauts were in their Apollo capsule going through a dry countdown with Launch Control. For a minute the audio was mundane, just the acronym-laden techno-talk that is part of any spacecraft checkout. Then one of the voices urgently cried, "There's a fire! Get us out of here!" NASA had designed the Apollo capsules to fly with pure oxygen atmospheres. Somewhere in Grissom's capsule a spark had set it ablaze. In seconds the cockpit was transformed into a furnace. The *Apollo I* crew was being burned to death. "We're burning up! Get us out of here!" Screams were cut off as the fire destroyed the communication system.

We sat in silence, listening to the echo of the tape playing in our consciousness. "*We're burning up!*" The motive of our teacher was clear. He was attempting to open our eyes to the reality of our new profession. It could kill us. It had killed in the past and held every potential to do so again. It was a lesson the civilian TFNGs in particular needed to be given. The military astronauts were well acquainted with the dangers of high-performance flight, but the post-docs and others were not. The instruments of their past careers, telescopes and microscopes, didn't kill people. I wondered if the other TFNGs would have an MCC tour guide who would play the tape for them. I hoped so. But, even if he did, it was too late. It should have been part of the astronaut interview process. Every interviewee should have had the opportunity to hear that tape so they could have made a fully informed decision as to whether or not they wanted to assume the risks of the business. No TFNG was going to quit now. How would they explain it . . . *I'm afraid*? We would all just have to pray that it wouldn't someday be our voices crying in terror as a space shuttle killed us.

CHAPTER 11

The F* * *ing New Guys

In spite of the sobering wake-up call delivered by the *Apollo I* tape, the first year of our TFNG indoctrination was one of euphoria. We didn't walk. We floated along the hallways in a weightless glory. You couldn't have beaten the smile from our faces with a stick. We slept with smiles. If we had been served shit sandwiches we would have gobbled them down through smiles. To the tourists who strolled the byways of Johnson Space Center we must have looked like village idiots. If any of us had been struck dead during those months, the mortician would never have been able to remove the smile from our face. It would have been part of our rigor mortis.

At summer's end the class hosted a party for the entire astronaut corps. The centerpiece of the entertainment was a skit that poked fun of the astronaut selection process, specifically the selection of the female and minority astronauts. The program starred Judy Resnik, Ron McNair, and some forgotten white guy. A bedsheet was hung from the ceiling in front of a chair. Judy was seated with just her face protruding through a hole cut in the sheet. Behind the sheet Ron stood at her right and extended his arm through another hole. The effect was that Ron's black arm appeared to be Judy's. Through a left-side hole, the white TFNG extended his excessively hairy arm as if it were also Judy's. Clothing was pinned to the sheet to give the appearance the mutation was dressed. And what a mutation—a woman with one black and one white arm, an affirmative action wet dream. The skit continued as an "astronaut selection board"—fellow TFNGs, of course—interviewed this androgynous creature. All this time, the arm and hand movements, comically uncoordinated, brought howls of laughter. The final question posed was "What makes you qualified to be an astronaut?" With ebony-and-ivory arms waving, Judy replied, "I have some rather *unique* qualifications." At that, the laughter hit max-q.

The skit obviously predated political correctness. For astronauts to perform such satire in today's America would have Jesse Jackson sprinting to the NASA administrator's office with a gaggle of lawyers in tow.

In fall 1978 we experienced our Astrodome welcome. Houston's professional soccer team, the Hurricanes, invited us and our spouses to be their guests for a game in the famed Houston landmark. We would be introduced to the crowd during a halftime ceremony. As Donna and I drove to the event, I couldn't help but imagine it would be like something out of *The Right Stuff*. When the seven Mercury astronauts had arrived in town they were welcomed with a Houston Coliseum BBQ. Thousands of cheering Texans filled the seats to catch a glimpse of their heroes. Battalions of Texas Rangers prevented them from being mobbed by the worshipers.

My first hint that TFNGs wouldn't have quite as many worshipers came as I pulled into the Dome's expansive parking lot. It was as empty as the Mojave. Had they canceled the game? Only after circling the lot did I finally see a clutch of cars, at least enough to have brought two soccer teams.

Donna and I rendezvoused with the other astronauts and spouses in our skybox. Skybox was an appropriate designation. We were in the stratosphere, perhaps even in the mesophere. Watching the game was like watching an ant farm from a block away. Most of us gravitated to the buffet at the back of the box and watched the match on TV.

Halftime arrived and we were escorted onto the field, where we formed a single line facing the crowd . . . if it could be called that. There were *tens* of Houstonians to greet us, most of whom were engaged with the beer or hotdog man. Obviously some things had changed since the days of the Mercury Seven.

A ridiculously enthusiastic commentator boomed our individual introductions.

"Please welcome astronaut James Buchli from Fargo, North Dakota!"

"Please welcome astronaut Michael Coats from Riverside, California!"

"Please welcome astronaut Dick Covey from Ft. Walton Beach, Florida!"

On it went. With each introduction I could barely hear a handful of claps over loud cries of "Beer here!" The applause reminded me of the clapping heard during the credit roll for the television show *Laugh-In.*

As each of us was introduced, we would step forward, wave to the empty seats, and receive a Houston Hurricanes T-shirt from one of the silicone-enhanced cheerleaders. At least she was clapping. Regardless of the vacant seats I still felt nervous to hear my name booming from those speakers. I couldn't wait for the voice of God to pass over me and go to the next in line. I noticed the other TFNGs appeared equally self-conscious and anxious to receive their shirts and melt back into the anonymity of the group, with one exception—Big Jon McBride from West by-God Virginia. Jon was a heavyset navy fighter pilot with sandy hair and a ruddy complexion. As his name was announced, he stepped forward just as the rest of us had. But here, all conformity ceased. Instead of a nervous wave and a quick step backward, Jon seized the startled cheerleader, swept her backward off her feet, and planted a kiss on her. Then he pulled on the Hurricanes T-shirt and waved a greeting to the crowd. Now there was applause. Even the Beer Man was cheering. Jon was a man for the masses. The rest of us exchanged wondering looks. Clearly Big Jon was cut from a different mold.

It came as no surprise to any TFNG when, in his retirement, Jon ran for governor of West Virginia. Unfortunately he lost in the Republican primary. If only his campaign had shown the video of him accepting the T-shirt from that cheerleader, he would have carried every county. After such a display of leadership, every good ol' boy in West Virginia would have voted for him.

For our one-year anniversary some of the class organized a celebration over July Fourth weekend. About a dozen of us went together to rent some stone cabins near Canyon Lake in the Texas hill country. We brought our wives and children and barbecue grills for fun in the sun. My daughters immediately fell in love with John "J. O." Creighton, a bachelor navy fighter pilot with a midnight blue Corvette and an awesome ski boat named *Sin Ship*. After rides in both, the kids ran to me shouting, "Dad, why can't you be like J.O.?" Apparently they were unimpressed by my choice of family car, an un-air-conditioned 1972 VW station wagon, powder blue in color except where the rust had rotted out a door panel. I silently prayed for the day J.O. would have six kids and be driving a Dodge.

After a day of swimming and waterskiing we adjourned to the cabin compound and fired up the grills and campfires. One of the physician TFNGs used a hypodermic syringe to inject vodka into an "adults only" watermelon. This fruit cocktail and an array of alcoholic drinks soon reduced mothering to an occasional, halfhearted warning to their broods: "Somebody is going to get hurt." A few of the kids were in a tree trying to remove Fisher's aluminum canoe that had previously been installed there by a group of intoxicated TFNGs.

Inside one of the kitchens the wives drank wine and chopped vegetables for a communal salad while outside the men flipped burgers and drank beer. We were just about to declare victory with the burgers when a loud pounding on the kitchen window caught our attention. We turned to see three pairs of naked breasts pressed against the glass. Three of the wives had pulled up their swimsuit tops and served us an hors d'oeuvre of six nipples under glass. We shouted and whistled our approval and lofted our bottles in a toast of their daring. The women dropped their tops into place and went back to the salad preparations. The TFNG wives were thoroughly enjoying their new roles as astronaut spouses. Ultimately they would pay for the title in crushing terror. But for now that was too distant to spoil the fun.

After dinner a load of illegal fireworks materialized from somebody's trunk. My kids suspected J.O. since he was so cool. Whatever the source, the astronauts were all over them like the eighth graders the alcohol had rendered us. Even "flaming hookers" didn't hold the promise of entertainment like drunken astronauts playing with fireworks. Soon the night was alight with sparklers, fountains, and assorted illegal devices normally seen only in combat firefights. Aerial bombs exploded over the campsite. Rockets swished into the black. If it hadn't been for the dampness left by an earlier thunderstorm, we would have burned down the surrounding forest. A couple of the wives were sober enough to shout at us, "For guys who depend on their eyes and hands for a living, you're sure taking chances," but we laughed away the warnings.

It was great fun until a particularly wicked aerial mortar fell off its stand. Balls of fire spewed into the crowd. There were shrieks of panic as mothers swept up children and hustled them behind the cabin walls. I flattened myself behind Fisher's canoe (finally extracted from the tree) as one ball whistled by my head. I was quickly joined by my son, Pat. With fear swimming in his eyes, he exclaimed, "Dad, don't you think this is kind of dangerous?" Even a ten-year-old could sense the idiocy of our play. We had become the kids. We were bulletproof. We were immortal. We were astronauts.

After the last bomb had exploded and the kids were asleep, the adults settled around a fire. We were growing close. Our competitiveness and the differences in personality (militant feminists to sexist pigs; propeller-headed scientists to Chuck Yeager clones) would ultimately strain relationships. It was impossible to throw thirty-five people together and not have some acrimony. But, like the fear the wives couldn't yet see, it was still too early for the enmity to get in the way of our fun.

As a sign of our closeness, we now had our class name: TFNGs. There was no official requirement that a new class of astronauts name themselves. It just happened. The *Mercury* 7 astronauts had become the "Original Seven." The class of 1984 would later become known as "Maggots," a play on the derogatory term that marine drill instructors used in reference to their new recruits. None of these names were ever formally put to a vote. Only through constant usage were they legitimized. For us, TFNG stuck. In polite company it translated to Thirty-Five New Guys. Not very creative, it would seem. However, it was actually a twist on an obscene military term. In every military unit a new person was a FNG, a "fucking new guy." You remained a FNG until someone newer showed up, then they became the FNG. While the public knew us as the Thirty-Five New Guys, we knew ourselves as The Fucking New Guys.

Deep in the heart of Texas, the fire crackled and glowing embers swirled skyward. More beers were popped. Brewster Shaw strummed his guitar to an Eagles tune as our talk turned, as it always did, to when we might fly in space. Like teenagers wishing for Saturday night to arrive, we wished for miracles to speed us to our launches. Our dreams were of the incredible things we would do. We would fly missions into polar orbits and fly jet packs on tetherless spacewalks. We would carry every science satellite, every military satellite, every communication satellite. We would use a robot arm to grapple satellites and repair them in orbit. We were going to do it all . . . The Fucking New Guys.

With the dream talk circling the fire I looked into the star-spangled night and felt supremely happy . . . but only for a moment. I was too seasoned not to know there would be tears on this journey. Some at this very campfire would die as astronauts. Perhaps I would, I thought. Perhaps in one of NASA's training jets. Perhaps on a space shuttle. It wasn't hearing the *Apollo I* voice tapes those many months ago that now brought on this melancholy. It was a much more intimate experience with death in the sky.

Christmas season, 1972. I was twenty-seven years old, stationed in England and flying in the backseat of RF-4Cs as part of the Allied Forces staring down the Russian threat. Jim Humphrey and Tom Carr were in the squadron planning area. We were kibitzing over coffee as they put the finishing touches on their training maps. I handed Jim my BX cigarette ration cards. I didn't smoke and he did. He thanked me. Then he and Tom headed for their plane. It was the last time I would see them alive. Shortly after takeoff their Phantom inexplicably nosedived into the earth at 400 miles per hour. There had been no distress call. The squadron commander came into the ready room and told us of the crash. "Stay off the phones," he ordered, then departed to pick up the chaplain and drive to inform the wives.

I worried for Donna. In a couple minutes she was going to see a staff car drive up to the apartment with the squadron commander and chaplain. Every apartment in the complex housed a flyer's family. Donna and I shared a wall with the Humphreys. Our entry sidewalks were fifteen feet apart. I could just imagine the two uniformed officers hesitating between those concrete ribbons, checking the address before choosing one. I could see Donna and Eurlene Humphrey watching in horror from their windows, wondering which one of them was the new widow.

Screw the commander's order, I thought. I grabbed a phone and called Donna. "There's been a crash. Jim Humphrey and Tom Carr are dead. The chaplain will be there soon. I wanted you to know it wasn't me. Go visit Eurlene as soon as they leave. Don't call anybody else." She was sobbing as I hung up the phone. Eurlene had two small children.

When the squadron commander returned, he appointed me as the casualty assistance officer for Jim's family and told me to visit the crash site. It was a muddy morass reeking of kerosene jet fuel. The vertical impact of the plane had left a crater about twenty feet deep. Shards of camouflage-painted aluminum littered the area. About thirty feet from the edge of the plane's crater was a smaller crater made by Tom Carr's body and ejection seat. Tom had ejected, but it had been far too late. His body, still strapped to the seat, had impacted the earth at the speed of the plane.

The flight surgeon was directing a group of hospital orderlies in the recovery of remains. Of Jim's there were none aboveground. The F-4 ejection sequence put the pilot out last. The twin craters made it obvious Tom, the backseater, had just cleared the rails when the plane hit, meaning Jim had to have still been in the front cockpit. Fifty thousand pounds of plane had been behind him at impact and had compressed his body deep into the Earth. Of Tom Carr there was nothing recognizable as human. Each orderly had a plastic bag and was picking up shards of his flesh—bright pink and red strings of it.

The saddest thing I yet had to do in my young life was to give Eurlene her husband's wedding band. I found it in his locker. Like most of us, he removed the ring prior to a flight to prevent it from snagging on a piece of aircraft equipment and causing injury.

Weeks later the squadron commander ordered me to have a brass plaque etched with an appropriate inscription memorializing Jim and Tom. The twin craters in the rolling hills of East Anglia, England, were truly their graves and the commander wanted the plaque on a nearby tree. I went to the site and screwed a brass plate into a majestic two-hundred-year-old oak.

Five months later Eurlene returned to England to visit her air force friends. The squadron commander invited her to attend a Memorial Day remembrance at the crash location. The day prior to the service he asked me to drive to the site and polish the plaque. I gathered Donna and our three children. The day was a rare one for England, sunny and warm. I wanted them to enjoy it. There was nothing at the crash site to identify it as such. A crop of wheat now covered the area. The scene was a postcard painting of English springtime tranquillity.

I began work on the plaque while Donna and the kids played with a farmer's dog that had followed them into the field. Moments later Donna screamed, "Mike, the dog has a hand in its mouth!"

I was sure I had misheard. "What?"

As she struggled to pry open the dog's jaws she screamed more urgently, "Oh God! It has a hand!"

I rushed to her, certain she was imagining things. She wasn't. Donna held a decomposed human hand. The presence of fingernails left no doubt about that. I was sure it was Tom Carr's remains. When his body hit the earth, it had exploded into countless pieces. The hand had been thrown into some nearby hedges and not discovered until that very moment by the wandering dog.

We wrapped the remains in the cloth I had been using on the plaque and drove to the base to give it to the flight surgeon. On the drive I thought of the many times I had clasped that hand in life. Tom and I had been classmates together in navigator training in 1968. When Donna gave birth to twins, he and several others in the class had gone together to buy two strollers for us. He had been a close friend. Now Donna cradled a piece of him in her lap.

At the Texas campfire I pulled Donna closer and prayed God would watch over all of us.

CHAPTER 12

Speed

Our TFNG freshman year also included an introduction to the ultimate astronaut perk—flying the NASA T-38 jets. Even fireworks, flaming hookers, and tits under glass couldn't put a smile on our faces like the ones we wore while flying the '38, a two-place, twin-engine, afterburning supersonic jet. Even its title, "'38," was the stuff of testosterone. It conjured up images of breasts and caliber. Originally designed in the late '50s by the USAF to serve as an advanced pilot training aircraft, NASA had acquired a squadron of them as proficiency trainers for the astronauts.

The years had been kind to the '38. Its sleek needle nose and exquisitely thin, card-table-size wings were timeless hallmarks of speed. In one hundred years people will look at the '38 in museums and still say, "What a sweet jet!" It was a crotch rocket that looked as if it had been especially built for astronauts. NASA's '38s were painted a brilliant white with a streak of blue, like a racing stripe, running the length of the fuselage. The agency's logo, a stylized script spelling "NASA," was emblazoned in red on the tail. It was an airplane and a paint scheme certain to turn heads on any airport tarmac.

The '38 served two functions: transportation to various meetings and proficiency training. NASA's simulators were great at preparing astronauts to fly the space shuttle, but they had one critical shortcoming. They lacked a fear factor. No matter how badly you screwed up, simulators couldn't kill you. But high-performance jet aircraft could. Flying the T-38s kept the pilots razor sharp in dealing with potentially deadly time-critical situations, something spaceflight had in abundance.

Before being cleared to fly the '38 we first had to complete water-survival training at Eglin AFB in the panhandle of Florida. We parasailed to several hundred feet altitude and then were released from the tow-boat to float into the water in a simulation of an aircraft ejection. Once in the water we had to release the parachute, climb into a one-man raft, signal a helicopter, then don a harness to be winched aboard in a simulated rescue. We were also towed behind fast-moving boats in a simulation of landing in a gale. We were dumped into the water and covered with the canopy of a parachute and taught how to escape before the nylon sank and became an anchor to pull us to our deaths. For the military flyers this training was a review. We had all been through it several times in our careers. But it was a first for the civilians. I carefully watched them, wondering if any would balk at some of the scarier training or need remedial instruction. In particular I watched the women for displays of fear or for a cry for help during the more physically demanding activities. But they performed as well as me or any of the other vets. I began to reassess my feeling of superiority over the post-docs and women. It would take several more years for the full transition of respect to occur, but this was my start.

The biggest surprise of our introduction to NASA's T-38 flight operations was the rules. There were none, or at least there were very few. In military flight operations, every phase, from engine start to engine shutdown, was usually part of a training program and closely monitored by superior officers. The ready rooms of military squadrons had credenzas filled with volumes of rules and regulations for operating the aircraft. NASA management, on the other hand, had the misguided belief that astronauts were professionals and didn't need big brother watching or a thick manual of rules to safely operate one of their jets. Sure . . . and a teenager being handed the keys to a 160-mile-per-hour Ferrari doesn't need any rules or supervision either.

We were the teenagers, and the skies over the nearby Gulf of Mexico were our back roads. After a radar-controlled exit from the Houston airspace, we would make that glorious call to Air Traffic Control (ATC), "Houston Center, NASA 904. Please cancel my IFR." Translation: "Houston, I'm off to play. Don't bother me. I'll call when I'm done."

On many occasions a cooperative TFNG pilot would say, "You've got it, Mike," and I would take control of the aircraft. (Since the '38 was designed as a trainer it had a full set of controls in the backseat.) When there were thunderstorms in the area I would send the plane twisting among their cauliflowered blossoms like a skier darting through the gates of a downhill slalom. Wisps of vapor would pass inches from the canopy, enhancing the sensation of speed. If there is orgasm outside of sex, this was it—speed and the unbound freedom of the sky.

We would flat-hat across the water, passing alongside container ships and super-tankers. That a seabird might come crashing through the windscreen like a cannon shot and kill us was a fear . . . but not much of one. We were intoxicated on velocity.

For thrills it didn't get much better than being in Fred Gregory's backseat. Fred, a USAF helicopter pilot, was one of the three African-American TFNGs. Apparently helicopter pilots believed they would get nosebleeds if they ever flew above a few feet altitude, or at least I got that impression from flying with Fred. We would depart Houston's Ellington Field and fly ATC control to the Amarillo City airfield in the panhandle of Texas. There we would refuel and then fly VFR (under our own control) at butt cheek–tightening low altitude to Kirtland AFB in Albuquerque, New Mexico. We would pass over the tops of windmills with just yards of clearance. The only thing that protected us from running into buzzards and hawks was that they had sense enough to cruise at higher altitudes. We streaked across the tips of 13,000-foot mountains and dove into canyons. The 600-foot-deep Rio Grande River Gorge in northern New Mexico was a favorite. I would look *up* to see the rim of that canyon. As power lines appeared, Fred would hop the jet across them and dive back on the other side. In what is truly a remarkable irony, many years later Fred was appointed NASA's associate administrator for safety. I guess we all eventually grow brains.

The most dangerous aerial play was "one-vee-one," or one-versus-one dogfighting. In a flight of two '38s we'd cruise a few miles over the water, then switch to company frequency, an unused frequency nobody would be monitoring. At least we *hoped* nobody would be monitoring it. Then each aircraft, flying in formation at the same speed and altitude, would simultaneously break 45 degrees in opposite directions. After flying for a minute on the new headings, we would turn into each other on a collision course. This maneuver ensured a neutral setup, one in which neither pilot had an advantage when the dogfighting started. There were obvious dangers in this arrangement. First, it put two virtually invisible objects on a head-on course at a combined speed in excess of 1,000 miles per hour. The other danger was more subtle. Pretending air-to-air combat with identical aircraft makes it difficult for either pilot to gain an advantage. Pilots are more tempted to push their vehicles to the edge of their performance envelopes to gain a simulated "kill." In my air force career there had been numerous incidents of dogfighting pilots crossing that edge, losing control, and having to eject—or dying when they didn't. It happened in my squadron in England. In fact, it happened so often worldwide the air force ultimately banned the practice of identical jets simulating a dogfight.

But in our Gulf of Mexico playground, the only rule was, "There are no rules," another witticism of Hoot Gibson. Pilots would make that final turn toward each other and slam the throttles into afterburner. It was a game of chicken and we strained to pick out the dot representing the competition. When the "tallyho" call was made, the game was on. Our jets would pass canopy to canopy, sometimes no more than a couple hundred feet apart, and the pilots would jerk their '38s into a vertical spiraling climb, keeping each other in sight and trying to maneuver for an advantage. Usually the first "vertical scissors" would end in a tie with both planes standing on their nozzles and the airspeed dropping lower and lower. When an out-of-control tail slide was imminent, the pilots would have no alternative but to pull the nose over. With the ocean steadily filling the windscreen, another scissoring dance to gain advantage would begin. Only after several of these up-down vertical helixes would one pilot finally gain a small advantage and a tail chase would begin. The pursued would twist in various escape maneuvers. The planes would shudder violently in high-speed turns that crushed us in our seats. Sweat would pour from our scalps and sting our eyes. Unintelligible grunts would fill the intercom as we strained to tighten our guts and prevent blood flow out of our brains. Unlike fighter pilots we did not wear anti-G suits, which added the danger of G-induced unconsciousness to the games. In a high-speed turn the G-forces could momentarily reach seven, which would pull the blood from our brains and bring on tunnel vision. Just a little harder pull and our vision would have gone to black . . . unconsciousness. Death at water impact would have followed. But we always managed to grunt our way through the yanking and banking to eventually hear the "*rat-a-tat-tat,* you're dead" call over the radio. A victor would be proclaimed and another game would begin.

How we survived this idiocy without an aircraft and/or crew loss, I have no idea. On several occasions the extreme maneuvering would lead to a flameout. A "break it off" call would be a certain indication the other crew was restarting a failed engine. It must have been the Almighty watching out after us. As I would later hear John Young say in reference to near disasters on early shuttle flights, "God watches out for babies, drunks, and astronauts." He certainly watched after dog-fighting TFNGs.

If only God would have watched out for me all the way to the chocks. On one occasion Brewster Shaw let me fly our jet to a landing. After touch-down, I made the mistake of lowering the nose too quickly. The tire impacted a barrier wire stretched across the approach end of the runway (used by tailhook-equipped aircraft in an emergency), which dented the wheel and caused the nose tire to go flat. In the vernacular of the military flyer, we had just "stepped on our dicks." One of the few rules in NASA's playbook was that backseaters didn't land the plane. Brewster attempted to cover our violation by telling the flight-line mechanics he had screwed up the landing. "I let the nose down too early." The maintenance chief seemed to accept this explanation and we thought we were home free . . . until the next Monday morning meeting. TFNG Dave Walker brought the flat tire into the conference room! He hefted it from behind the table and said, "Brewster, you want to explain this? The incident report says you forgot to hold the nose up for aero-braking." Dave had recently been appointed the TFNG safety rep for flight operations, so it was not surprising he had heard about the flat tire.

Brewster, a short, wiry, reticent air force pilot, shot Walker a look that read, "After this meeting is over, I'm going to personally shove that freakin' tire up your ass and then reinflate it!" We had been caught, given up by one of our own. There was no way anybody in that room was going to believe Brewster, a test pilot, had forgotten to hold up the jet's nose, any more than they would have believed Nolan Ryan had forgotten how to throw a fastball. John Young, in particular, was looking for an explanation.

An hour later we were in his office giving him one . . . the truth. "I let Mike do the landing, John." After the confession it was painfully obvious we were going to need new assholes. Young gave us a well-deserved reaming. "I'm constantly fighting headquarters to keep these '38s and stunts like this jeopardize it for all of us. MSes aren't pilots. Letting Mullane land was the dumbest thing I've heard in a long time."

During our grilling, I was struck by how uncomfortable Young appeared to be with command. As with his welcome speech, he couldn't make eye contact. He looked at his shoes. He looked at papers on his desk. He looked out the window. He looked everywhere but into our eyes. In every prior ass chewing I had ever received from my military commander, and there had been a few, their eyes had been the worst. They had drilled into my very soul and filled me with dread. I recalled my last commander, Colonel Jim Glenn, haranguing me and my pilot for having disobeyed a checklist procedure to jettison some hot ordnance before making an emergency landing. Colonel Glenn had stared at us with the intensity of a cobra. I was embarrassed for Young and his dancing eyes. His unease was palpable. But we were finally dismissed to return to our offices and worry about whether we'd ever fly in space.

The flat tire incident didn't hurt Brewster's career. In fact he flew before Dave Walker, on STS-9 as—get this—John Young's copilot! I can only assume that Young never really knew it was Brewster who had screwed up because he had never looked him in the face.

There was one aspect of the T-38 flying I wondered about. How would the civilians handle the issue of airsickness? Some of them were going to be affected, of that, I was sure. I had been in my early air force flying career. When I made my transition to the backseat of the F-4 Phantom, I vomited enough for a squadron of men. Would any of the civilians have a similar experience? Would any of them give up? There were whispers of some being as tormented as I had been. The office grapevine had it that Rhea Seddon was struggling and Hoot Gibson was taking her on flights to help her adjust. She didn't quit. None of the civilians did. I admired them for it. It was a shared experience and another lesson for me that the civilian TFNGs were not the wimps I had imagined they were.

CHAPTER 13

Training

The training we had all been anxiously anticipating—how to operate and fly the space shuttle—began in earnest in 1979. It was a training program that would last our entire careers. The shuttle cockpit has more than a thousand hardware and software switches, controls, instruments, and circuit breakers. Before our first ride, we would have to know the function of all of them.

The heart of NASA's training was simulation. Anything associated with spaceflight that could be simulated, was. There were part-task trainers in which individual systems were simulated: the hydraulic system, the electrical system, the environmental control system, the main engine system, the attitude control system, the orbital maneuvering system, and all the other systems that made up the space shuttle. We were scheduled in these trainers again and again until we had a working knowledge of each switch and computer display for that particular system. Then we would go through the emergency procedures for each system.

There were no tests in our training. We had a motivation far more compelling than any written test . . . ourselves. The military flyer's creed said it best, "Better dead than look bad." Nobody wanted to look bad in front of their peers. So we attacked the training as if something more important than our lives depended upon it, since something more important did . . . our egos.

We graduated from the part-task simulators to the Shuttle Mission Simulator (SMS). NASA had two of these machines, both featuring exact replicas of the space shuttle cockpit. The "fixed-base" SMS was, as the name implied, fixed; it didn't move. It was used for orbit simulations. The "motion-base" SMS was used for ascent and entry training. Perched atop six hydraulic legs that could tilt, pitch, and shake the cockpit to simulate launch and landing maneuvers, it looked like a giant mutant insect from a sci-fi movie. In both SMSes, computer-generated graphics appeared in the windows to provide representations of the cargo bay, robot arm, payloads, rendezvous targets, and runways. The simulators could be electronically linked to Mission Control for "integrated" simulations in which the astronaut crews would fly missions with the same MCC team that would watch over them during the actual mission.

The SMS training was orchestrated by a Simulator Supervisor (Sim Sup, pronounced "sim soup") and his/her team. Sitting at computer consoles in back rooms, these engineers could input malfunctions and watch the responses of the crew and MCC. Sim Sups were virtuosos from hell. Astronauts joked that simulation supervisors intentionally remained celibate for weeks prior to a simulation, wore shoes a size too small, and starched their underwear just to be frustrated and mean.

Within seconds after a simulated liftoff, a Sim Sup would introduce malfunctions that would have the crew scrambling to respond to a failed engine, an overheated hydraulic pump, a leaking reaction control system, and a shorted electrical system. Astronauts scheduled for "Ascent Skills" training jokingly referred to it as "Ascent Kills." It was an exaggeration. The simulation objective wasn't to kill the crew. Any mission that ended in a crash was considered poorly written or poorly executed. Instead, the Sim Sups and their teams designed missions that stressed the astronauts and MCC to their absolute limits. And their genius and dedication showed in the missions. No astronaut crew has ever been lost in flight because they were not adequately trained. No mission has ever failed to achieve its objectives because of a deficiency in training.

In my first SMS session I had a flashback to my arrival day at West Point. Then, an upperclassman had told me to relax and turn around to take in the magnificence of the campus. "Mr. Mullane, have a good look. Over there is Trophy Point and the beautiful Hudson River, and up there is the famous Protestant Chapel. Take it all in . . . because this will be the last time for eleven months that you'll see it. You have now died and gone to hell. GET YOUR NECK IN AND YOUR EYES STRAIGHT FORWARD, MISTER!"

My first SMS experience was similar. The Sim Sup let us enjoy a perfectly nominal ascent into orbit. We felt the simulated rumble of engine ignition (delivered through the hydraulic legs), saw the computer-generated image of the gantry speed past the window, felt the Gs rise (simulated by increasing the tilt of the cockpit), experienced the bang-flash of SRB separation, and enjoyed the ride all the way to Main Engine Cutoff (MECO). Then, as he was resetting his computers, Sim Sup said, "I hope you enjoyed it. It's the last one you'll ever see." He was almost right. In my twelve-year NASA career, I saw only three more malfunction-free SMS ascents. Each of those came just before departing to Kennedy Space Center for each of my three missions. In spite of rumors to the contrary, Sim Sups had hearts. They wanted us to leave for our missions on a high note, to see what we'd hopefully see during our real ascent a few days later . . . a completely nominal ride into space.

There were other simulators besides the SMSes. Astronauts trained for spacewalks in an enormous indoor swimming pool, the Weightless Environment Training Facility (WETF). The pool contained replicas of the shuttle airlock, cargo bay, and payloads. We dressed in 300-pound spacesuits and were craned into the water, where scuba divers ballasted us with lead until we floated at a fixed depth. This "neutral" buoyancy provided a fair replication of what occurred on real spacewalks, where a push on a tool would result in an equal and opposite reaction. The WETF facilitated the design of tools, handholds, and foot restraints for spacewalkers, all necessary to complete weightless work.

There was also a simulator for MSes to acquire skills with the Canadian-built Remote Manipulator System (RMS). The Manipulator Development Facility (MDF) contained a full-scale mock-up of the shuttle cargo bay (60 feet long and 15 feet in diameter) and a fully functional 50-foot-long robot arm. Huge helium-filled balloons served as weightless payloads. MSes would stand in a replica of the shuttle's aft cockpit, look through the aft windows, and operate the robot arm controls. We would lift the balloons from the cargo bay and/or stow them in the bay in simulations of orbit activities.

Robot arm operations were challenging. A camera at the end of the arm transmitted images to a screen in the cockpit. MSes would look at these images and simultaneously use two hand controls to bring the arm's business end to a successful grapple with the target. Using these hand controls while tracking a moving target on a display screen (how we would grapple a free-flying satellite) was like patting your head and rubbing your stomach at the same time. It required lots of practice. To assist us in developing tracking skills, the engineers provided a moving target that hung from the ceiling of the building.

In one of my MDF sessions, I employed my newly acquired tracking skills to tease Judy Resnik. I knew from the schedule that she was next up for the training and watched for her to enter the building. When she did, I maneuvered the end of the robot arm so as to track her with the camera at its tip. She glanced up and saw the huge boom dipping and swaying and twisting to her every turn and knew exactly what I was doing. She stopped, and I flew the arm outward as if it were reaching for her, then slowly tilted the wrist joint so the camera scanned her body from head to toe. When she entered the cockpit, she smiled and said, "You're a pig, Mullane." I smiled back and pretended not to understand, but of course she was right.

While I eagerly looked forward to SMS, WETF, and MDF simulations, there was one simulator I could have done without. . . . NASA's zero-G plane, nicknamed "the Vomit Comet." This was a modified Boeing 707 aircraft. Large sections of seats had been removed and the interior surfaces padded. After taking off from Ellington Field, the pilot would steer for the Gulf of Mexico, where he would fly the craft in a roller-coaster trajectory. While climbing toward the top of each "hill," he would push forward on the controls so the trajectory of the plane exactly matched the pull of gravity. The result was a thirty-second free fall in which everything in the plane was weightless. Unrestrained astronauts in the back would float in their padded chamber. At the end of the dive, the pilot would perform a 2-G pullout that would smash everybody to the padded floor. He would then advance the throttles, climb back to 33,000 feet, and start all over. On a typical mission the process would be repeated about fifty times.

It took only one flight in the jet to understand why it was named the Vomit Comet. The plane was a barf factory. Just climbing aboard, the nose would detect a faint odor of bile. Like cigarette smoke that cannot be removed from the drapes of a two-pack-a-day addict, the smell of stomach fluid had permeated the very aluminum structure of the machine. Even when its aged bones are someday sold for scrap and melted down, the recycled aluminum will still bear the aroma of our stomach acid.

I quickly learned that the videos NASA released to the public of Vomit Comet–borne astronauts laughing and tumbling were recorded on the first couple dives because by the tenth weightless parabola someone would have already retreated into his or her seat and be vomiting copiously. Like the chain reaction of a nuclear explosion, the odor of fresh barf would drift through the cabin and send a few more over the edge. Those new smells would combine to affect yet more people. Even those who tried to block the smell by breathing through their mouths could not shield their senses, for the guttural sounds of the damned would fill the volume like a pack of barking German shepherds. By the twentieth parabola there were few smiles remaining. By the thirtieth parabola, some would be wishing the flight controls would freeze and the plane would smash into the sea at 600 knots and put them out of their misery. But through it all there would be the lucky minority, the immune who would smile and whoop and tumble and ask for more. I hated them.

I never barfed on any of my Vomit Comet rides, but I had wanted to on all of them. From the fifth dive onward my gorge was continually at the back of my throat and only by a super-human effort was I able to keep it there. I sucked on Life Savers by the gross, hoping the constant swallowing reflex they generated would keep my stomach where it belonged. I knew I would have felt better had I periodically retreated to the rear of the plane and vomited, but that would have been a sign of weakness and a violation of rule number one: Better dead than look bad. Besides, there were female TFNGs unaffected by the maneuvers. The image of me strapped into the back with my head in a barf bag while Anna Fisher and Judy Resnik did loop-de-loops was too much for my testicles to take. So I faked it. When Judy suggested we do simultaneous somersaults I smiled through gritted teeth and nodded agreement, all the while cursing my balls for their bravado.

Of all of NASA's simulators, none was more memorable than the toilet trainer. It occupied a room next to the fixed-base SMS so astronauts could practice on it when they were in those training sessions. And practice was certainly needed.

The shuttle toilet was basically a vacuum cleaner. (Do not try this at home.) The urinal was a suction hose with attachable funnels to accommodate male and female users. Because of its strong suction (one marine proposed marriage), the toilet checklist contained a warning for males not to allow the most cherished part of their anatomy to get too deep into the funnel. If an inattentive astronaut's appendage got sucked into the hose, he would find himself qualified for a second career as a circus freak working under a banner heralding, "See the world's longest and skinniest penis!"

Urine was collected in a holding tank and dumped into space every few days. I would later find these urine dumps spectacular to watch. The fluid would freeze into thousands of ice crystals and shoot into space like tracer bullets.

The toilet solid waste collection feature also used airflow as a flush medium. A plastic toilet seat sat atop a "transport tube" approximately four inches in diameter and a couple inches in length. Users attached themselves to the seat with padded thigh clamps then pulled a lever to open the transport tube cover and turn on the steering air jets. The waste would be directed into a large bulbous container directly beneath the user. Astronaut solid waste is not dumped outside but is retained in the toilet, no doubt to the great relief of the rest of humanity. If solid waste is ever dumped into space, it will give new meaning to the phenomenon *meteor shower*.

One feature of the toilet made it particularly difficult to use . . . the narrow opening of the solid waste transport tube. This was an engineering necessity to achieve an effective downward airflow, but it made transport tube "aim" critical to waste collection success. A user not perfectly aligned in the center of the tube could find their feces stuck to the sides of the tube and smeared over their rear end. To help the astronauts find their a-holes, NASA installed a camera at the bottom of the toilet simulator transport tube. A light inside the trainer provided illumination to a part of the body that normally didn't get a lot of sunshine. A monitor was placed directly in front of the trainer with a helpful crosshair marker to designate the exact center of the transport tube. In our training we would clamp ourselves to this toilet and wiggle around until we were looking at a perfect bull's-eye. When that was achieved we would memorize the position of our thighs and buttocks in relation to the clamps and other seat landmarks. By duplicating the same position on a space mission we could be assured of a perfect "shack" (fighter pilot lingo for a perfect bomb drop). Needless to say, this training took a lot of the glamour out of being an astronaut.

The toilet design was essentially complete by the time TFNGs were undergoing waste management training, but an Edwards AFB Vomit Comet pilot told me of some of the early development efforts. These included female nurse volunteers who flew hundreds of weightless parabolas. They drank gallons of iced tea and during the thirty-second weightless falls would void into various toilet designs. Volunteers for the solid waste collection tests included a USAF lieutenant. The Vomit Comet would be parked near a taxiway with all the ground support equipment attached and ready to go, just like a Cold War nuclear bomber. And just like those bomber crews, the Vomit Comet pilots made sure they were ready for the scramble call . . . not from the president of the United States but rather from the bowel-distressed lieutenant screaming, "I've got to go!" At that, everybody would run to the plane, fire up the engines, and roar skyward. The weightless parabolas would begin and the test subject would have multiple thirty-second intervals to try a bowel movement. Where do we get such men?

Urine collection for spacewalking females proved to be a particularly challenging engineering problem. Catheterization was quickly eliminated—too dangerous and uncomfortable. Diapers were messy. The most bizarre design was the brainchild of a gynecologist. He proposed that a mold of the inside of a woman's vagina could be used as an alignment tool for urine collection. Before dressing in the spacesuit, the woman would insert her personal mold into her body, which would bring the exterior-mounted urine collector into a seal around the urethra. Urine could then be cleanly collected as it left the body. A test subject was needed to try the design and a call went out for volunteers. Kandy answered.

Kandy was a free-spirited Ellington Field flight operations secretary with a wonderful sense of humor. She easily tolerated the AD astronauts, as when she pulled up a chair to join a group of us waiting for the fog to lift so we could fly our '38s. Several of the navy astronauts were telling "beat this" stories about bizarre tattoos they had seen. One pilot recalled a photograph of a man's crotch in the window of a Filipino tattoo parlor. Tattooed on the thighs of both his legs were huge elephant ears that gave the man's penis the appearance of the animal's trunk. (Who says we men aren't in touch with our inner feelings?) Kandy joined in our laughter.

It was later in my TFNG life when, at a party, she recounted being the volunteer for the vaginal-insert urine collection design. The gynecologist had made the mold and she had tried it, but with limited success. Eventually the design was rejected and diapers were adopted as the best solution. Kandy finished her story: "I've got the mold sitting on my coffee table at home." Upon hearing that, I choked, shooting beer out my nose in the process. I had an instantaneous vision of a guest at Kandy's home picking up the object and asking, "What's this unique knickknack?" I told Kandy NASA should have given her a medal, or at least mounted the device on a plaque signed by the NASA administrator with an inscription, *For service above and beyond the call of duty.*

NASA is filled with thousands of men and women who have labored in anonymity to put astronauts in space and make our lives somewhat comfortable once we get there. As I once heard an astronaut say, "We stand on their shoulders to get into orbit." In the case of Kandy and those other toilet testers, we stood on other parts of their bodies.

In our space-wardrobe fitting sessions, we encountered one other waste collection detail, which included a man's worst nightmare. These sessions were conducted by white-smocked young ladies armed with tape measures, calipers, and clipboards. They measured our skulls, hands, limbs, and feet for helmets, gloves, and spacesuits. During my session I was as witty and charming as Burt Reynolds. I was a brand-new astronaut being fitted for a spacesuit. A bottle of tequila couldn't have gotten me higher.

At the end of the session a particularly sweet little custard walked me to a corner of the room that was screened from the rest of the facility. "Step inside and tell me what size fits you."

I pulled back the curtain and boldly walked forward, expecting to find a fitting room for underwear. But I was wrong. I had stepped into male hell. Forget about blowing up on a space shuttle. This was *real* fear. On a table, laid out like indictments, were four different-size condoms.

I would learn an open-ended condom was part of the male urine collection system worn under the pressure-suit cooling garment. One end of the latex slipped over the penis, the other end was connected to a waist-worn nylon bladder. Urine could pass through the condom, through a one-way valve, and into the nylon bladder. After a launch, landing, or spacewalk (the three times when the toilet was inaccessible) the bladder/condom combination, known as a Urine Collection Device (UCD), could be stripped from the body and thrown away. In a really cruel joke, God created different-size penises, so NASA provided different-size condoms. The cute little filly on the other side of the curtain needed my stud size on her clipboard so the correct condom could be loaded in my personal locker when I finally flew in space.

With all the enthusiasm of a prisoner walking to the gallows I dropped my pants. Until this moment in my life I had worn a condom only during brief periods in my marriage when my wife had stopped her birth control pills. On those occasions there had been a sense of urgency and enthusiasm about donning the one-size-fits-all latex scabbard. Not now. I looked down at an appendage that was in the process of renouncing circumcision and finding some heretofore unknown foreskin to hide behind.

I reached for the largest condom. Astronauts are the most competitive people in the world. From supplying an autograph to fitting a rubber, we're out to be the best, the fastest, the smartest . . . the *biggest.* If there had been a hula hoop on that table, male astronauts would have seized it with hope in their souls.

I grabbed my cowering little friend and began work. "Don't you have anything bigger?" I nervously joked to the cutie on the other side of the curtain. I'm sure she had never heard that one before.

Why didn't they have a man collecting this information? I wondered. Then, I thought, *That would be even worse.*

Putting a flaccid penis in a condom is like shoving toothpaste back in the tube. I finally managed to corral the beast and did a few jiggles to see if the rubber would stay on. It fell to the floor. My testicles might as well have joined it. I had been emasculated. Clearly, I wasn't going to place first in this competition. Of course I could have lied and said I needed the *annihilator* size, but to do so would have been to invite disaster during a spacewalk. If the condom didn't fit, it would leak or even come off altogether, in which case the cooling garment would become a urine sponge. As uncomfortable as that sounds, it would be the least of the victim's problems. An astronaut would never outlive the teasing.

I finally made a fit and gave the technician my size, wanting to add, "I'll have you know I've fathered three children with this!"

Many years later astronauts were outraged when a pilot's medical records were compromised to the press. Some in the media were questioning his suitability to command an important shuttle mission since he had been treated for kidney stones. Astronauts were livid that the flight surgeon's office had somehow leaked this private medical information. As the brouhaha raged, I told a fellow TFNG, "I don't care if they publish my medical records in the *New York Times*. I just hope the record of my condom size is locked up in a vault in Cheyenne Mountain." He understood. There are worse things to read about in the paper than the fact that you have passed a kidney stone.

CHAPTER 14

Adventures in Public Speaking

With the astronaut title came two duties few of us had ever performed in our past careers: giving public speeches and press interviews. While NASA didn't force astronauts onto the speaking circuit, they did expect everybody to voluntarily take about a dozen trips a year to represent the agency at the head tables of America. The astronaut office received hundreds of requests a month for speakers, so there were plenty of events to pick from.

Like the majority of people, most astronauts fear public speaking more than death. As the joke goes, "Most people would rather be in the casket than delivering the eulogy." I witnessed this hierarchy of terrors one dark and stormy night in the backseat of a T-38. My pilot was Blaine Hammond (class of 1984). After finishing a day of practice shuttle approaches at the White Sands shuttle runway, we were making a night takeoff from El Paso's airport for our return to Houston. Our eastward departure was sending us into an ink black sky over a similarly darkened desert. Just as Blaine pulled the nose from the runway, I noticed a yellow flickering in the cockpit rearview mirrors and was about to comment on it when the El Paso tower interrupted. "Departing NASA jet, you're on fire. There's a flame trailing from your aircraft." We were already airborne and well beyond our maximum abort speed. We had no choice but to continue our climb. I quickly informed Blaine of the flickering yellow in the mirrors. Clearly, our jet was burning behind us. Blaine yanked the engines out of afterburner (AB) and declared an emergency. El Paso tower immediately cleared us to land on any runway we could make. My thoughts were on ejection. The checklist was clear: In bold lettering it read, "Confirmed Fire—Eject." You don't get better confirmation than having the tower tell you you're riding a meteor. I cinched my harness to the point of pain and placed my hands on the ejection handles and mentally reviewed the bailout procedures. As I was doing so, I continued to watch the engine instruments. The nozzle position on the left engine was the only off-nominal indication. At the power setting of the throttle the nozzle should have been more closed than what was indicated. There were no firelights and the fire-warning circuitry checked okay. I snatched my mask from my face and breathed the ambient air. There was no odor of smoke. The tower was telling us we were on fire, but there was no indication of it in the cockpit.

"Something's wrong with the left engine. I'm going to keep it at idle and make a single-engine approach." Blaine stated his intention and immediately banked the plane toward the nearest runway.

I challenged the decision. "That's not what the checklist says we should be doing." I didn't have to say the word *eject.* Blaine knew the emergency procedures as well as I did.

"I know, but she's flying fine." I could tell in his voice Blaine was as frightened as I was about our predicament. The plane *was* flying fine and neither of us wanted to leave the security of his cockpit for the black outside. The thought of pulling those handles was absolutely terrifying. But, by staying with the plane, we were in clear violation of the emergency procedures.

I heard the tower wave off an airliner to give us every option for landing. We had the field to ourselves. I wondered if we would soon be putting on a fireworks display for a planeload of TWA passengers.

As the runway lights came into view, I followed Blaine through the landing checklist, including making a computation of our touchdown speed . . . nearly 180 knots. We were full of gas, making a high-speed, single-engine landing on a high-elevation runway. To complicate things a thunderstorm had just passed over the field. The runways were sodden. *This could get ugly* was my thought. Assuming we made the runway, there was an excellent chance we were going to blow some tires. To run off the runway would probably result in death. That assumed we made the runway at all, which was far from guaranteed. The fire had already damaged the engine nozzle positioning system, which put it in the vicinity of the tail-control surfaces. If those failed while we were deep into our landing attempt, we would probably die. The ejection seat wouldn't be able to save us in an out-of-control situation close to the ground.

The annals of military aviation are filled with stories of aircrews that died doing exactly what we were about to do . . . ignore the "Fire—Eject" rule, play the heroes, and attempt an emergency landing. "Crewmember death occurred when ejection was attempted out of the ejection seat envelope." I had read that conclusion in accident reports a hundred times in my career. I could imagine the comments of our peers at the next Monday meeting: "If they had followed the checklist, they would be alive today."

Eject? Stay? Eject? Stay? The runway lights were looming and I was in the agony of indecision. Finally I decided that I would stay. I put the checklist aside and resumed my death grip on the ejection handles. If the master caution light illuminated or there was any other indication of ongoing fire damage, I was gone. It might be too late by then, but that was my decision.

Throughout this period, which had been less than two minutes, I could hear Blaine's breathing through the intercom. He had the respiration of a marathoner. He was stressed to the max.

We touched down and within seconds the right-side tire blew and the aircraft started a drift to the right. In correcting our trajectory Blaine blew the left tire. We were riding on shredded tires but at least we were skidding straight down the runway. Fire engines followed us.

It was soon a story for the ready room. We came to a safe stop. The firemen used handheld extinguishers to spray the smoking wheels. Blaine and I climbed from the cockpit and immediately walked to the back of the plane. Sure enough, we had been on fire. There was a hole burned in the bottom of the fuselage near the left engine nozzle. Later we learned a piece of the afterburner plumbing had failed and had served as a blowtorch when the throttle had been in AB. The problem was far enough aft to be out of the range of our fire sensors, which explained the lack of any firelights. When Blaine had pulled the left throttle from afterburner, the fuel source for the fire had been isolated. It was the remaining fuel in the engine compartment that had been burning as we made our landing. We had ignored the checklist and lived to tell the tale.

As we were being driven back to the operations office, I was thinking of what a great job Blaine had done. It wasn't the stuff of legends, but it was a fine testament to his piloting abilities. He had handled a serious threat with confidence and poise. But then, that was to be expected. He was an astronaut and test pilot and he had merely been dueling with death. A far greater menace was about to leap from the shadows.

The radio scanner at a local TV station had picked up the words "NASA jet on fire" and a reporter had been dispatched to the scene. As we walked into operations, we were blindsided by the lights of a camera. A microphone was shoved in our faces. There was to be film at ten and we were to be the stars. Speaking into the lens of a camera was the most fearful form of public speaking. Fumbling for words in front of a Rotary Club didn't compare with having your deer-in-the-headlights, fear-twisted face and bumbling dialogue transmitted into the living rooms of tens of thousands.

I quickly extricated myself. "I was the backseater," I told the reporter. "Blaine was the pilot. He landed the plane. He's the one to talk to." The reporter fell on him like a hyena on a wildebeest carcass. I scuttled off camera.

With the reporter in his face Blaine became living proof that fear of public speaking far exceeds fear of death. In a span of twenty minutes he had faced both and it was the blazing camera spotlight that was killing him. His eyes dilated in fear. His nostrils flared open and closed like a bellows. Everything in his body language screamed, "Eject! Get me out of here!"

Blaine wasn't unique. Most of us were equally terrified of TV cameras and public audiences. And NASA was no help. There was nothing in our TFNG training to prepare us for the great unknowns of the press and the public spotlight, an astounding oversight given the fact that astronauts were the most visible ambassadors of NASA. Apparently the agency thought our talents with machines extended to the lectern. They did not.

One of the most egregious examples of an astronaut abusing the microphone occurred when a pilot, who was renowned for a sense of humor even Howard Stern would have found offensive, attempted to hide his nervousness by opening his speech with a joke. With a hushed and expectant crowd of hundreds awaiting pearls of inspiration from one of American's finest sons, he threw out the following:

A golfer walks into the clubhouse with a severe injury to his neck. He can barely talk. His buddies rush to him: "Bill, what happened?" Bill goes on to explain. "I teed off on number eight and sliced my shot into the rough. As I was looking for it, I noticed this woman searching for her ball in the same area. When I couldn't find mine, I walked up to a cow grazing nearby thinking the ball might have ended up between its legs. But again, it wasn't there. Finally out of frustration, I lifted up the cow's tail to see if maybe it had hit there. Sure enough, a golf ball was stuck in its rear end. I looked closely and noticed it was a Titleist. Since I was hitting a Top-Flite, I knew the ball wasn't mine. So, with the tail of the cow upraised in one hand and my other hand pointing at the animal's ass, I shouted at this woman, "Hey, lady, does this look like yours?" That's when she hit me across the throat with a seven-iron."

The joke might have been appropriate for a group of golfers or military pilots or any similar crowd of crotch-scratching, crude, and coarse males. Unfortunately, that wasn't the audience. The astronaut in question delivered this joke to open a high school commencement address! Only if it had been delivered at a NOW convention could it have generated more outrage. One can only imagine the horror on the faces of parents and faculty, the snickers of the students, and the subsequent crucifixion of the person who had suggested, "Let's get one of America's finest to speak at graduation. Let's get an astronaut. It'll be a commencement address to remember." Indeed, it was.

NASA got what it was looking for in this astronaut's presentation, a lot of visibility with the grassroots taxpayer. Unfortunately that visibility was, well, a little negative. Cards and letters rolled into NASA. The general message was something along the lines of, "Where did you get this bozo?" The answer was simple. NASA had plucked him from Planet AD.

Most of the military astronauts had no idea what constituted an appropriate sense of humor in a public setting. I once attended a dinner with a marine fighter pilot (not an astronaut) who rose from his seat with glass in hand and offered this toast to the ladies and gentlemen present: "Here's to gunpowder and here's to pussy. One I kill with, the other I'll die for, but I love the smell of both." You would think even the most AD-affected of the military TFNGs would probably have concluded such a toast would be inappropriate at a Shriners' dinner, but I wouldn't have put any money on it.

Lacking any other real-life experience, military males just assumed everybody had our perverted sense of humor. I certainly did. At one of my very early public appearances, I showed a slide of the six TFNG females intending to make a statement about the diversity of the new NASA class. But instead my alcohol-lubricated words came out as "pigs in space," a reference to a popular Jim Hensen Muppets' skit of the same title. Actually, I didn't say, "Pigs in space." Rather, I mimicked the Muppet announcer's overly enthusiastic call: "Piiiiiiiigs innnnnnnnnn spaaaaaaaaaaaaaaace!" The only reason NASA didn't get protests from my performance was that my audience was a U.S. Army "Dining Out," a black-tie gathering of army officers and their spouses. Most of them had similar disturbed senses of humor. The audience loved my wit.

At another military formal dinner, Rhea Seddon and I were cospeakers. In my comments I used the word *girls* in reference to the female astronauts. I had done so without malice. It was just as natural as breathing for me to refer to the women as *girls* or *gals*. Afterward, a wife from the audience approached me with a smile that would have chilled Hannibal Lecter. She asked, "Do they call you a *boy* astronaut?" I was baffled by the comment . . . but not for long. She enlightened me while tearing me a new fundamental orifice. "How dare you refer to Dr.

Seddon as a girl! Where is your PhD? Are you a surgeon? She has better credentials than you." She stormed off. It was one of my earliest lessons in political correctness.

Besides contracting with Miss Manners, Toastmasters, and NOW for remedial training, NASA should have also reviewed with its astronauts the various songs they might be asked to sing during a public appearance. Many of the requests for astronaut speakers came from organizations planning patriotic-themed events. Nothing was bound to excite more pride in the American soul than a trim, square-jawed, shorthaired, steely-eyed war-veteran astronaut poised next to Old Glory leading the audience in the singing of a patriotic song. Every Rotary Club, VFA, and Elks Club in America wanted that Norman Rockwell scene on their stage. But that assumed the astronaut knew the song in question.

At one of my appearances I was blindsided by a request to lead the audience in the singing of "America the Beautiful." I was prepared for my speech. I had it on my notecards. What I didn't have on my cards was "America the Beautiful." As the master of ceremonies beckoned me to the podium I could feel my bowels liquefying. I held on to his handshake just to keep from collapsing. My brain was logjammed with every patriotic lyric I had ever heard: *for-purple-mountains-majesty-our-flag-was-still-there-the-caissons-go-rolling-along*. Retrieving "America the Beautiful" from that mess was going to take a miracle.

The MC handed me the microphone. I wished it had been a gun so I could have blown out my scrambled brains. They were all looking at me, hands on hearts. Hundreds of them. Only a lone cough disturbed the silence. *It doesn't get any worse than this,* I thought. But I was wrong. As a courtesy to a group of hearing impaired who were sitting in the front row, there was a signer at the edge of the stage staring right at my lips. Her hands were poised to record my every utterance. How I didn't wet myself (or worse), I'll never know.

I placed my hand on my heart and turned to face the flag. I could feel my pulse through my suit pocket. The MC punched "play" on a boom box and the first strains of the melody flowed into the room. I sang the only words I was absolutely certain of, "Oh beautiful . . ."

Those words proved enough. Everybody joined in and my voice was lost. Actually, I lowered the microphone from my mouth so my incoherent babbling couldn't be heard. I had pulled it off. Or so I thought. Then, the signer caught my eye. She was focused on my mumbling lips with the precision of a laser. Not a syllable was getting by her. If I could have read sign language, I knew what those flying fingers would have been saying. "Hey, everybody! This guy is a fraud. He doesn't know 'America the Beautiful.'"

I wasn't the only astronaut to be surprised on the way to a stage. Hoot Gibson once served as a last-minute replacement speaker for Judy Resnik at a women's event. The MC began the introduction by reading Judy's entire biography. Hoot was dumbstruck. Judy wasn't there. Everybody in the audience knew he was to be the substitute speaker, yet the MC droned on with Judy's bio as if she were going to step out of the wings to give the program. Only after it was completely rendered did Hoot realize the MC's purpose in reading it. It was to establish Judy's irreplaceable importance to NASA. The MC went on with Hoot's introduction in words that loosely translated, "Judy is so important to NASA there was no way she could be spared to come to speak at today's event. But NASA could easily do without this useless dirt bag of a man so they sent him. We'll just have to be disappointed and listen to his forgettable comments." Then, after Hoot's speech, the MC presented him with a plaque inscribed to Judy.

As my NASA career continued, I discovered new land mines to step on while in front of the public. In the Q&A that followed one of my speeches, a woman asked, "Have you seen any aliens?"

I answered, "No, but I believe there is alien life elsewhere in the universe. There are so many trillions of stars it's easy for me to believe there will be planets around some of those stars that harbor intelligent life." I should have quit right there, but like a fool, I continued. "However, I don't believe any UFOs have landed on earth. Why," I rhetorically asked the audience, "would an advanced civilization go to the trouble of building an interstellar craft, fly to earth to find it teeming with life, and then only hover over lonely women and beer-drinking men?" The crowd laughed. The woman asking the question did not. If looks could kill, I was a dead man.

The next week I received an anonymous letter postmarked Salt Lake City, Utah, viciously attacking my position on aliens. It was clear the writer believed *the truth is out there* and that I was part of the cover-up. I suspect the letter was from the woman who had asked the alien question.

This question was just one of many that could turn a public appearance into a gut-wrenching torture. "What happens when you fart in a spacesuit?" or "Do women have periods in space?" were the easy ones to answer. But questions like "Are there gay and lesbian astronauts?"

and "Has there been sex in space?" had the potential to put a TFNG's name in a Johnny Carson monologue.

The prizewinner in the category of fielding the most difficult question was Don Peterson (class of 1969). After one of his speeches, several members of the audience came to him with their questions. One asked, "Is there privacy on the shuttle to masturbate?" Don was immediately thrown into a panic. It was like being asked, "Do you feel better since you've stopped beating your wife?" It was impossible to answer. He considered saying no, but that implied astronauts had searched for such privacy. He imagined his face on a supermarket tabloid under the headline "Astronaut Complains: No Privacy to Spank the Monkey." A yes reply held equally embarrassing possibilities: "Astronaut Admits to Five-Knuckle Shuffle in Space." He mumbled an incomprehensible answer, praying whatever it was it wouldn't come back to haunt him in the *National Enquirer.*

As Blaine Hammond learned in the El Paso flight operations office, the most dreaded form of public speaking was a TV interview. A streak of antiaircraft fire passing your wing doesn't get your heart rate up like looking into a black camera lens and hearing, "Three . . . two . . . one . . . you're live." For me, it was a cadence that always brought on nausea. Once, as I was listening to this on-the-air countdown, the anchor leaned in to me and said, "It's just like a shuttle launch. When you hit zero, there's no going back." He was right. Hearing, "You're live," was just like hearing the rumble of SRB ignition. You were flying. The camera was scattering your image and words into the living rooms of America and there would be no do-overs. I was sure my Adam's apple was dancing like a bobblehead on a dashboard and my fear-widened eyes were darting like minnows. I imagined people at their breakfast tables laughing as I choked, trying to respond to a simple question like, "What's your name?"

Live interviews could be made even more torturous by the AD antics of other astronauts. Several of us were in a Houston bar one evening when the TV caught our eye. A local station was airing a call-in interview with Ed Gibson (class of 1965) and TFNG Kathy Sullivan. One of our group immediately asked the bartender to borrow the phone and called in his questions: for Kathy, "How do girls pee in the toilet?" and for Ed, "What does Mrs. Gibson think of Mr. Gibson flying single women around the country in a NASA jet on overnight business trips?" We all hooted and hollered as the victims struggled with their answers.

Interviews with the print press were much more relaxing but still held the potential to screw an astronaut. During one interview I explained to the reporter my feelings of boundless joy and visceral fear while being driven to the pad for my first launch. I said, "To see the xenon-lighted *Discovery* and know it was my shuttle, that I was only hours from the culmination of a lifetime dream come true, nearly had me crying with joy." But I was quoted as having said, "Astronauts cry from *fear* as they are driven to the launchpad." The story was picked up by Paul Harvey and repeated to a huge national audience on his radio show. I was outraged and excruciatingly embarrassed.

Experiences like this explained why the astronaut office bulletin board occasionally displayed news articles in which an offending quote was circled with "I didn't say this" written next to it by a pissed-off astronaut.

On August 31, 1979, Chris Kraft came to the astronaut office to tell us NASA was dropping the *candidate* suffix from our titles. Apparently we had impressed the agency enough for them to designate us *astronauts* nearly a year earlier than originally planned. We were no longer Ascans. I was happy to hear it. Even though I wouldn't consider myself an astronaut until I got into space, I was tired of having to explain the title on PR trips and watching the crestfallen faces of event planners as they realized I wasn't the *real* astronaut they had been expecting. At our next office party we were each given silver astronaut pins to go with our new title. These were lapel pins fashioned in the shape of the official astronaut symbol, a three-rayed shooting star passing through an ellipse. When we finally flew in space, we would be given gold pins. Actually, we would then be allowed to purchase, at a cost of $400, a gold astronaut pin. (The silver pins were paid for out of the office coffee fund.)

After returning from the party, I took my pin off, put it in a drawer, and never wore it again. To me it was a meaningless token, like the plastic pilot wings that stewardesses give to children. Those Delta Airline wings weren't going to make a child a pilot and a silver pin and title weren't going to make me an astronaut. Only a ride into space could do that.

CHAPTER 15

Columbia

Columbia was less than a year from launch, and, when it flew, it would mark NASA's first manned spaceflight in six years. That was a concern for the NASA safety office. A six-year hiatus in manned operations provided a fertile environment for complacency. In defense, the office sent astronauts to various factories and shuttle support facilities to refocus the workers. We wanted to put a face on manned spaceflight, to reacquaint people with the deadly consequences of making a mistake on the job. Teams of astronauts were dispatched around the country and around the globe to give speeches, shake hands, and pass out NASA safety posters. We astronauts referred to these appearances as "widows and orphans" visits. While we never said, "Don't fuck up or you could kill us and make widows of our wives," that was exactly the message we hoped to impart by just standing there in our blue flight suits.

Steve Hawley and I were tapped to travel to Madrid, Spain, and the Seychelles Islands to deliver that message to the NASA and air force contingents who manned the shuttle tracking sites at those locations. NASA did not yet have its own communication relay satellites in orbit, so we depended upon an earth-girdling network of ground sites to communicate with orbiting astronauts. Other TFNGs were sent to Australia, England, Guam, Ascension Island, and the other overseas sites that completed this global tracking system.

To go to the Seychelles is to die and go to heaven. The nation is a collection of islands a thousand miles east of Africa just south of the equator in the bath-warm waters of the Indian Ocean. The beaches are white, the surf is turquoise, and both are filled with topless vacationing Scandinavian women. As if that isn't enough of a temptation, many of the local island women are beautiful manhunters. Their preferred quarry are American men, as they represent a means of escape to the land of the Big BX (the USA). At a party hosted by the tracking site commander, Hawley and I learned just how aggressive they could be. A young and exceptionally beautiful woman came to us and requested our autographs. "Sure, we'd be happy to sign something for you," I replied. I was expecting her to hand over one of the space shuttle photos we had previously distributed but, instead, she pulled up her skirt, thrust a cheek of her ass in my face, and asked me to sign her panties. I searched my memory but couldn't remember "ass signing" being covered in our JSC training. I looked at Hawley and suggested, "To refuse could cause an international incident." We had been cautioned by the resident state department official not to alienate the locals, as the United States was in sensitive negotiations with the island's current Dictator for Life. Steve concurred: "It's our NASA *duty* to fulfill her request." That settled it. I turned my pen to the silky fabric, only to be struck by the limited real estate. Her petite posterior didn't give me a lot to work with. But astronauts love a challenge. In a font so tiny I could have penned the Declaration of Independence on a grain of rice, I leisurely inscribed on the side of her underwear, *Richard Michael Mullane, Major, United States Air Force, Astronaut, National Aeronautical and Space Administration.* I was thinking of adding, *In the Year of Our Lord One Thousand Nine Hundred and Eighty* and the date, but Hawley was getting impatient. If anybody's hand should be on that heinie, it was his. He was the bachelor of our duo, a fact that had spread across the island on the coconut telegraph as fast as a trade wind. This young woman had probably set her sights on him when we stepped off the plane. I finally capped my pen and she immediately presented her other cheek for Steve and he began his treatise. The things we do for our country. There ought to be a medal awarded to men who return from the Seychelles. Bachelors, like Hawley, should get the Order of *I Walked Away from Heaven* with accouterments of oak leaf clusters, laurel wreaths, dangles, bobbles, flames, and shooting stars.

As if the local women weren't enough, Hawley and I also discovered the vacationing Dereks . . . as in John and Bo Derek. Even among the hard-bodied, oil-smeared Danish pastries decorating the beach, Bo stood out. To say she was a "Ten" didn't do her justice. She made a Stepford Wife look like a hag. Unfortunately, she wasn't topless. Nor was she jogging down the beach in slow motion. But, like Dudley Moore's famous character, I had an active imagination.

Hawley and I debated whether or not to approach the star, a debate that lasted about as long as it takes a quark to decay. We were at her side in a flash, mumbling and stuttering like Dumb and Dumber. I think I blurted out, "I want to have your baby!"

We made sure to include the title "astronaut" in our introduction. John, at least, was impressed by that and asked us several questions about the upcoming launch of STS-1, including some technical questions about landing speeds and glide path angles. Bo didn't ask us anything. In fact, she didn't say much at all. Maybe it was the way Hawley leered at her. Surely it couldn't have been me. We learned the couple was taking a break before the filming of that celluloid classic *Tarzan, the Ape Man.* I said to Bo, "Me Tarzan. You Jane." John looked at my 145-pound frame and said, "I don't think so."

"How about Cheetah?" I certainly had the chimp ears to qualify. But, once again, I was rejected.

Hawley and I posed with Bo for some photos and said our good-byes. (Or maybe John said he was going to call the island police if we didn't leave. I can't recall.) I couldn't wait to get back and phone every male I had ever met in my entire life beginning with my high school classmates and scream, "Eat your hearts out! Guess who I met?"

Back in Houston, when Judy Resnik heard our story, she began to call me Tarzan. For the rest of her short life, she never again called me Mike. Always Tarzan.

By the second year of our TFNG careers the bloom had begun to fade on our management: George Abbey and John Young. George had chaired the dozen-man astronaut selection committee. If the office vets were to be believed, the "committee" title was a joke. George didn't operate by committee any more than Josef Stalin had. His was the only vote that counted in the TFNG selection process. George was a pear-shaped man with silver-tinted buzz-cut hair, a permanent five o'clock shadow, and sleepy, basset hound eyes. The word *enigmatic* was coined to describe a man like George. His heavy face revealed nothing. His rare smiles were hardly more than grimaces. I never saw him in a teeth-showing laugh. I never heard him raise his voice in anger. I never saw him animated in any way. When he spoke, which wasn't much, it was in low mumbles. He was as unreadable as a marble bust.

George's parents had obviously expected great things from their son, christening him George Washington Sherman Abbey at his birth in 1932. It was a handle that earned him the acronym GWSA from us TFNGs. George met the challenge of his name. He graduated from Annapolis in 1954, took a commission in the USAF, and accumulated more than four thousand hours of flying time as an air force pilot. He earned a master's degree in electrical engineering from the Air Force Institute of Technology. In 1967 he resigned from the air force and began his NASA career as an MCC engineer (he wasn't an astronaut). For his work on the *Apollo 13* Mission Operations Team, he was awarded the Medal of Freedom, the nation's highest civilian award.

Every TFNG walked into NASA a slavishly loyal subject of King George, and we competed in pathetic attempts to brown-nose him. The nineteen new astronauts of the class of 1980 did the same, so there was real crowd around George's backside. Two of the 1980 newbies made an exceptionally flamboyant attempt to put their names in front of George. On Abbey's birthday Guy Gardner and Jim Bagian called the JSC security police pretending to be employees of a window-cleaning service needing access to the ninth-floor windows of the JSC HQ building. After the police unlocked the windows and departed, Bagian dressed in a Superman costume, dropped a rope to the ground, and repelled to Abbey's eighth-floor office. There, he pounded on the glass to gain George's attention and sang "Happy Birthday." Mission complete, he continued to the ground. Gardner freed the rope of its anchor, closed the window, and disappeared.

It didn't take long for word of the prank to reach the security police and for its chief to be pounding on Center Director Chris Kraft's office door with an angry complaint about astronauts duping his people and conducting a dangerous stunt. In a classic demonstration of the old adage "Shit flows downhill," it didn't take long for the turds being shoveled onto Kraft's ninth-floor office desk by the chief of security police to find their way to Abbey's eighth-floor desk and thence to John Young's third-floor desk in Building 4. In effect, Dr. Kraft's message to John was "Johnson Space Center isn't a private playground for your astronauts." So Guy Gardner and Jim Bagian picked up some early, if not exactly *positive,* visibility with George.

But even as we TFNGs and the class of 1980 were doing our best to gain George's favor we were also developing serious doubts about our leader. While he frequented our social functions, he rarely made appearances in the astronaut office. In particular he offered no insight into the one thing that mattered most to us, the shuttle flight assignment process. Initially, we believed that John Young would be making shuttle crew assignments. Since he bore the title *chief of astronauts,* how could it be otherwise? But the older astronauts were certain Abbey would be assigning crews independent of Young. In our rookie naïveté we found that hard to believe. Young was in a much better position to know our capabilities, limitations, and interpersonal compatibilities. Abbey's office was in a separate building. How could he know what crew composition would be best for a particular mission? We could understand why Abbey wanted crew assignment authority, since it represented considerable power, but we could not understand why Young would have rolled over and allowed him to take it. While NASA's management hierarchy did put Young under Abbey's authority, it seemed to us Young could have easily insisted on having a big say in crew assignments without the slightest risk to his career. John was a living legend. He was a four-time veteran of spaceflights—two Gemini missions and two Apollo missions. He had walked on the moon. There was no way a midlevel bureaucrat like Abbey could have ever prevailed against him if Young had told Chris Kraft, "These are my astronauts. I know them. I want to have a hand in crew assignments. I'll consider HQ's inputs, your inputs, and Abbey's inputs, but I want a significant say in the matter because I will have to bear the ultimate responsibility if there are any mistakes made by crews." But the vets in the office were adamant in their opinion that Abbey was a rapacious power monger who had taken all flight assignment responsibility from Young. Why Young would have ever accepted such an office-neutering arrangement would remain a mystery throughout my astronaut life.

There were occasional hints that Abbey's rule over astronauts *was* absolute, as when Jerry Ross (class of 1980) returned from Chris Kraft's welcome for his class. Jerry told us he had been shocked when Kraft had implied he didn't understand why their class had even been selected. He thought there were enough astronauts as it was. (As Jerry said, it was a strange way of welcoming them.) Jerry's story implied Abbey had selected a new class over Kraft's objections. Did even Dr. Kraft answer to Abbey on the subject of astronauts? Nobody knew. Kraft, Abbey, and Young never said a word about their responsibilities. Everything about the most important aspect of our career—flight assignments—was as unknown to us as the dark matter of space was to astrophysicists. Who made assignments? Who approved them? Who had veto power over them? Would there be a rotation system? Would our preferences for a mission be considered? Would military astronauts fly only military missions? Abbey said nothing. Nor did he ever provide the slightest performance feedback—positive or negative. If he had an agenda, that was never revealed either. I have never worked in any organization where there was such a complete lack of communication from above. The result of this information vacuum was predictable. FEAR. The line into space was long and nobody wanted to be at its end, or worse, be banished from it altogether. We were all terrified of doing something that might cross our king. We lived by rumor and innuendo because that was all there was. An early instance was a warning to Steve Nagel from Don Peterson (class of 1969) to stop work on a shuttle autopilot improvement project, "because rumor has it Abbey hates that project." Nagel was stunned. He had been *assigned* the work by another office vet. It wasn't something he had initiated. Yet he was being told he was jeopardizing his career by doing his assigned job. Shannon

Lucid and I had a similar experience. Moon walker Al Bean directed us to prepare a report justifying why nonpilot MS astronauts should be trained as pilots. Later we heard from another office vet that Abbey was vehemently opposed to such a program. Shannon and I dropped the work as if it were radioactive waste. Everybody was constantly second-guessing their actions. It was a poisonous situation.

If John Young had been more involved in our professional lives, things might have been better, but he was also an absentee leader. He was consumed with training for STS-1. His interaction with the rank and file was mostly limited to the weekly one-hour Monday meetings, and at those he had an irritating and morale-eroding habit of publicly rebuking us when we failed to win battles on shuttle issues at the various NASA review panels. I recall one meeting in which Bill Fisher (class of 1980) leaned over to me and sarcastically whispered, "That's it, John, yell at *us*." Fisher's implication was obvious to all within earshot: John should have been at the panel meeting in question using his vast experience as a veteran spaceman to defend his position instead of expecting one of us rookies to carry the day.

Many TFNGs would grow to loathe the Abbey-Young duopoly and its black hole of communication.

In our second year at JSC we received our first real astronaut job assignments. Because we lacked any other information on the flight assignment process, we quickly constructed a belief system in which these early jobs portended our place in the line into space. To draw an "STS-1 Support" job was thought to be indicative of a position at the head of the TFNG line because of the overarching importance of that first shuttle flight. My name wasn't under "STS-1 Support." Next were jobs supporting STS-2, -3, and -4. Again, it was assumed TFNGs assigned to support those missions must be impressing Abbey and be in line for an early space mission. My name was absent from those assignments. And neither was my name typed next to jobs supporting spacewalk, robot arm, and payload development. I finally found "Mullane" next to "Spacelab Support." This was at the rock bottom of TFNG job preferences. I felt as if I were back in high school after baseball tryouts seeing my name penciled next to *B-squad backup right fielder.*

Spacelab was a cylindrical module that would be installed in the cargo bay of a shuttle and connected to the cockpit by a pressurized tunnel. Since Spacelab flights would be science missions, I had assumed the post-docs would fly those missions. But it was my name on the jobs list next to "Spacelab Support," not theirs. Over lunch in the cafeteria I got to listen to Pinky Nelson and Sally Ride and the others excitedly discuss their work of validating robot arm malfunction procedures, developing spacewalk procedures in the WETF swimming pool, and getting down and dirty with STS-1 issues. I averted my eyes, praying nobody would ask me about my days of listening to science briefings on upper-atmospheric gases and the Earth's magnosphere. I was crushed. I now had the scent of Spacelab on me. I had to believe I was at the end of the flight assignment line, and, most maddening, I had no idea how I had gotten there or how I might recover. But, as I had done throughout my career, I resolved to set aside my disappointment and do my best at my new job. I also resolved to do a better job at getting my nose up George Abbey's behind.

My disappointment at my Spacelab job was mitigated when, in late 1980, I was assigned to be part of the STS-1 "chase" team. When *Columbia* came streaking to a landing, NASA wanted a T-38 chase crew on her wing to warn Young and Crippen if anything looked amiss, if there was evidence of leaking fluids or fire or damaged flight controls or the landing gear didn't extend properly. Thermal protection engineers also wanted the backseater in the chase aircraft to photograph *Columbia*'s mosaic of ceramic belly heat tiles before she landed. There was some concern those tiles could be damaged by fragments of the Edwards AFB dry lakebed runway being hurled backward by the tires. Prelanding photos would enable engineers to determine whether a tile sustained damage during the mission or during landing. Several chase crews were formed and I was assigned to fellow TFNG Dave Walker's backseat. During STS-1's launch we were to be positioned at El Paso's airport in case *Columbia* had a problem that necessitated an Abort Once Around the Earth (AOA) with a landing at the nearby White Sands Missile Range runway. If that happened we would scramble to do the rendezvous and I would take the photos.

Dave was known by his navy call sign, Red Flash, bestowed on him for his red hair. (Air force flyers of my era did not have personal call signs, as those in the navy did.) Over several months Red Flash and I, along with the other TFNG chase crews, practiced shuttle rendezvous with ground-based radar personnel. One T-38 would simulate the landing shuttle while the others would be vectored toward it by the radar controllers, as would be the case if *Columbia* made an emergency White Sands landing.

During this training, trajectory engineers in Houston asked us to examine other dry lakebeds in southern New Mexico and Texas as potential emergency landing sites for *Columbia*. They wanted to cover every contingency, including "low energy" trajectory errors that might prevent the shuttle from reaching the White Sands runway or "high energy" errors that would result in the shuttle overflying White Sands completely. Their fears were well founded. When *Columbia* came to Earth, it would do so as a powerless glider. It had no engines the pilots could use to fly around and search for a runway. If there wasn't a suitable landing site within reach, Young and Crippen would have to eject and *Columbia* would crash.

Day after day, Red Flash and I would take off from El Paso and search the Chihuahuan desert for twelve thousand feet of straight, flat, firm earth. And day after day, I would return to El Paso with my butt cheeks fatigued from an hour of ass-clinching fear. It wasn't that Dave was a bad pilot. Rather he was too cocky, the type of pilot who thinks he's bulletproof even when he's sober. (All fighter pilots think they're bulletproof when they're intoxicated.) He was the pilot that backseaters had in mind when they had coined this joke:

Question: "What are the last words a dead backseater ever hears from his pilot?"
Answer: "Watch this."

I was living that grim joke in Dave's backseat. When we spied a likely playa from altitude, Dave would say, "Watch this," and dive for the sand. To my left or right I would see our plane's shadow paralleling us at 300 knots. It would porpoise over hill and dale, quickly drawing closer and closer as Dave dropped lower and lower, until it finally disappeared underneath us. If our jet had had the curb-feelers of a '59 Edsel, I would have heard them scratching a warning into the desert a foot underneath us. Our engine exhaust had to be frying lizards, snakes, prairie dogs, and other ground-hugging fauna. And while I was white with fear, Dave was jotting observations about the condition of the terrain on his knee board.

In the early morning hours of April 12, 1981, Dave and I and the rest of the chase team were in the El Paso airport flight operations office, gathered around a TV watching the final moments of *Columbia*'s countdown. The previous night I had slept lightly and each time I awoke I would pray for Young and Crippen. I had a strong sense of dread about the mission. When *Columbia*'s hold-down bolts blew, her crew would be irrevocably committed to a flight that was more experimental than any manned flight in history. Forget Alan Shepard, John Glenn, or Neil Armstrong as astronauts who had taken unequaled gambles. Their Redstone, Atlas, Titan, and Saturn rockets had all been proven before they had ever climbed on board. Young and Crippen would be making history by riding a rocket on its very first launch. They weren't doing so reluctantly. The astronaut office had no problems with this decision, even though it would have been relatively easy to modify the vehicle to fly a first test mission unmanned. (In 1988 the Russians successfully flew, *unmanned,* the first and only mission of their space shuttle. It made two orbits of the Earth then flew under autopilot control to a perfect touchdown.) While the manned/unmanned debate over *Columbia*'s maiden flight had occurred before TFNGs had arrived at NASA, I could easily guess how long astronauts had discussed the topic before concluding a manned flight was the way to go . . . about five seconds. Astronauts will always be ready to jump into a cockpit. Any cockpit. Any time. There wasn't a single TFNG who wouldn't have volunteered to be ballast aboard *Columbia.*

But Young and Crippen would be taking an enormous risk and I feared for their lives. The only thing that had been positively demonstrated about the shuttle design was that it would glide from 25,000 feet to a landing. Four flight tests off the back of a 747 carrier aircraft had proven that. The solid rocket boosters and SSMEs had been ground-tested multiple times but had never actually flown in space. In fact, the SRBs had never even been tested vertically. In each of their firings, the rocket had been in a horizontal position, a fact that made many of us doubt the tests were really duplicating the stresses and strains of a vertical launch. The massive gas tank had never experienced the shake, rattle, and roll of a launch. There had been no full-scale flight tests of the 24,000-heat-tile mosaic that was glued to *Columbia*'s belly. How would it do in the 17,000-mile-per-hour, 3,000-degree wind of reentry? And no spacecraft had ever glided 12,000 miles to a "one-chance" landing—but that's exactly what *Columbia* was going to have to do. And the unknowns weren't just in the STS hardware. The shuttle's computer system contained hundreds of thousands of lines of code. Billions of dollars and years of labor had been spent to validate that software but there were still thousands of permutations that had not been tested and that could contain fatal flaws. Would an engine failure at precisely T+1:13 in conjunction with an unexpected wind sheer at 65,000-feet altitude cause a software switch in some black box to flip to a different polarity and send *Columbia* out of control? To an extent never before seen in spaceflight, the space shuttle was certified to carry astronauts based upon the wizardry of computer modeling. For a decade, engineers conducted thousands of ground tests in every imaginable engineering specialty: aeronautical, electrical, chemical, mechanical, hypersonic flight dynamics, cryogenic fluid dynamics, propulsion, flutter dynamics, aeroelasticity, and a hundred others. They digitized data gleaned from wind tunnel tests, engine tests, hydraulic tests, heat tile tests, and flight control tests and dumped the results into computers humming with the equations of Max Planck, Bernoulli, and Fourier. When the thousands of answers were finally assembled and examined, the engineers cheered. Computer models said their new shuttle system would work, that its twin SRBs and three SSMEs burning 4 million pounds of propellant in 8½ minutes would propel a quarter-million-pound winged orbiter to a speed of nearly 5 miles per second and an altitude of 200 miles. These same models also assured their brainy authors the orbiter would be able to make a powerless hemispheric-long glide to a 15,000-foot-long strip of runway. Of course, many of these same engineers had done the same thing in the development of the Redstone, Atlas, Titan, and Saturn rockets of the past manned programs, but in those programs, after all their testing and modeling were complete and the answer was "This rocket will fly," they had still walked cautiously. "We could be wrong in this model, or maybe in this model, or in this one," they had said. "We better test this puppy unmanned a couple times before we strap astronauts to it. And when we do, we better give the crew a way of surviving a booster failure through *all* of ascent." And they had. But not with the space shuttle. Its cherry flight was going to be manned. The engineers had foreseen the possibility of catastrophe and had included SR-71 Blackbird ejection seats for the two-man crew, but those were only usable during the first two minutes of launch; after that, the shuttle would be too high and too fast for an ejection seat bailout. The seats wouldn't again be usable until the shuttle was below about 100,000 feet and Mach 3.0, about ten minutes prior to landing. During the rest of the ride Crippen and Young would have zero chance of escape. Actually, the two-minute launch envelope of the ejection seat was even suspect. Many felt there was a good chance an ejection during launch would send them through the 5,000-degree plume of the SRBs. They would be vaporized. There was no doubt about it. Young and Crippen were human guinea pigs like no other astronauts before. It was another manifestation of Apollo hubris. Mere mortals might not be able to certify a rocket as *man-ready* with computers, but the gods of Apollo could.

I watched on TV as *Columbia*'s SSMEs came to life and a cloud of steam billowed from the flame bucket. When the SRBs ignited and *Columbia* was airborne, I almost pissed my pants. We jumped from our seats cheering. It was an act duplicated around televisions at Johnson

Space Center and Marshall Spaceflight Center and on the floors of countless aerospace factories and in millions of living rooms around the country. The TV showed a man at the Kennedy Space Center jumping up and down and punching his fist into the sky like a Little Leaguer celebrating a home run flying over the centerfield fence. Another view showed a man standing on top of an RV wildly waving an American flag as he watched *Columbia*'s smoke trail arc to the east. Another camera caught a woman dabbing at her tearing eyes. Everywhere the cameras captured a frenzied public. It was Woodstock, a NASCAR race, and a Virgin Mary appearance all wrapped into one overpowering, soul-capturing Happening.

And it only got better. At T+2 minutes and 12 seconds a flash of fire and smoke signaled the separation of the boosters. It was another computer-modeled milestone successfully passed. *Columbia* rapidly diminished to just a blue-white star and then disappeared completely. But we didn't need to see her to know how things were going. We could tell by the abort boundary calls coming from MCC: Negative return, Two Engine TAL, Single-Engine TAL, Press to MECO. It was gobbledygook to most of America, but for astronauts it was the sweet song of nominal flight. At Young's call of "MECO!" we all cheered again. *Columbia* had given her crew a perfect ride. I knew our celebration was premature. There was still a lot that could go wrong before *Columbia* was safely back on Earth. But, like the Apostle Thomas, I had seen with my own eyes and now I believed. If those gods of Apollo could put her into orbit with their computer models, they could certainly bring her safely home on the wings of their computer models.

On the flight back to Houston I couldn't relax the smile on my face. It was giving me a headache. But I didn't care. It had been nearly three years since I had entered NASA and this was the first time I really felt I had a chance of becoming an astronaut in anything but name only. Until I heard Young's MECO call, I hadn't truly believed it could happen. I had been convinced *Columbia* was going to end up on the bottom of the Atlantic and the closest I would ever get to space would be in a T-38. And I wasn't the only doubter. I would later hear that Pinky Nelson, upon the MECO call, had jumped from his seat and shouted, "Now I can put in a swimming pool!" Pinky had been a heretic, too. He hadn't truly believed in the gods of Apollo, and he had put off a decision to build a swimming pool until he knew he had a real job. In *Columbia*'s 8½ minute ascent, his dream of spaceflight, *all* of our dreams of spaceflight, had taken a giant leap toward reality. Mine was no longer the diaphanous mirage I had been following for twenty-five years. The gods of Apollo had fashioned a machine that could turn my astronaut pin to gold.

CHAPTER 16

Pecking Order

April 19, 1982, effectively marked the end of the TFNG brotherhood. It was on that day George Abbey assembled us to announce, "We've made some crew assignments." Like Hollywood stars hearing, "Can I have the envelope, please," we held our breath at Abbey's words. For four years, in hundreds of Outpost Tavern happy hours, on thousands of T-38 flights, around countless supper tables, we had asked the question of one another, of ourselves, of our spouses, of God: *When would we be assigned to a shuttle mission?* The room was space-silent as Abbey read the names. "The STS-7 crew will be Crippen, Hauck, Fabian, and Ride. STS-8 will have Truly, Brandenstein, Bluford, and Gardner. STS-9 will be Young, Shaw, Garriott, Parker, and two payload specialists. Hopefully we'll get more people assigned soon." That was it. God walked from the room.

Poof. With Abbey's words TFNG camaraderie vaporized. I don't believe there was ever again a social gathering of all TFNGs. As a group wallowing in a common uncertainty and united in a common distrust of our management, it had been easy to share a beer at the Outpost. Now we had been cleaved into haves and have-nots. There was a pecking order; some of us were better than others. I tried my best to be rational—somebody *had* to be first. We couldn't all be. But I couldn't accept that rationale and I doubted any of the others could either. We were too competitive. It was *The Right Stuff* syndrome as described by Tom Wolfe. The seven flight-assigned TFNGs had more of that *stuff* than the rest of us. We, the *unassigned,* had been left behind. I would later see first-flight assignments have the same effect on every astronaut class. Their all-for-one and one-for-all camaraderie would end just as abruptly as ours had. The effect could have been somewhat assuaged if Young and Abbey had been open about the flight assignment process, but all Abbey left us with was "Hopefully we'll get more people assigned soon." That wasn't a lot to hang on to. Abbey's and Young's silence on the mechanics and calendar of flight assignments was earning them a growing enmity.

With George's announcement still echoing in my brain, I wished for a hole in the earth to open and swallow me. I wanted to nurse my wounded ego in private, but that wasn't an option. Like the also-rans at the Academy Awards, I had to don a fake smile and shake the hands of the winners. They were incandescent. You could feel the heat from their faces. Several of the blessed tried to mollify us with comments like, "You'll be getting a flight soon" and "Your day is coming, too." I was being pitied. I didn't think I could feel lower. But I was wrong. I heard Sally comment, "George told us of the assignments a week ago, but he wanted us to keep it quiet until the press release." I wondered how many times in the past week I had been eating lunch in the cafeteria with Rick Hauck or John Fabian and whining about the delay in flight assignments, and all the while he had been silently celebrating his mission appointment. God, I felt so pathetic.

As I drifted from the room, I heard Fred Gregory's sotto voce growl, "This is bullshit!" His head and shoulders slumped in depression. Another casualty. Then it dawned on me. He had not just been passed over for an early flight assignment. He was black. He had just been passed over as the first African American in space. Guy Bluford would seize that title on STS-8. I was just a white guy. My name would never be on anybody's Trivial Pursuit card regardless of when I flew. But Guy Bluford would be history. And Sally Ride, as the first American woman in space, would become an icon. Some had lost more than just a mission assignment with Abbey's announcement. Some had lost history and the payday that came with celebrity. Sally Ride, in particular, had just been handed a free ticket through life. As the first American woman in space she could look forward to book deals, speech honorariums, corporate board seats, and consulting fees that could earn her millions.

As the seismic wave of Abbey's announcement was tearing apart the TFNGs, we were blissfully ignorant of another 9.0 wave moving through the system. Five months earlier, one of the eight O-rings on STS-2's recovered right-side booster had shown heat damage. This discovery had shocked the SRB engineers. Since the boosters were twelve feet in diameter, had a hollow center, and burned from the inside toward the outside, the perimeter-installed O-rings were far from the 5,000-degree gas throughout most of the burn. The unburned propellant served as an insulator. (In the final seconds of the burn, other insulation material at the walls kept the heat from the O-rings.) The O-rings should never show heat damage. And in seven ground tests and one mission (STS-1), involving a total of sixty-four primary and sixty-four backup O-rings, no heat damage had ever been recorded. The fact that an O-ring inside STS-2's right-side booster had been damaged was an indication that, at some point in flight, it had not held the nearly 1,000-pounds-per-square-inch pressure inside the tube and a finger of fire had worked between the segment facings to touch it. This suggested a serious problem with the joint design. But no consideration was given to stopping shuttle flights and conducting more ground tests. The NASA PR machine had promised Congress and the American public a rapid expansion of the shuttle flight rate with a vehicle turnaround time measured in a few weeks. *Schedule* had become the 800-pound gorilla in shuttle operations. Nobody wanted to wrestle with it. Instead, engineers at Thiokol and NASA searched for a way to continue operations in spite of the STS-2 anomaly. So they intentionally damaged an O-ring to a much greater degree than the damage they had observed on STS-2, put it in a laboratory test article, and pressurized it to three times the pressure developed by a burning SRB. The damaged O-ring held the pressure. Armed with these impressive results the Thiokol engineers endorsed their product as flight worthy. Lost in this process, however, was the fact something never expected and not completely understood had been accepted.

No astronaut was aware of the SRB O-ring problem. In fact, most of us were ignorant of the entire SRB design. There was only a single indication of SRB performance available in the cockpit of a launching shuttle. As the tube pressure fell to less than 50 pounds per square inch, a message flashed on the computer screens giving a warning that burnout and separation were near. Since we had little insight and no control over a burning SRB, we didn't waste our time in studying its design. We had too many other things into which we did have insight and over which we did have control (the liquid-fueled engines, hydraulics, electrical system, etc.). We devoted our time to learning the design and operation of these systems. We were convinced the SRBs were just big, dumb skyrockets, as safe and reliable as a hobby store model rocket. It was the SSMEs, which periodically blew up in ground tests, that we feared most.

The Thiokol and NASA SRB engineers were buoyed when STS-3's boosters returned with no O-rings damaged. It was full speed ahead with the shuttle program.

And the program shifted into overdrive on July 4, 1982. It was then that President Ronald Reagan and the First Lady celebrated Independence Day at Edwards AFB by personally welcoming Ken Mattingly and Hank Hartsfield back from space after their successful STS-4 mission. Reagan called attention to the latest orbiter to join the shuttle fleet, *Challenger.* Fresh from the nearby Rockwell factory, that vehicle was mounted atop its 747 carrier aircraft ready to take off for Florida as soon as the president finished his comments. It was an incredibly intoxicating sight. *Columbia* sat on the cracked dirt of the lakebed looking every bit the veteran of four spaceflights, with her nose and fuselage streaked with soot from four blazing reentries. *Challenger* sparkled in her virgin newness. It was the perfect backdrop as the president continued his speech and declared the space shuttle program "operational."

That label had never really been defined, but it was easy to sense how most of NASA and all of the public interpreted it. *Operational* meant the shuttle was nothing more than a very high-flying airliner. I doubt there was a single military aviator astronaut who believed that. Fighter jets of far less complexity than the shuttle routinely suffered malfunctions resulting in crashes. We were certain one awaited the shuttle, too, and when it happened, it would mean death for her crew. While the operational label was nebulous, it did contain one certainty—all future shuttle missions would be flown in vehicles with no in-flight escape system. There were no ejection seats in *Challenger*'s cockpit and the two in *Columbia* would soon be removed. That had been the plan from the very beginning. President Reagan's "operational" declaration was merely photo-op tensile. But contained in it was a shuttle design feature that would condemn some of us to death.

With flight crews named to all the planned missions through 1983, I knew I would not be getting a flight assignment for many months, perhaps even a year or more. But at least my purgatory of Spacelab support had ended. I was now assigned to shuttle software checkout in the Shuttle Avionics Integration Laboratory (SAIL). My frequent partner in that facility was the now very pregnant Rhea Seddon. She and Hoot Gibson had married in 1981 and their first child was due in July. In the SAIL cockpit I would watch Rhea's nine-month distended belly crowd the control stick as she flew simulations to perfect landings. It was a sight certain to have sent some of the old Mercury astronauts fumbling for their nitro pills. Rhea would ultimately give birth to a son, one of the rare boys born to astronauts. We had long noticed a propensity for astronauts to sire daughters and wondered if the G-forces of our jet-jockey training were pushing male sperm to the end of the line. As Hoot and Rhea were being congratulated at a Monday meeting, one pilot shouted, "This proves Hoot isn't an astronaut." I answered, "No. It proves Hoot isn't the father." Rhea had a good laugh at that.

I enjoyed Rhea immensely. Like Judy, she was a smart and capable beauty with a limitless tolerance for us AD males. She frequently parried our sexist BS with biting humor. I once saw Hoot, our AD King, skewered with it. One of the men chosen to sit on an upcoming astronaut interview board had ducked his head into our office and asked for inputs on the selection criteria for the new class of astronaut candidates. Hoot gave Rhea a body-appraising scan and answered, "Yeah, how about selecting some women with big breasts and small asses instead of the other way around." Rhea smiled wickedly at her husband and replied, "Robert, some night while you're asleep, I'm going to amputate your penis [she was a surgeon] and graft it to your forehead, and when you come to work people are going to think it's a zit." Hoot had married perhaps the only woman on the planet who was his equal. When they were together it was a laugh a minute. I loved them both.

By 1982, like the other AD men, I had learned my boundaries around the six females. Rhea's and Judy's were the widest. Sally's were the tightest. Though I repeatedly warned myself to watch my mouth around Sally, I would have relapses, as when I once observed, "The female cosmonauts are sure ugly." Sally snapped, "Have you ever thought they might be good at their job?!"

Alcohol always held the potential to wreck my resolve. One evening, as Donna and I walked from a local restaurant (after a dinner that included more than a few beers), a friend stopped Donna and they fell into conversation. As I dallied, I noted Sally and Steve Hawley at another table dining with an attractive woman I didn't recognize. At the time, Steve was dating Sally so there was nothing surprising about seeing them together. With my wife engaged I walked over and said, "Hey, Stevie, are you getting cookie recipes from these girls?" Sally glared at me like I was something growing in her bathroom grout. Hawley cringed as if he had taken a bullet to the gut and shot Sally a glance that said, "I don't know this guy." There was an awkward silence during which the unidentified woman examined me as if I were whale shit, the lowest thing on the planet. Finally, I bid a good-bye and escaped back to my wife, my hands discreetly checking the zipper of my fly as I walked. The threesome's rude reaction made me wonder if I had forgotten to zip up after my last visit to the urinal. Nope, everything was secure.

As I returned, Donna's friend gushed, "You know her?!"

Of course, I assumed she was referring to Sally.

"Sure, that's Sally Ride."

"No, not her. The other woman."

"No. I wasn't introduced." I was still puzzled by that table's hostility toward me. Was it something I said?

"That's Jane Pauley."

I shrugged. The name was a mystery to me. "Who's Jane Pauley?"

Donna's friend nearly had a seizure. "Who's Jane Pauley!? You don't know? She's the NBC *Today* show newswoman."

I honestly didn't know. I didn't watch much TV. I certainly didn't watch those chatty morning shows. If she wasn't in *Aviation Week & Space Technology* magazine, I wouldn't know her.

With this new bit of knowledge, it slowly dawned on me why I had been stonewalled at Sally's table. No doubt Ms. Pauley was talking to her about her recent flight selection. I could just imagine how my cookie recipe comment must have played with those two pioneering females. I made Hugh Hefner look like a beacon of enlightenment. I guess it's no surprise I was never invited to the *Today* show.

On October 5, 1982, three more TFNGs were named to a flight, STS-10 (later to be designated STS-41B).* I wasn't among them.

I put on another happy face and congratulated the winners. A few weeks later Norm Thagard became the eleventh TFNG to draw an

*Initially the STS numbering system was straightforward, STS-1, -2, -3, etc. After STS-9, NASA instituted a new number/letter system to provide more information in the mission designator. But there was another reason for the change—superstition. Astronauts and engineers aren't immune from it any more than the rest of the population. NASA did not want to have the bad luck number 13 hanging on a shuttle mission, particularly given the near disaster of *Apollo 13*. So Gerald Griffin, the JSC director, came up with a new STS mission designation system to shortstop an STS-13 label. The first number would designate the calendar year in which the mission was planned to launch. The second number would be a 1 or 2—1 designating a KSC launch and 2 indicating a launch from Vandenberg AFB in California. The letter designation would show the intended launch sequence in the calendar year, with A, B, C, D, etc., translating to the first, second, third, fourth, etc., calendar year launch. So the STS-41G label shows the mission was planned to be the seventh (G) mission of 1984 (4) and would be launched from KSC (1). It was hoped that this code would sufficiently blind the god of bad luck to the fact that STS-41G was actually the thirteenth shuttle mission. Labels were fixed at the birth of a mission on the planning manifest, years before flight. They were not later changed as the flight schedule changed due to shuttle and payload problems, so they are not indicative of the sequence of actual launches. For example, because of schedule changes, STS-51A, the first planned KSC flight of 1985 when it first appeared on the manifest, actually flew as the last mission in 1984. STS-51L, the fated *Challenger* flight, was supposed to be the last flight of 1985, but flew as the second mission of 1986. After *Challenger,* NASA reverted to the straight numerical designation, beginning with STS-26.

assignment when he was retroactively assigned to STS-7. NASA was growing concerned about the incidence of space sickness and wanted Thagard, a physician, to run some experiments on what was being officially labeled Space Adaptation Syndrome (SAS). SAS had impacted the recently landed STS-5 mission in a very big way. One of the two spacewalkers on that flight had been so stricken with vomiting the crew had asked MCC for permission to delay their EVA (Extra-Vehicular Activity, i.e., a spacewalk) to give him time to recover. Vomiting inside a spacesuit could kill an astronaut. The emesis could smear the inside of the helmet visor and blind the spacewalker, making it impossible to respond to a suit emergency. Also, because there was no way to remove the fluid, the astronaut could inhale it and choke to death or it could clog the oxygen circulation system and suffocate the victim. The STS-5 spacewalk, which was to have been the first from a shuttle, had just been a demonstration exercise (and was ultimately canceled for a suit malfunction), but in future missions spacewalks would be essential for mission success. Norm Thagard would be the first of many physicians sent into space to determine the cause of SAS. He, like all who would follow, would have their studies seriously hampered by astronaut paranoia. Spacewalking was the most sought-after prize for MSes. It filled a powerful need to be in ultimate control. The pilots had their shuttle landings to fulfill them. Their hands and eyes delivered a 200,000-pound orbiter to a runway. It was the same with a rendezvous mission. A pilot's personal skill brought two 17,300-mile-per-hour objects together 200 miles above the earth. It was heroic work. On the other hand, much MS work was mundane—throwing a switch to release a satellite, drawing blood, changing a data tape on some scientist's experiment. Spacewalks and, to a somewhat lesser degree, robot arm operations were the exception in MS jobs. Like a pilot feeling the kiss of the runway on the space shuttle wheels, MSes could enjoy a powerful sense of being in control as they assembled structures or repaired satellites or performed other hands-on spacewalking tasks.

So physicians studying SAS, like Thagard, were hamstrung. Astronauts didn't want to admit to an episode of vomiting out of fear that it would eliminate them from consideration for future spacewalk missions. As a result many astronauts were less than truthful about their symptoms. Some blatantly lied. We would hear stories of crewmembers who were seriously sick, yet the data would never appear on the flight surgeon's bar charts. SAS was considered an individual health issue and was therefore privileged information between the astronaut and flight surgeon. If an astronaut didn't tell the flight surgeons the truth, the doctors were not going to hear it from anybody else.

To be SAS-free was considered so important, many astronauts attempted inoculations. When it was first assumed the problem was related to Earth-based motion sickness (later disproved), astronauts would perform stomach-churning acrobatics in T-38 jets in the days prior to a launch. I was flying in Story Musgrave's backseat when he decided to prep his body for an upcoming mission. He asked ATC for a block of altitude and then went into a series of spiraling rolls and violent maneuvers that alternately had me slammed into my seat at 4-Gs and lifted from it in negative Gs. My head snapped back and forth like a palm tree in a hurricane. Within a minute I was ready to blow my last meal (and perhaps a few before that) and had to plead with him to stop.

Another equally ineffective attempt at SAS inoculation was to sleep on an incline with your head lower than your feet. This became popular when the flight surgeons hypothesized that the fluid shift of weightlessness might be causing the inner ear to be disturbed, inducing vomiting. All astronauts experience an uncomfortable eye-popping fullness in the head during weightlessness because of an equalization of body fluid. By sleeping in a bed with bricks under the foot posts to tilt the head down, it was thought the resulting fluid shift to the upper body would somehow prepare it for weightlessness and eliminate SAS. It didn't. Some of those practicing head-down sleep still got sick in space, suggesting that those head-downers who didn't vomit had probably been immune anyway. To this day doctors are baffled by the cause of SAS and it continues to affect nearly 50 percent of astronauts.

As the calendar turned to 1983, my fifth year as a TFNG, I was suffering from something far worse than SAS—the depression of being an unassigned astronaut. That status had me doubting everything about myself—my abilities, my personality, even my astronaut friends. *Were they on Abbey's shit list and, by association, was I too?* I wondered if others had already been told of a mission assignment and were keeping it secret until the press release. Might an office mate already be assigned? Every few days a new rumor on flight assignments would sweep the office like a pandemic flu. Some of this scuttlebutt would have my name assigned to a mission. Before the press release for STS-10 appeared, one such rumor had me on that flight. But it was a lie. We all searched for any indication that another round of flight assignments was in the offing. We watched from our office windows for groups of our peers walking to Building 1, Abbey's lair. Were they on their way to be told of a mission assignment? One astronaut kept a pair of binoculars on his desk to better observe that traffic (as well as hard-bodied, halter-topped female tourists). Unassigned TFNGs were ready to explode in frustration. At parties I could see the tension had infected our spouses. *There is no rank among wives* was an old military proverb. Yeah, and the Easter Bunny is an astronaut. Every wife of one of the unassigned, mine included, knew her position had changed. The wives of the assigned were working with the NASA PR people to schedule TV and magazine interviews while the spouses of the unassigned were wiping the baby's ass. These Queens for a Day would soon be boarding NASA Gulfstream jets to zoom to Florida as VIPs. There was no doubt some marriages were suffering from the new reality of assigned and unassigned TFNGs. Mine certainly was. When Donna commented at a party, "George Abbey couldn't lead a pack of Boy Scouts" (something I said every night), I pulled her aside and snapped, "Goddammit, don't bad-mouth Abbey with others around! There's no telling what gets back to him." It wasn't a fluke outburst. My frustration was a loose cannon and Donna was frequently in the line of fire. I was an asshole.

I continued my drab life. I would pull into the Building 4 parking lot by 7:30 A.M. so I could fight for a parking space (cringing at the sight of the assigned TFNGs pulling into their reserved parking places), attend some SAIL-related meetings, go to the mail room to sign autographs (wondering why anybody would want mine), go to the gym to exercise, eat lunch in the cafeteria (to catch up on the latest rumors), attend more meetings or study shuttle-training schematics, perhaps take a T-38 flight (if the assigned crews had left any), then go home. On my SAIL days I would pull one of the eight-hour shifts of its 24/7 operation. If I was lucky, I would be called to the SMS for some real shuttle training as a substitute crewmember. When Guy Bluford was absent for an STS-8 simulation I received such a call and eagerly jumped on it.

From my perspective the racial integration of the astronaut office with Bluford, Gregory, McNair (all African American), and El Onizuka (Asian American) had occurred seamlessly. The entire astronaut corps seemed color-blind. I certainly was. My family upbringing, so abysmally lacking when it had come to the topic of females, had been radically progressive on the subject of race. "When you're in a foxhole and the damn Japs are shooting at you, you don't care about the color of the American at your side," was my dad's version of an "I Have a Dream" speech. And my religion, so medieval in its attitude about women, was enlightened in its preaching on race. Jesus Christ had said, "Love your fellow man." He hadn't added any footnotes on color. Hell's fire awaited the racist, just as it did for boys imagining the cheerleaders naked. I never gave a second thought to the skin color of the minority astronauts and I got the impression the other palefaces in the astronaut office didn't either.

While there was no racism in the astronaut office, the topic of race, like that of gender, religion, sexual orientation, the pope, motherhood, apple pie, and just about anything else, was fair game for the office humorists. Nothing was sacred to them. While substituting for Guy Bluford in the SMS I had a ringside seat to some of this humor.

During the course of the sim, we received a call from the Sim Sup, "I want you guys to come up with a medical problem and call the surgeon about it." We were used to such requests. There was a "surgeon" console in the MCC manned by a NASA physician and the Sim Sup wanted to ensure he had a crew health problem to work. In the cockpit we put our heads together and the suggestions flowed.

"Let's tell him Dan has a sharp pain in his stomach. It might be appendicitis."

"Let's tell him Dick has flulike symptoms."

"Let's tell him Dale has a bad toothache."

We were mulling over these and a few other ideas when Dale Gardner snapped his attention to me, the Guy Bluford substitute, and exclaimed, "No! I've got it. Let's tell them that Guy has turned WHITE!" NASA HQ had been in orgasmic ecstasy over the impending flight of America's first black astronaut. Knowing this, the suggestion was outrageously funny.

Somebody mimicked the call with an *Apollo 13* header, "Houston, we've got a problem. Guy has turned white!" It would be a call certain to turn a few people in HQ *white*.

Dick Truly looked at us and said, "If you guys make that call, the closest you'll get to space is the ninth floor of Building 1 while Kraft fires you."

We all understood. Even color-blind astronauts couldn't publicly joke about race. It was career suicide in America. We had to settle for Dan's stomachache.

During another simulation in which I was a substitute crewmember, I thought my career had ended and the circumstances had nothing to do with race. During a break another crewmember and I climbed down to the mid-deck and made ourselves some lunch. Since every aspect of a real shuttle mission was being simulated, our food was space food—sandwich spreads and dehydrated food. Bread wasn't on the menu. It crumbled too easily in weightlessness. Instead, tortillas were used. We made peanut butter sandwiches from these and then cut into a package of dried fruit. My lunch partner held up a dehydrated pear. "Mullane, check it out, it looks like a [part of the female anatomy]." Only this TFNG didn't say "part of the female anatomy." He used a popular planet AD euphemism. I laughed. "You're right. It does look like a ——" I repeated the word. The butterflied pears had dried into an X-rated art form.

A moment later Dale Gardner, who had been absent on a toilet break, reentered the mid-deck. The vein on his head looked ready to burst. "Jesus Christ, what did you guys say on the intercom! Some woman working on the Sim Sup console heard you guys say something about a dehydrated pear and was totally grossed out. She stormed off to Kraft's office to lodge a complaint."

"Oh God," I said. "She must have heard us through an open mic."

This wasn't going to look good . . . some young woman leaning over Dr. Kraft's desk and screaming, "Some of your wonder-boy astronauts just saw a part of the female anatomy in a dried pear!" God, why couldn't we have seen the Virgin Mary in a tortilla instead?

Dale again asked, "What did you say to piss her off?"

My compatriot hung his head like a six-year-old in front of an irate parent and mumbled, "I think I said ——"

"You said THAT? Jesus . . . you're toast. Kraft is going to crucify you." Then he climbed to the flight deck shaking his head at our idiocy.

We threw our food in the garbage, including those offending pears. Our appetites were gone. Our careers would soon be in that garbage can, too, I thought. We climbed back to the flight deck and went through the motions of being an astronaut. We were hooded victims tied to a post waiting for the bullet to be fired from Kraft's office: "Get your asses over here . . . and clean out your desks on the way!" But hours passed and no call came. In fact the sim proceeded to completion and still there was no call.

As we sulked back to our offices expecting to find messages on our desks, Dale came to our sides and said, "Hey, guys, that was pretty funny, wasn't it?"

We looked at him. "What was funny?"

"That joke I pulled on you about the woman hearing your pear comment."

"That was a joke?"

"Yeah, I was standing outside the mid-deck and heard you guys talking about it. I thought I'd rattle your cage."

I was ready to rattle his cage with both hands on his throat. "You bastard!"

A few days later a note did appear on my desk requesting my presence in Building 1. It was from George Abbey.

CHAPTER 17

Prime Crew

I knew there were binoculars focused on us as we made the trek to Building 1. I was in the company of four others who had also been called for the same appointment: Hank Hartsfield, a veteran of STS-4, and fellow TFNGs Mike Coats, Steve Hawley, and Judy Resnik. There was no mistaking the meaning of this gaggle. It screamed *flight assignment.* It was all I could do not to break into a sprint for the headquarters building.

It might have been the very first time in my five years as a TFNG that I had been to Abbey's office. As befitting a deity, it was a large corner office on the eighth floor that looked out on our home, Building 4. None of us believed that sight line was an accident. George wanted his shadow to fall over his subjects at all times.

The secretary waved us through and we entered to find him standing behind his expansive desk. He wore a coat and tie, the coat unbuttoned and his belly prominent. Though it was midmorning his jowls were already darkened with a faint beard. Several documents and an overflowing in or out box (I couldn't tell which) littered the mahogany. I had a momentary wonder, *What was all the paper about?* It was a standing TFNG joke that Abbey never left a paper trail. Few could recall ever seeing his signature on any document.

"Have a seat." He motioned us to a ring of chairs.

"We've been looking at the mission manifest . . ." As was his custom, George never made significant eye contact and he mumbled his words. I could sense everybody leaning forward to gain another decibel. ". . . and think it's time to assign some more crews. I was wondering if you would be interested in STS-41D? It would be the first flight of *Discovery.*" It was beyond what I had prayed for. Not only was I being offered a mission, I was being offered a position on the first flight of the orbiter *Discovery.* Aviators live for the day they might be the first to take a new jet into the air, and we were being offered the first flight of a space shuttle. Like the Stockholm Syndrome hostages we were, we all groveled in thanks.

George asked us not to publicly mention the assignment until the press release was made in a couple days. Mike Coats whispered to me, "When I get home, I'm calling everybody in my Rolodex." There was no way I was going to withhold this information from my family, either.

On the walk back to Building 4, I considered my incredible fortune. Besides getting the first flight of *Discovery,* I was also part of a great crew. At age forty-nine, snowy-haired Hank Hartsfield was the old man. He had come to NASA in 1969 as an air force test pilot with more than seven thousand hours of fighter jet flying time. STS-41D would be his second space mission and first as a commander. Unlike a number of his peers who were so anal that even the rest of us anal-retentives noticed, Hank was easygoing and quick with a laugh. He was an Alabama boy who had retained the drawling speech and the political ideals of the Deep South. Hank was so far right on the political spectrum he made even the John Birch Society look like a collection of hankie-wringing, pantywaist liberals.

Mike Coats would be our PLT. He was a former Navy A-7E attack pilot with movie-star good looks. My daughter hadn't been wrong when she likened him to Christopher Reeve's Superman character. A curl of black hair would periodically fall across his forehead, making the Clark Kent appearance complete. Mike was a quiet family man, devoted to his wife, Diane, and their two children. I never heard him swear or bring a disgusting joke to the table or get intoxicated or look twice at any of the beautiful women we met on our trips. Mike was a rare exception to the rule that military aviators are root-bound on Planet AD.

I was also happy to be crewed with Judy. It wasn't just because she was so AD tolerant, although that was a big reason. She was smart, hardworking, and dependable, all the things you would want in a fellow crewmember. Of course the male in me also appreciated her beauty. The five years since our arrival at JSC had been kind to JR (her nickname). As most of us had done (under the hammer of astronaut competitiveness) she had taken to jogging and dropped her weight. Her ravishingly curly anthracite black hair now framed a leaner, bronze tan face.

Steve Hawley was a gift to all of us. More than any other post-doc, baby-faced Hawley was the source of my conversion from doubter to believer in the capabilities of the TFNG scientists. A Kansan with a PhD in astrophysics, Hawley was living proof an advanced civilization of aliens have visited Earth. He was one of them. No human had a comparable brain. In just a glance, Steve could commit the most complex shuttle schematics to permanent memory. Every checklist, including the two-volume malfunction massif, was in a virtual file drawer in his brain, ready for instant retrieval at the sound of a cockpit warning tone. He was a maestro in simulations, directing responses to ten different system failures simultaneously. John Creighton once commented that to have Hawley in the cockpit was to have a sixth GPC aboard, a play on the fact that the shuttle computer system consisted of only five IBM General Purpose Computers (GPC). Steve had so much brainpower, the space shuttle was hardly enough to occupy him. He found additional challenges. One was to attend professional baseball umpire camp over his vacation (he was a sports addict). It wouldn't have surprised me to have learned he was also ghostwriting Stephen Hawking's books. Hawley was held in such high regard by the military TFNGs that the pilots christened him "Attack Astronomer." Since pilots were particular about their own titles—fighter pilot, attack pilot, gunship pilot—Hawley's title was an honorific. Steve was recently married to Sally Ride and for the sake of that marriage he was trying to distance himself from us AD bottom feeders. But it was a struggle. Hoot kept pulling him back to the dark side.

There would also be a sixth crewmember aboard, a McDonnell Douglas employee, Charlie Walker, flying as a payload specialist (PS) to operate a company experiment.

That night, February 3, 1983, Donna and I celebrated over dinner and later in bed. As she slept beside me, I stared at the ceiling and thanked God for passage through one more gate on my journey to space. I finally had a mission. Space was looking closer than ever. But there were six other shuttle flights in front of me. A lot could yet go wrong. I prayed for the crews of those missions. I prayed for their safety and success. I prayed more fiercely than they were praying for themselves. I wanted them out of the way.

The official NASA press release followed within a couple days and we bought the beer at an Outpost happy hour. George had also named the STS-41C crew so there was a total of seven TFNGs who had jumped to the sunlit side of the unassigned/assigned fence. I was now the one being congratulated and I ached for the others who showed me their fake smiles.

During the party I heard one frustrated astronaut redefine TFNG—

thanks for Nothing, George. I would never understand George Abbey. Some interpreted his dictatorial style as megalomania, but I never saw him seek a spotlight. In fact, a recent newspaper photo had appeared on the astronaut B-board showing Abbey shaking hands with a shuttle crew. It was captioned, "Unidentified NASA official welcomes astronauts." We all laughed at that. It was as if a photo of the pope had appeared in a newspaper over the caption, "Unidentified papal official welcomes pilgrims." But it was an indication of how invisible Abbey was. Though his management of the astronaut corps provided many opportunities for him to be in the press and on TV, he was never featured in either. Abbey was no megalomaniac. I don't think anybody had a clue what he was. Hoot Gibson would later offer me his best guess . . . that Abbey loved us, but, like a stern parent, he didn't care how we felt. He knew what was best for us and would give it to us at the time and place of his choosing. The problem was, we weren't children. We were freakin' astronauts who would have gladly taken whatever he thought was best for us, if he would have just told us what that was. Abbey's secretive leadership style was a cancer on astronaut morale.

Our 41D crew was at the end of the training line for the JSC simulators, but there were still opportunities for payload training and we traveled to Seattle to learn the intricacies of the 25,000-pound Boeing-built Inertial Upper Stage (IUS) rocket booster that would be our primary cargo. After we deployed it from *Discovery*'s cargo bay, it would lift a large communication satellite to geosynchronous orbit 22,300 miles above the equator. At the contractors' factories, we also did some widows and orphans appearances, passing out "Maiden Voyage of *Discovery*" safety posters to the workers. Judy had me laughing when she whispered, "There are no *maidens* on this flight."

Maiden or not, Judy was the center of attention wherever we traveled. At one contractor event a young engineer went, quite literally, mad for her. Throughout a daylong factory visit, he was constantly at her side trying to anticipate her needs. When she had none, he created some, bringing her water, soft drinks, and snacks. When we sat for briefings he would stand in a corner and stare at her like a Labrador waiting for the Frisbee to fly. On our factory tour he would rush ahead to hold a door until she walked past, then sprint ahead to the next. If there had been a puddle anywhere on our route, I was certain he would have flung himself face first into it, offering his back as a bridge. I expected the senior contractor official to tell his drooling puppy to get lost, but he turned a blind eye. I could tell Judy was seriously upset by the attention, but she was too much of a lady to say what needed to be said—"Fuck off!" We all breathed a sigh of relief when we were finally in our cars and on the way to the sanctuary of our T-38s. Those were parked at the gate-guarded military apron of Los Angeles International Airport. Then, to our astonishment, as we sat in our cockpits with the engines running, Judy's want-to-be paramour appeared out of nowhere, rushed under her jet, and pulled the chocks!

Back in Houston, over an Outpost beer, we laughed off the incident as a one-time case of extreme infatuation. It wasn't. It proved to be Jody Foster–John Hinckley creepy. Judy began to receive letters, poems ("your raven hair and eyes"), and gifts. JSC security was notified and they promised to call the wacko's employer and have them discipline the man. I thought that was the end of it until one night I received a panicked call from Judy. "Tarzan, can you come over right away? I just got home and there was a package at my door from that engineer. It doesn't have any postage on it." The implication was obvious: It had been hand delivered. He was in town. The guy was a stalker and Judy his prey.

By the time I arrived, Judy had already called the JSC security people and they had sent a car to patrol around her house through the night. They had also promised to call the man's employer . . . again. But, again, whatever warnings were delivered didn't take. A few weeks later the man walked into our office! I could only assume he was there on official contractor business because he wore the proper JSC badges. He went immediately to Judy's desk and asked her to autograph one of his poems. She refused. He begged her to write him one letter a year. She refused. He begged her to come to dinner with him. She refused. As this was going on, I was moving to Judy's side, watching the man like a secret service agent watching the crowd at a presidential event. He didn't look violent, but if he reached into his briefcase I was going to tackle him. Grabbing JR by the arm I said, "We've got a meeting to attend," and escorted her from the room. We called security and, this time, whatever they did apparently had the desired effect—the stalking ended. Beauty and celebrity had their downside, as Judy was learning.

Our crew soon acquired nicknames. Tarzan stuck on me. Judy christened Hawley, my Bo Derek salivating cohort, Cheetah. I'm sure Steve would have preferred the more macho handle Attack Astronomer, but Cheetah stuck. In keeping with the Ape Man theme, I branded Judy *Jane,* asking her as I did so, "Would you like to swing on my vine?" She replied, "Sure, Tarzan. But first I'll have to tie a knot in it so I have something to hold on to." Judy always had a comeback for my AD bullshit. Mike Coats maintained his Superman call sign. Upon hearing these titles the office secretaries began to refer to STS-41D as the "Zoo Crew" and Hank Hartsfield naturally became the "Zookeeper."

God apparently didn't hear my prayers to watch over every mission in front of us. STS-6 returned home safely but the IUS booster rocket it deployed, identical to the one we would carry on *Discovery,* malfunctioned. Its communication satellite was released into an unusable orbit. It was going to take as much as a year for the Boeing engineers to fix the IUS, meaning several IUS missions—including ours—had just lost their payloads. I was miserable. Any ripples in the flight schedule could generate changes in flight assignments. But after many tense weeks of worry, we acquired a new payload of two smaller communication satellites with different booster rockets. Best of all, we still retained the first flight of *Discovery.*

We set to work on our crew patch design. Since *Discovery* was named after one of Captain Cook's eighteenth-century ships, we included a sailing vessel morphing into the space shuttle *Discovery.* We also teased Judy about adding the symbol for the female gender to the patch ♀. There was precedent for this: The STS-7 crew had included a da Vinci Vitruvian Man–like representation on their patch. Four "male" symbols, arrows radiating outward, formed the head, arms, and one leg, while a lone female symbol—obviously representing Sally Ride—formed the other leg. When the patch appeared, Mike Coats observed, "Sally wears her gender like a chip on her shoulder." I jokingly suggested to Judy we add something similar to our STS-41D patch and penciled an idea. It had the + of the female symbol as the *center* of the creature and five male arrows pointed inward at it. It would have been interesting to see how HQ would have reacted to that design.

STS-7 and -8 flew into history and I prayed my hallelujahs.

In November our crew celebrated Hank's fiftieth birthday at the Monday meeting. Because he wore his political leanings on his sleeve, he was an easy target to lampoon. We presented him outrageously satirical gifts, including a copy of *Ms.* magazine dedicated and autographed to him by Gloria Steinem, "In recognition of your support of the feminist movement." (Sally Ride, a friend of Ms. Steinem, had secured the magazine and her autograph, a one-of-the-guys act that shocked me.) We read fake congratulatory messages from Hank's supporters, including the ACLU, Jane Fonda, and the Nuclear Freeze Movement. There was also a congratulatory card from Yuri Andropov thanking Hank for "promoting global communism," as well as a card from Senator Ted Kennedy thanking him for his recent donation to the Democratic party. A final gift, a box of Ayds diet candy, was from the gay rights political caucus acknowledging Hank's support for their cause. The gift card read, *Here are some AIDS for you.* Nothing was out of bounds when it came to astronaut humor.

On December 8, 1983, my dream of spaceflight, not to mention the entire shuttle program, almost ended when STS-9 landed on fire. During the final moments of *Columbia*'s approach, one of its hydraulic pumps experienced a propellant leak that dumped hydrazine, a particularly wicked fuel, into the aft engine compartment. The resulting fire quickly spread to a second hydraulic system and both systems failed shortly after touchdown. Had the fire started a moment earlier, it probably would have caused all three hydraulic systems to fail while *Columbia* was still airborne. Like a car losing power steering, the controls would have frozen. *Columbia* would have rolled out of control and crashed into the desert. John Young and his crew missed death by a handful of seconds. As I later examined photos of the fire damage, I thought of John's earlier pronouncement, "God watches out for babies, drunks, and astronauts."

The failure mode that caused the hydrazine leak was quickly identified and corrected. The shuttle program rolled on and my spirits soared . . . and then, just as quickly, came crashing back to Earth. On the very next mission, STS-41B, both of its deployed satellites failed to reach their intended orbit due to booster rocket malfunctions. I was thrown back into hell. We had the identical booster rocket attached to one of our two communication satellites. It was unlikely NASA would launch *Discovery* with only a single satellite as freight. For weeks we fretted and sweated while NASA HQ shuffled payloads and, for a second time, we survived with *Discovery.* One mission to go.

We were now practically living in the simulators—one session was fifty-six hours in duration. Hank and Mike were spending most of their time practicing launch aborts and landings. Steve, Judy, and I were consumed with payload training. There was also spacewalk training. None was planned for our mission but every shuttle crew included two astronauts who were prepared for an emergency spacewalk. This was to provide one more line of defense against things that could kill a crew, like not being able to close the Payload Bay Doors (PLBD). To attempt reentry with those open would be certain death. This was why every component associated with the door closing and locking systems was redundant. Redundant motors were powered by redundant electrical systems through a myriad of redundant black boxes and redundant wiring. One failure of anything would not prevent an astronaut crew from closing and locking the doors. But that wasn't good enough for

NASA. They wanted to back up even this redundancy with astronaut spacewalkers who could manually string a lanyard to a door edge and winch it closed, then hand-install and manually tighten locks. Other contingencies also had to be considered. The shuttle's high-gain antenna and the robot arm were both mechanisms that could become stuck outside the PLBD envelope and interfere with door closure. The two-man contingency spacewalkers were trained to muscle these devices inside the bay and tie them down. Hank designated Hawley and me as the EVA (spacewalk) crewmembers and Judy as our Intra-Vehicular Activity (IVA) crewmember. It would be her job to help us dress in the 300-pound Extra-vehicular Mobility Unit (EMU), i.e., a spacesuit, assist us in the suit checkout, and follow us in the EVA checklists to ensure we didn't make a mistake. A spacewalking crewmember entered a whole new arena of risks. When pressurized, the suits became as hard as a steel-belted radial, severely impairing movement and tactile feel. In this condition a mistake was possible, perhaps a deadly one. If Hawley and I had to do a spacewalk, Judy would be our omnipresent guardian angel, watching us from inside *Discovery* and making certain we followed every procedure exactly.

The primary facility for practicing spacewalks was the WETF swimming pool. Few experiences in life are more claustrophobic than being lowered underwater while dressed in an EMU. One astronaut confided in me that he repeatedly exhibited excessively high blood pressure when the flight surgeons conducted their pre-WETF vital sign checks. In fact, the docs became so suspicious, they required him to do multiple blood pressure checks at the flight clinic to ensure it wasn't a more chronic problem. I never had a problem with these checks. I could always put myself in a happy place while the cuff was on. After it was off, my veins got an elastic workout.

Training in the WETF pool was the most physically demanding of all astronaut training. As it would be in space, the suit was pressurized to the consistency of iron. The simple task of opening and closing one's hand rapidly fatigued those muscles. And while the suit was neutrally buoyant, the body inside wasn't. When I was in an upright position, I was standing inside it. But when the spacewalk practice sessions required me to work upside down, which was frequently the case, I would "fall" an inch or two inside the suit until my entire body weight was borne by the top of my shoulders pressing into the EMU neck ring. That was torture. Moving against the stiffness of the suit also resulted in abrasions on the arms and wear marks on other parts of the body. But, in spite of the pain and occasional nibbles of claustrophobia, I loved the WETF sessions. Like the robot arm training, the work was more personally challenging and yielded a greater sense of accomplishment than learning how to throw a switch to release a satellite. I prayed that someday I would get to do a spacewalk . . . a *planned* spacewalk. I never wanted to hear the word *contingency* on any of my missions.

Our final EVA training session afforded me a unique insight into the burden of feminism on the TFNG females. The lesson involved a soup-to-nuts spacewalk dress-out conducted in an exact replica of the shuttle cockpit/airlock. The shuttle mid-deck, though the largest volume in the two-deck cockpit, was small, measuring about 7 feet fore-to-aft, 10 feet wide, and 7 feet floor-to-ceiling. Astronauts liked to impress the public with the gee-whiz fact that the Texas prison system allocated more space for one inmate than the shuttle provided for crews of six people. The airlock was even smaller, a cylinder 7 feet tall and 4 feet in diameter. Most of this space was filled with the two wall-mounted EMUs.

Under the critical eye of our instructor, Judy entered the airlock, disassembled our suits, and passed the helmets and pants into the mid-deck. The torso portion of the suit that contained the electronics, oxygen bottles, water supply, and controls—and weighing nearly 200 pounds—would remain on the airlock wall. While she was busy with this task, Steve and I retrieved our condom UCDs, bio-data attachments, and Liquid Cooling Garments (LCG) from the EVA lockers. The LCGs were a netlike long underwear. Weaved into the material were small tubes that carried chilled water next to the skin to prevent spacewalkers from overheating.

The first task of donning a spacesuit is to get naked and put on the UCD. I had assumed Judy would step out of the cockpit during this intimacy, but I had failed to appreciate how feminism had complicated our situation. No male IVA crewmember would have left the mock-up while other males rolled on their condoms, so Judy knew she could not either. To do so would be a violation of the feminist cause, to send a message that women were different from men. As Judy made no attempt to leave, I shot Steve a nervous *You go first* glance, only to see his eyes answer, *Screw you. You go first.* This was going to be interesting, I thought. At least if I was going to get naked around a nonwife woman, I had a ripped body and the ass of a Greek god to impress her. But, even for me, a guy with few inhibitions, the thought of rolling on a condom while JR was standing in front of me discussing the checklist was, well, inhibiting. Certainly she didn't intend to make sure we did *that* task correctly?

I need not have worried. As Steve and I started with our shirt buttons, Judy sat on the edge of the hatchway, put her head down, and faked reading the checklist. She gave us as much privacy as possible while still holding on to her feminist sensibilities.

As fast as I could, I sheathed myself in latex, Velcroed the nylon bladder around my waist, then slipped into the LCG. Hawley did the same. We were once again presentable.

Before zipping our LCGs fully closed, we attached biosensors to our chests. It was only on spacewalks that MCC monitored astronauts' heart rates. The data was a measure of how much the spacewalker was exerting him- or herself. Astronauts also suspected flight surgeons wanted to be able to remotely pronounce a spacewalker dead in the event of a suit malfunction.

Judy referred to a torso photo in the checklist to ensure we had positioned the sensors correctly. The photo was of a man's chest, his nipples being the landmarks used in positioning the sensors. Those of us from Planet AD had jokingly complained of the sexism represented in the photo. There were female spacewalkers, too, we argued. The checklist should also have a photo of a woman's naked chest showing the sensors properly applied. One AD male cut out a *Playboy* model's photo, drew in the biosensors on her naked breasts, and pasted it into an EVA checklist. He said he was going to clandestinely substitute it for the actual checklist in his next training session with a female spacewalker, but I never heard about the prank being executed, so I suspect he chickened out.

We continued with the rest of the dress-out. We pulled on our pants, waddled into the airlock, and squatted under the wall-mounted torso/arm pack. While Judy held the suit arms vertical, I drove my head and arms upward and into the torso part of the suit. The squeeze through the neck ring ripped at my ears and made my eyes tear. Judy locked the pant waist to the torso, then dropped my helmet into place and locked that down. She next pushed on my gloves and locked those to the wrist rings. It had taken the better part of two hours, but I was now fully dressed. The training session would go no further. If I released myself from the wall mounts, as I would have to do on a real spacewalk, I would collapse under the 300-pound weight.

Judy finished dressing Hawley but before she could get out of the airlock, I encircled her and pulled her into my front in a writhing hug. She laughed. The embrace was as sensual as a fair maiden hugging an iron-suited knight.

Hawley and I had graduated. We were ready for a spacewalk. And as much as each of us wanted to do one, we both prayed it wouldn't happen. If we were on a spacewalk, it would be to save our lives.

Friday, April, 13, 1984, proved to be a very lucky day for the "Zoo Crew." STS-41C landed at Edwards AFB. We were next. We were Prime Crew. With that title came top priority for simulators and T-38s. For me, it also brought Prime Crew night terrors. Until this moment every time my soul tried to deal with the fear and joy of what was fast approaching, I disallowed it. STS-6's IUS failure, STS-9's hydraulic fire, and STS-41B's twin satellite booster failures had made me a skeptic. There was too much in front of us that could jeopardize our mission. Even now, with the horizon clear of any other shuttle missions, I kept my emotions on a very short leash. Until the hold-down bolts blew there were no guarantees, I told myself. An engine could blow up in a ground test and stop the program. My health could become an issue. The payload contractors could find something seriously wrong with their machines. There were thousands of unknowns in this business. *Don't even think about flying in space,* I ordered myself. And during my waking hours I obeyed that order. I had plenty of distractions. However, in sleep, the reality of being Prime Crew would creep past my defenses. I would bolt awake with my heart wildly drumming and my brain overwhelmingly aware that I would be next off the planet. Every fear I had ever harbored about death aboard a space shuttle, every doubt I had ever held about my competence to do the mission, every joy I had ever celebrated at the thought of flying into space would flash through my consciousness in a wild, chaotic fury and vaporize any hope of further sleep. I would get up and go for a walk or run.

By this time "Zoo Crew" had been together for fourteen months and we'd be together a few more. Because of delays in earlier missions, *Discovery*'s launch had slipped to June. In our thousands of hours of training Judy and I had become close friends and I would be a liar if I said I hadn't thought about expanding our relationship beyond the study of payload checklists. That thought was certainly nibbling at me as our T-38s landed on a warm spring Sunday at the KSC shuttle landing strip. Judy and I were there, alone, to support some payload tests that would begin the following day. We jumped into a rental car for the drive to the KSC crew quarters. Wearing Prime Crew smiles, sitting in a convertible (top down, of course), dressed in our blue flight suits, the wind in our hair, the sun on our face, we were everybody's image of the Right Stuff.

Judy parked the car and we grabbed our luggage and headed for the elevator. The crew quarters occupied a small portion of the third floor of a huge Apollo-era rocket checkout building. The facility included a fully equipped kitchen, a small gym with weights and stationary bicycles, some conference rooms, ten or so bedrooms, and a handful of unisex bathrooms. NASA must have consulted with Benedictine monks on the decor of the bedrooms: They were monastery spartan, containing a bed, desk, telephone, lamp, and chair. No TV. To ensure no outside noises would disturb a sleeping crew, the quarters were located on the interior of the floor. There were no windows.

Judy and I found the facility deserted. *Come on, Satan, give me a break,* I thought. I was going to be in sixteen hours of solitary confinement with a beautiful woman and idle hands, those instruments of the devil.

"Hey, JR," I shouted down the hall, "let's check out the old Cape Canaveral launch facilities." On multiple trips to KSC I had tried to fit in such a tour but the schedule had not allowed it. Now was a propitious time. My brain was screaming, *Don't do something stupid. Get out of here!*

"Sure, Tarzan," she called back.

It was too warm for flight suits so we changed into our NASA gym wear. I grabbed the NASA phone book, which included a map, and jumped in the car, letting Judy drive while I navigated. The early launch pads had been preserved as part of the Air Force Space and Missile Museum. The centerpiece of the museum was the concrete blockhouse that had served as the control center for the 1958 launch of America's first satellite. An outdoor display of a couple dozen rockets had been added to the area. The orange-painted latticed gantry of Launch Complex 26 speared the sky a mere four hundred feet east of the blockhouse.

It was late in the afternoon, long after tour hours. The facility was as deserted as the crew quarters. Judy looked at the rocket displays. "How many of these can you identify?"

I did a quick survey. "All of them."

"Bullshit, Tarzan. I'll bet you a six-pack you can't identify all of these."

"Judy, I lived and breathed rockets from the age of twelve. Photos of these things wallpapered my bedroom. You're challenging a rocket geek. You're going to lose that bet."

Her smile said, "No way," and she rushed ahead to look at a placard. "What's this one?"

"The Navajo. It was the world's first supersonic cruise missile. Range fifteen hundred miles."

"Lucky guess." She walked to the next display. "This one?"

"Bomarc. A ramjet-powered supersonic antiaircraft missile."

I could see she was beginning to believe my rocket identification powers might not have been exaggerated.

"This one?"

"Easy. Firebird, an early air-to-air missile. By the way, make it a six-pack of Moosehead."

"You haven't won yet."

But I did. After correctly answering several more of Judy's challenges, she capitulated in front of a Skybolt missile.

"Tarzan, did you do anything as a kid besides memorize rockets, like go to rock concerts or dances?"

"I have one autograph in my high school yearbook. Does that answer the question?"

She laughed. "Yeah, I guess it does."

I was worried an air force security officer would arrive at any moment to lock the blockhouse, so I suggested we take a quick tour of it. For me, stepping inside was a spiritually moving moment. I had never been to this place before, yet I was connected to it. As a child in Albuquerque, I had watched TV scenes of this building and the gantry beyond as the earliest satellites and monkey-nauts, Able and Baker, had ridden pillars of fire into the sky. Werner von Braun had stood where I now stood and directed America's first steps in the space race. I touched a lifeless control panel and felt even closer to him and the history he and his team had written. My fingers brushed across the blockhouse periscope and archaic lights and switches and oscilloscopes. *God,* I thought, *what I wouldn't give to go back to January 31, 1958, and be standing at this very spot as the final seconds clicked off the countdown clock for* Explorer I*'s launch.*

"Be careful, Tarzan. You'll launch one of those rockets."

Judy interrupted my reverie. Her obvious indifference to the history of the site prompted a question that had been on my mind since I had first stood on the stage with her at our TFNG introduction. "JR, when did you first want to be an astronaut?"

"In 1977, when I saw the announcement on the company bulletin board."

She answered as I had expected. I had already heard several of the other females say the same thing in various press interviews. Only Shannon Lucid had a different answer. She had a copy of a letter she wrote to *Time* magazine in 1960 challenging NASA's male-only astronaut corps. She had dreamed of spaceflight as a child, as I had. Only recently had I matured enough to give Judy, Sally, and the others some slack for their lack of lifelong zeal for the astronaut title. If I had been raised in a society that told me I could never be an astronaut because of my gender (or color), would that dream have ever taken root in my soul? Probably not. How, I asked myself, could I hold it against this woman if she had not carried the dream from her childhood? I could not. Judy and the other women were teaching me the meaning and consequences of discrimination.

We returned to the car, Judy still behind the wheel. "Let's go to the beach house," she suggested. "I'll buy you a beer there." It was a destination certain to test the male animal in me. The beach house was as isolated as Mars, situated just behind the dune line only a couple miles from the shuttle launchpads. The house was a relic of the 1950s, before the days of the great space race. Then, the Cape Canaveral area was just one more place for snowbirders to build their winter retreats, and private homes had dotted the landscape. But the *beep-beep* of Sputnik had wrought a great change in this part of America. The newly formed space agency needed a place to launch its rockets and Cape Canaveral was ideal. Exercising its right of eminent domain, Uncle Sam acquired the land and began its spaceport renovations. Only one of the existing structures survived demolition, saved by some enlightened bureaucrat who had decided it would be the perfect retreat for the early press-hounded astronauts. The building selected was well into government property, so privacy was absolute. Even Jehovah's Witnesses wouldn't have been able to find this address. While the press no longer pursued astronauts as they had the Mercury Seven, the building was still used as an astronaut retreat.

On the drive I tried to keep my eyes forward but could not. They kept going to Judy's smile, to her wind-flagged hair, to her golden legs. *Danger, Will Robinson! Danger!* There's never a good robot around when you need one.

Judy turned the car onto a shell-covered driveway and parked. The house wasn't exactly Frank Lloyd Wright. It was something the Unabomber might have cobbled together: small, boxy, utilitarian. The downstairs was concrete and comprised a garage and storage area. The flat-roofed, wood-framed upper story contained a living area of two small bedrooms, a bath, and a kitchen/living area that opened onto an elevated wooden deck. NASA had done little to the structure over the decades. The exterior wood finish was sandblasted and warped, the weather stripping shredded, the concrete walkways uneven and crumbling. The interior furnishings were similarly old and worn.

I stayed outside while Judy walked upstairs to the kitchen with a handful of bills for the honor cash box. While she had been a model of professionalism and had done nothing to suggest there was more to this beach visit than watching the waves and having a beer, every molecule of testosterone in my body was busy suggesting otherwise. I could no longer see her as a fellow astronaut and crewmember. I could only see her as the beautiful woman she was. She came out with a six-pack of Coors hooked on a finger, stood with her hip cocked to the side, and smiled. "It's not Moosehead, Tarzan, but here're your winnings." She tossed the package to me. *God help me,* I prayed.

We walked over to the dunes and sat in the sand. I extracted beers for both of us and for a moment we were silent, just enjoying a perfect beach evening. A thunderstorm lashed the distant ocean at our front, its anvil head glowing orange in the dying sun. There was just enough of a breeze in our faces to keep the bugs away.

"Here's to Prime Crew, Tarzan." Judy held out her beer and I touched it with mine. Her face was illuminated by the reflections from the cloud and I could see her expansive smile. The Prime Crew title did that to astronauts. We'd all be wearing those smiles until Hank's call of "Wheel stop."

We fell into conversation about our training and new issues on our communication satellite deployment procedures. I joined in halfheartedly. She was the only astronaut present on that beach. My mind was busy dealing with the scent of her shampoo, the feel of her body heat radiating across the gap between us, and the voices in my head. Those whispered that it would be different with Judy and me. Mortals had dirty, sinful, sordid affairs. But we weren't mortals. We were astronauts. We were demigod and -goddess, alpha male and female. The rules didn't apply to us. Not on this beach, they didn't. This was a place separate from Earth where vows and social conventions and the Sixth Commandment didn't apply. Of course, it was all testosteronic bullshit, but I listened to those voices with the same intensity I listened to MCC in our simulations.

As I was popping another beer, Judy left the subject of communication satellites. "Tarzan, I want to thank you and Donna for including me in your family." I was glad to hear Donna's name. It was a reminder of what I was . . . married. My thumb folded onto my wedding band.

"You're welcome. We're always glad to have you over." Donna and I had extended numerous invitations to Judy for dinner or other social get-togethers.

"Well, you guys are the only ones. Most of the wives hate me." She was right. Many of the wives did not like Judy. They were threatened by her. Their husbands jetted across the country with her in private planes to train in factories and go jogging together and perhaps, as I was doing at that very moment, to sit on a beach together enjoying a sunset and a beer. *And what happened then?* the wives wondered. At one TFNG party I saw a visibly shaken JR leave early. Later I heard the reason. One of the child-widened wives had taken her aside and screamed at her, "Stay away from my husband!" Though I had not heard a single rumor connecting JR to the woman's spouse and strongly suspected the accusation was completely groundless, I could understand where the woman was coming from. She understood what female youth and beauty did to men (and was doing to *me* on that beach) and she was dropping a preemptive nuke. Judy was feared by most of the wives. At one of our earliest TFNG get-togethers, she arrived wearing formfitting jeans and a white knit T-shirt. Every head, male and female, turned and the hubbub of the party diminished noticeably. A few of the wives looked into their man's eyes to read his thoughts. Some stepped closer to their husbands. There was nothing trashy about Judy's dress. I had seen many of the wives similarly dressed at various casual functions. It was just that Judy looked like one of those impossibly curvaceous mannequins in a boutique window. It was a fact of life that wives never looked like that.

The last thing I wanted to talk about was Judy's beauty and its effect on men and their wives. I mumbled some bullshit about her misreading the women and attempted to switch the conversation back to our training. A segue of "How 'bout them Astros?" would have been more sincere. Judy knew I was lying and I sensed she was hurt by the rejection of the wives. She ignored my training question and continued to talk about relationships, this time about her childhood relationship with her mother. I was shocked by this intimacy. I had never heard her talk about her past. She had been a TFNG for years before someone discovered she had been previously married and was divorced. Now she was opening her soul to me as if I were some harmless confidant. I could only assume she was trying to justify a heated telephone argument with her brother that she had recently conducted in my presence. The topic had been invitations for the Resnik family to attend *Discovery*'s launch. In that conversation it was obvious there was some tension in the family.

I wasn't trained for this. I ached to return to the subject of communication satellite deployment. But Judy continued. She revealed a deep bitterness with her family's demands that she date only Jewish boys and other aspects of teen oppression in the name of religion. (Gee, and I thought only us Catholics were screwed up.) I had more in common with this woman than I had previously thought. But Judy's coming-of-age trauma had been significantly greater than mine. It had led to an estrangement from her mother. I couldn't imagine a daughter dealing with that.

This was the only time Judy ever gave me a glimpse into her past. And, while I'm no Dr. Phil, I sensed she was a deeply wounded and lonely woman. Of course I also considered her vulnerability at this moment. She wasn't crying, but I had never seen her more emotional. It would have been so easy to reach across and offer a consolation hug. I had a couple beers in me. My inhibitions were as feeble as the starlight. But I didn't. I didn't make any physical contact. Not a hand squeeze. Not a pat on the back. Not a hug. Nothing. My resistance to temptation was nothing short of miraculous. Those moments on that sand had been my Garden of Gethsemane. I offered Judy only conciliatory words about how things might change in the future for her and her mom. It was a prophetic comment. Things did change. Twenty-one months later Judy would die a few miles from where we now sat.

I rose from the sand. "We better get back to the crew quarters. Things are going to start early tomorrow."

CHAPTER 18

Donna

A month prior to our June 25, 1984, launch date another milestone was passed. It wasn't noted in any press release but it was significant all the same. At a crew dinner the wives selected two astronauts to be their family escorts. The expanded training hours in the homestretch to launch made all prime crewmembers absentee spouses and parents, so the astronaut office had created the family escort role to take some of the load off the families. They helped spouses deal with the logistics of traveling to KSC and the landing site. They helped with airline, rental car, and condo reservations and, in general, served as 24/7 contacts for spouses seeking help on any mission issue. Some of NASA's rules on family travel necessitated this escort help. While NASA carried the spouses to launch and landing at government expense aboard the agency's Gulfstream jets, children were not allowed on those aircraft. Their travel arrangements (and expenses) were the responsibility of the families. So, as they departed for the most stressful week of their lives, spouses had to pass their children to grandparents or other family members serving as travel escorts and deal with the coordination of getting them from the Orlando airport to their condos. The spouses were also required to arrange their own lodging. This could be a big headache if the mission slipped, particularly during the prime Florida tourist season. Some spouses of earlier missions had found themselves begging with condo reservationists not to be evicted. The escorts could be an enormous help.

There were no formal criteria for selection of family escorts. Crew spouses usually threw out a few names to consider and quickly settled on two. Our spouses picked TFNG Dick Covey and Bryan O'Connor (class of 1980) as their escorts. Unspoken in their deliberations was another duty for which the family escorts were being selected: If *Discovery* killed us, they would become casualty assistance officers. I suspected every wife knew this. Even if their husbands were negligent in not telling them, they probably heard from other wives. I had told Donna years earlier. NASA required her and the kids to watch my launches with the family escorts from the roof of the Launch Control Center. It wasn't the view NASA had in mind: NASA wanted to isolate the families from the press in the event of disaster. In that case the family escorts, turned casualty assistance officers, would drive them to KSC flight operations, where a NASA jet would whisk them back to Houston.

That evening, on the ride back from the party, Donna turned to me and said, "This is a strange business when you have to preselect an escort into widowhood." She was enduring a lot for my dream.

I was selfishly consumed by the flight, and it weighed on the entire family. Why Donna didn't just walk away from me in the final weeks was a miracle. On one occasion I arrived home to news that Pat had strep throat. "The flight surgeon wants you to come in for a throat swab, too." It was no surprise that Donna had sought medical help at the surgeon's office. The doctors also served as astronaut family physicians. But I was furious with her. Though I was feeling fine, I had no idea what a throat culture would reveal. Visions of *Apollo 13* and Ken Mattingly's removal from that mission because of an exposure to German measles aroused my paranoia to insane levels. I raged at her, "Goddammit, Donna, I'm ten days from leaving for KSC! This could screw me!" I made no inquiry of Pat's condition. Donna had never met the man who was now in her face. Tears streamed down her cheeks. I ordered her, "Until I launch, don't go back to the surgeon's office for anything! Nothing! Find a civilian doctor." I would later apologize to her, but I will always carry the memory of this failure as a husband and father. There are some things you can't take back. I ignored the flight surgeon's request, but he badgered me at my office until I finally submitted. The swab results were negative.

At T-7 days to launch I moved into the temporary trailer complex that served as the JSC crew quarters. This was a requirement of flight medicine's mandatory health quarantine, a program designed to minimize the chances of an ill family member infecting a Prime Crew astronaut in the homestretch to a mission—just what I had feared in the case of Pat's strep throat. From this point onward, everybody, including our wives and all NASA employees whose duties put them in contact with us, would have to first be checked by the flight surgeon before they could be in our company. School-age children were forbidden any contact.

I said good-bye to the kids. Pat and Amy were now sixteen, Laura thirteen. I had always been open with them about the dangers of spaceflight, so they understood the significance of this parting, that it might be the last time they would ever see me. Pat and Laura were composed and quiet, while I detected a nervous intensity in Amy's eyes and voice.

I invited Donna to every crew quarters' supper and, after a quick exam by the surgeon, she was allowed to attend. On the last evening before our departure to Florida, we went to my room and slowly and quietly enjoyed ourselves under the sheets (*very* slowly and *very* quietly, for it was a trailer). In the dark I whispered in her ear, "The next time we do this, I'll be radioactive." Neither of us mentioned the other possibility . . . that there might not be a next time.

On June 22, 1984, "Zoo Crew" departed for Florida in a flight of four T-38s. In a routine that had long been perfected by NASA PR, our wives had preceded us. The press liked this human touch of the women waiting to greet their men and NASA was happy to oblige them. As we entered KSC airspace we took a turn around *Discovery,* then slipped into a fingertip formation and entered the "break" over the shuttle landing facility. Hank waited until every plane was landed and we taxied to the apron together. We cut the engines, popped our canopies, and climbed from our jets. Our spousal embraces were captured by a clutch of news photographers. It was a *Life* magazine moment. We were the heroic knights, come to joust with the forces of death . . . fire and speed and altitude . . . and our fair maidens were there to bid us adieu.

The final two days before a launch were designed to be relaxing. There were no simulations. We studied our checklists, flew in T-38s, and enjoyed suppers with our wives. But relaxed? Not a chance. I was hours from achieving a lifelong dream. Pure adrenaline was surging in my veins. Sleep was a struggle. The night terrors were ready to awaken me at the instant of unconsciousness.

We said the final good-bye to our wives at an L–1 luncheon at the astronaut beach house.* The last time I had been there was with Judy. As Donna stepped from the van, I was glad I had no regrets about that night. The NASA-catered lunch was attended by our wives, the family escorts, and key launch personnel. The gathering was informal. There were no speeches, no toasts. Everyone helped themselves from a table set with sandwich fixings and chips. We filled our plates, found a place to park a beer, and enjoyed ourselves.

After lunch, all but our significant others departed and we were left alone to say farewell. Diane Coats took it hard. She was a naval aviator's wife. She knew the danger. Hank's wife, Fran, seemed composed. This was her second time through the drama and that probably helped. Or maybe she was dying inside but hid it for the benefit of the younger wives. After all, she was the commander's spouse and had to set the example. Our payload specialist, Charlie Walker, and his wife were struggling. Judy was spared the spouse separation issue. If Sally Ride, Steve Hawley's wife, was experiencing any fear, it wasn't on display.

Donna and I walked to the beach and turned north. The day was a furnace and the surf splashing on our legs was welcome. Just a few miles away was Pad 39A and *Discovery*. In our stroll we joined the end of a line of astronauts and their spouses, stretching two decades into the past, who had made this same walk in the shadow of their machines: Redstones, Atlases, Titans, and Saturns. A river of tears had been shed on these sands as couples struggled to come to grips with their tomorrows and the potential for glory or death. Now it was our turn.

I was ill-equipped to deal with this moment. When it came to emotions, I was my mother's son. I once teased Mom about her seeming lack of emotion and she replied, "You'll never know what a Pettigrew [her maiden name] is feeling. It's just the way God made us. We keep it all inside." At the DNA exchange of my conception I had been stamped a Pettigrew. It's not that I don't deeply feel things or that I'm afraid of unmanly labels if I reveal them. It's that I can't. What I feel in my soul and how those feelings are verbalized are two entirely different things.

*NASA uses the term L– (pronounced "L minus") to indicate the days remaining to launch. Within twenty-four hours of launch, the term T– (pronounced "T minus") is used to indicate the hours, minutes, and seconds to launch.

The good-bye could not be delayed and Donna finally brought it to the surface. I could hear her sniffling. She stopped and embraced me. "Mike, hold me." As I had always done in poignant moments in my life, I now tried to hide behind humor. "We could go back to the beach house bedroom and do more than hold each other." Always the joker, that was me.

"Just shut up, Mike, and hold me. It's not funny."

She sobbed into my neck and I felt like shit. I tried to calm her with comments about the shuttle's critical component redundancy, but that went over about as well as my "Let's make whoopee" comment. I wasn't going to talk her out of her fear. This was a woman who knew the cost of high-performance flight. She had held the decomposed hand of a friend at an aircraft crash site. She had seen the squadron commander and chaplain step from a car and walk to the door of a neighbor to deliver the "Your husband is dead" message. She had comforted the widows and children of how many friends? I could not guess. She was the woman who had seen through the NASA euphemisms to identify the astronaut family escorts as "escorts into widowhood." Nothing I could say was going to bury Donna's fears.

We sat for a while and just listened to the waves and watched pelicans kamikazeing after their meals. Donna broke the silence. "It's been a lot of water under the bridge to get here."

"Yes, it has."

"I can see you right now as a teenager launching your rockets from the desert. It's amazing where it led."

"I've got a rocket right here you can launch." There it was again, my shield of crude humor.

"Mike, can't you be serious?"

I forced myself to be the man she wanted at this moment. "Okay. I'm sorry. I will be serious. Whatever happens tomorrow," I felt her tense at the implication of the word *whatever,* "I'll be living a dream. It wouldn't have happened without you." It sounded corny but it was the truth.

We embraced and kissed. It wasn't *From Here to Eternity* passion but it was sufficient for the moment. It was easier for me to convey my feelings in this physical contact than it was through words. I could taste the salt on her cheeks . . . tears, not ocean.

I thought of how many random, seemingly inconsequential events steered me through life. If my mom and dad hadn't ignored those "Danger, Unimproved Road" signs, where would life have taken me? If they hadn't settled in Albuquerque, where the sky had captured me, what path would I have journeyed? If I had married a different woman seventeen years before, would I now be sitting on this beach?

In a remarkable coincidence, Donna was born on the identical day and year of my own birth, September 10, 1945, in Albuquerque, New Mexico. She was just a few hours older than me. (In her childhood, my youngest daughter was convinced that men and women had to marry someone who shared their birth date.) Donna's mom and dad, Amy Franchini and Joseph Sei, were first-generation Americans, born of Italian immigrants. Both spoke fluent Italian, argued incessantly, and smoked like forest fires. When Donna was born, the couple already had one child, a boy, ten years old. They had been trying to have a second child for nearly a decade, praying to a pantheon of saints for a daughter. Amy Sei was thirty-five years old when she finally conceived. In their minds Donna was a miracle and she quickly became the center of the couple's existence.

Donna's life was the polar opposite of mine, root-bound. She never moved. Throughout her youth she lived in the same home, only a few blocks from one of the major pathways of adventure for the Mullane clan, fabled Route 66. As a little boy, I had passed within a few hundred yards of the little girl I would one day marry.

We were high school students when we first met. She attended the downtown Catholic high school, St. Mary's, while I was a student at the uptown school, St. Pius X. Her cousin was my classmate, and through this family connection, Donna and I were introduced in 1961 in our sophomore year. This was a horribly insecure time in my life. My blemished face would have repulsed the Elephant Man. It looked as if I had lost a paint ball game in which the other side had been using Clearasil bullets. And, of course, there were my radar-dish ears to horrify the ladies. I could not imagine any girl finding anything attractive about me. When Donna was introduced I said "hi" and ran to hang with the guys. Destiny would have to wait for another four years.

I continued my tortured journey through high school, occasionally running into Donna at various teen functions, but never talking to her, much less asking her on a date. She wasn't beautiful. Attractive, with a bubbly personality, would be an honest description.

In May 1963, I graduated from St. Pius and several weeks later departed for the hellish rigors of West Point. Donna was now two thousand miles away and nowhere in my mind. I was fighting to survive. Upperclassmen were taking numbers to get in line to scream in my face. Even after putting plebe year and its hazing behind me, the pressure did not diminish. The academic workload was overwhelming. I couldn't imagine any other nineteen-year-old in America having it worse than I did. I was wrong.

In faraway Albuquerque, Donna was ill, suffering periodic bouts of nausea and vomiting. Since she had previously experienced a kidney infection, her mother assumed a reoccurrence and took her to the doctor. The blood test results came. With her mom sitting primly at her side the doctor delivered the diagnosis . . . pregnancy. The father was another teenager.

Donna's parents were destroyed. This was 1964 and it was a minor scandal for even a Hollywood starlet to be pregnant and unwed. For a traditional Italian-Catholic family to have a pregnant, unwed daughter was worse than a diagnosis of a terminal disease. For the first time in her life, Donna got to see her father cry.

It was her brother who organized a face-saving escape. Donna would stay in Albuquerque as long as possible. Before her belly could betray her, she would travel to an out-of-state Catholic home for unwed mothers. The baby would be given up for adoption. Extended family and friends would be fed the lie she had left town for college, but few would believe it. The daughters of traditional Italian-Catholic families did not leave home until they were married. But a priest, who was a family friend, taught at a nearby Catholic university so he willingly joined the conspiracy, prepared to cover for Donna if anybody inquired after her.

While I was thinking my life had ended at West Point, Donna was certain hers was also over. She had far greater reasons to feel condemned. She was making the first trip of her life away from her parents . . . as a social and family outcast. Joe and Amy made it clear she had shamed them. "Don't you dare look at that baby when it's born" had been her mother's send-off warning. "I don't want you getting attached to it."

For months Donna cried herself to sleep in a Catholic geriatric-care center operated by the Sisters of Charity. A portion of one floor of the facility had been converted into a dormitory that Donna shared with two dozen other scarlet-lettered women. In exchange for their room and board, the girls helped the nuns with the care and feeding of the aged. They carried trays through rooms and hallways scented with urine, feces, and death. Depression hung on them like a shroud. The mother superior proved to be the quintessential witch, treating them as morally flawed beings. There was no counseling, no trips, and few phone calls to or from loved ones. Donna wrote letters home, putting coded notes on the envelope to designate which mail could be shared with the extended family. In those she created a life at college. In the others she begged for forgiveness.

The girls found comfort only among themselves, but even that succor was transient. As quickly as friendships blossomed, they would end. Girls would give birth and move on to uncertain futures. Unlike other traumatic events that bond people for life, living in a home for unwed mothers was naturally terminal for friendship. None of the girls wanted continued communication for fear of discovery of their sinful secret.

Donna's moment arrived in the summer of 1964. When the baby came there were no exclamations of joy, no rush to take photos for grandparents, no happy tears. Instead, the child was immediately taken away.

Yeah. I had it tough at West Point.

Donna returned home to distrustful parents who watched her like wardens. She had no future but what her mom and dad would allow.

Meanwhile, I had become adept at shooting an M-14 with laserlike precision, getting across a ten-foot-deep pool in full combat gear, and enduring the shit being pounded out of me in boxing class. But none of it helped in my quest to attract a girl. I retained the romantic IQ of a snail. On second thought, snails have no problem being attractive to other snails. I was something else, maybe an evolutionary dead end. My genes would never go forward. I was alone and unwanted.

On January 3, 1965, destiny decided to reintroduce Mike Mullane and Donna Sei. We were partying with family and friends at Donna's cousin's home. My yearling (sophomore) Christmas leave was ending and I had a plane to catch back to West Point. There is nothing more depressing than returning to West Point from a leave, particularly a Christmas leave. It's akin to going back to prison or perhaps dying and going to hell, except this hell is cold and gray and more depressing than anything Beelzebub could ever dream up. To top it off, my girlfriend had dumped me earlier that day. When I say "girlfriend," I exaggerate. I met her in my senior year of high school and throughout plebe year had pined for her. It was a one-way infatuation. To be "dumped" implies there was something that ended. There was not. It was more like she threatened to get a restraining order.

In my despair, I resorted to that cure of the ages, alcohol. There was plenty at the party and I drank to forget . . . to forget being alone and to forget a flight back into the ninth circle of hell. As the moment of departure approached, I walked outside to get away from the fun. I wasn't having any and it was depressing to be around people who were.

Donna observed my exit and minutes later followed me. We walked for a while making small talk about our friends and our new lives. Romance was nowhere on my mind—it was Donna who took the lead. She leaned into me and kissed me . . . on the lips, no less. And it was all her doing! I didn't have to beg or plot. It was as if the sun had risen, West Point had slid into the Hudson River, and I was on infinite leave! I was in love . . . well, lust maybe, but it would do. Never in my young life had a girl shown any romantic interest in me. Never. I found heaven in Donna. She was a life preserver in the sea of my muddled adolescence and I grabbed her and held on for dear life.

Donna drove me to the airport, as I was in no condition to do so myself. As we parted, she kissed me again. It was all I could do not to propose marriage. She asked me for something to write her address on. SHE ASKED ME! Again, I didn't have to beg. She *wanted* me to write. It was truly a night of firsts. I fumbled in my wallet for a piece of paper and found my Army Code of Conduct card, a card that detailed how a soldier was to act if captured by the enemy: "If I am captured, I will continue to resist by all means available. I will make every effort to escape and aid others to escape, etc., etc." Why I was carrying this, I have no idea, but if there was ever a signature of what a nerd I was, this was it—giving an Army Code of Conduct card to a girl to write her address on. Donna should have known right then and there what a doofus she was hooking up with.

How quickly one's heart can change. Now I couldn't wait to get back to West Point. I couldn't wait to send a letter. I flew to Colorado Springs, where I connected with an Air National Guard flight to a field near West Point. The plane was filled with returning cadets who were slumped in their seats in near suicidal depression. But not me. As the C-97 droned eastward, I wore a permanent smile, the dopey smile of young love. Other cadets stared at me, certain I had lost it, certain at any moment I would rush the door and leap to my death. No sane cadet smiled while returning to the granite asylum.

I penned my first letter within an hour of arriving in my room. As I sealed the envelope, I stared at the photo of my imaginary girlfriend. I thought of how long I had defined happiness as getting this girl to love me, how long I had prayed she would send me letters (none ever came). As Garth Brooks sings, some of God's greatest gifts are unanswered prayers. I tossed the photo in the garbage.

Donna's and my relationship continued through the mail. I sent out more letters than Publishers Clearing House. I needed continual assurance she was still there, that she wasn't as imaginary as my prior girlfriend. Her letters arrived by the truckload. We had "known" each other for a total of two hours, yet in our correspondence, we each professed our everlasting love.

Donna's entry into my life probably saved me from expulsion from the academy. While my academic grades were satisfactory, my "military bearing" was seriously lacking. I had loathed the hazing of plebe year, never understanding how it could possibly contribute to the development of a leader. (I still don't.) Upperclassmen sensed my contempt for the tradition and I was rewarded with a steady stream of demerits for various infractions, such as scuffed shoes, unpolished brass, and failure to satisfactorily render plebe knowledge. As a yearling I was rated near the bottom of my class in leadership skills by senior cadets. I was certain some of those cadets interpreted my lack of zeal at enforcing the plebe system on the latest class as further evidence of my disdain for it. Before Christmas leave, I was warned by my tactical officer that I could be terminated if my attitude didn't improve. My parents received a letter from that same officer, saying that I was floundering, and my dad called in an attempt to rally me. But I remained indifferent to the warnings. I was rudderless, not sure I even wanted to stay at West Point. Then Donna stepped into my life. In her I found clarity and focus. I had to succeed—not for myself, but for her. Almost overnight my attitude and behavior changed. While I'm certain my superiors thought it was their great leadership that had turned me around, it was really Donna. I still had discipline relapses, such as when I was caught skipping a senior class's graduation ceremony, a transgression that earned me another tactical officer rebuke and forty-four hours of "walking the area" with a shouldered rifle, plus a two-month confinement to my barracks. But I was on the road to graduation, guided unerringly by a star two thousand miles away—Donna.

In February 1965, I sent her an "A-pin" (A for *Army*), which was West Point's version of a fraternity pin. It was another blitzkrieg escalation of our relationship.

In March 1965, I flew home for a three-day spring leave. Donna and I were inseparable. We grew more emotionally—and physically—intimate. At age nineteen, at the Silver Dollar drive-in theater in the backseat of a 1954 Chevy Bel Air, I finally got to second base with a girl. It was also in this passion pit that Donna told me of her dark secret, that she had had a baby. I didn't care. It didn't change anything between us, I said. This might sound mature and noble except, at the time, I had my hand in her bra. She could have told me she was a whorehouse madam and it wouldn't have mattered.

Then, as the dialogue of some forgotten film squealed and popped through the window-mounted speaker, I proposed marriage and Donna accepted. There was no ring, no romantic dinner, no months of wonderful anticipation. It was as spontaneous as a heartbeat. I was mad to legitimize my claim to this woman and *mad* was the correct word.

Much later in life Donna and I would recount a PG-13 version of this story to our teenage children and warn them that if they ever did what we did, we would kill them. It was insanity. We were engaged to be married after knowing each other for a total of three days and a hundred letters. I was marrying for sex. Donna was marrying to escape her parents. *Oh, yeah. This is gonna last.*

For her birthday in 1965 I mailed Donna an engagement ring. That's correct . . . I *mailed* it. I couldn't wait until we were together again. This woman had become my life. I couldn't let her escape. But marriage was going to have to wait until after my graduation, two long years away. West Point cadets were forbidden to be married.

We were able to wait for marriage, but not the honeymoon. On my summer leave of 1966—in my twentieth year of life—I finally slid into home plate with a girl. It happened in Donna's bedroom. Her parents were away for a few hours, which established opportunity. Motive had long been raging. Two hours later Donna and I were in the confessional admitting our sin to the tobacco-breath shadow behind the curtain. The priest reminded me that having premarital sex was a violation of the temple of God (our bodies) and I would burn in everlasting fire if I didn't change my ways. (I guess it was okay to smoke in God's temple.) Donna and I shared the same kneeler as we prayed our penance and promised God that in the future we'd keep our hands and the rest of our bodies to ourselves. Even under penalty of losing our immortal souls, we couldn't keep that promise. On every leave we'd end up in that Chevy, parked in a drive-in theater or the wilds of the desert, the windows steamed over and our sacred "temples" in rhythmic collision. The next day we'd be in the confessional hearing more dire warnings of hellfire ahead. I have no doubt we frustrated that priest into a three-pack-a-day habit.

At my graduation from West Point I took a commission in the USAF, something I was permitted to do because my dad was a retired USAF NCO. But I was not released to the commissioning ceremony until my tactical officer made one last effort to get me to pledge my life to the U.S. Army. "Mr. Mullane, going into the air force is the dumbest thing you could ever do. Your background is all army. You'll never get far in the air force." Thank God I tuned him out.

Donna and I married one week later in the Kirtland AFB chapel in Albuquerque. She made a lovely bride. In high school she had never worn the tiara of the homecoming queen or the uniform of a cheerleader or played the lead in the senior class play. She didn't possess the beauty of girls who typically captured those honors. But seen through the lens of my young love, she was the most beautiful woman in the world.

Three of my West Point classmates served as groomsmen. We were all in uniform—they in their army dress blues and I in my black-tie air force livery. Military weddings are timeless. With the carefree smiles of youth and the lights glistening from our polished brass, the scene could have been lifted from WWII, or even a Civil War daguerreotype. We were still too intoxicated by our recent release from West Point to hear the guns of our war . . . Vietnam. But they were waiting for us. Mike Parr, one of my groomsmen, would be killed in action seventeen months later.

Donna and I took a honeymoon to someplace. I hardly recall where. We never left the sheets. I only remember that the rented room had hardwood floors and the bed was on castors. If there had been an odometer on the bed frame, the instrument would have recorded a couple thousand miles during our short stay. By the time we returned to Albuquerque, Donna was already morning sick, pregnant with twins. (This was before the days of ultrasound. We wouldn't know she was carrying twins until two weeks prior to her delivery.) As we had done everything else, we had children spontaneously. There had been no real thought or discussion. We were Catholic. You got married and had kids. What was there to discuss?

In July 1967, we drove from Albuquerque to begin our life as military nomads. In that car was a social retard . . . me. It is true what cadets say about West Point: "It takes eighteen-year-old men and turns them into twenty-one-year-old boys." Did it ever. I had learned to drive tanks and fire a howitzer and field-strip a machine gun, but I had never used a Laundromat or cooked a meal. I couldn't dance. I had never written a check. I had never made a stock investment or shopped for a car or clothes or groceries. I had no clue about home ownership.

God only knew what this woman at my side saw in me. But throughout my journey toward the prize of spaceflight, Donna never wavered in her support. Even ten minutes into that drive from Albuquerque, she was there for me. I was still trying to come to grips with the fact that my bad eyesight had blocked me from pilot training and thrown me into navigator training instead. To be an astronaut I would have to be a test pilot and that wasn't going to happen if I couldn't get into pilot training. Donna knew how bitterly disappointed I was and gave me her shoulder to cry on. "It'll all work out for the best, Mike. God has a plan. You'll see." That was Donna's hallmark, the faith of the pope. She would turn every house we would ever occupy into a mini-Lourdes, with wall-mounted crucifixes and Virgin Mary statuary everywhere. In our bedroom she always had candles burning for one saint or another. She would send cash gifts to various orders of nuns and ask them to pray for us. Priests would get checks asking that they say Masses on our behalf. If the Mike Mullane family had a connection to God, it was certainly through Donna, not me.

A small cinder-block house on Mather AFB in Sacramento, California, was our first home. While we waited for our few possessions to catch up, we enjoyed Uncle Sam's furniture. We sat on metal folding chairs and ate our meals off a card table and made love on a one-man canvas cot. It was the richest we've ever been. We had each other and that was all we needed.

My immediate goal was to graduate from navigator training into the backseat of F-4 Phantoms, so I worked like a Trojan to finish high in my class. It wasn't going to be easy. Flight assignments were given in order of class rank, and the group was filled with Air Force Academy wizards who had been through much of the coursework during their academy years. But Donna was there for me. I would hang a sextant from a neighbor's child's swing and practice shooting three-star "fixes." She would be at my side, teeth chattering in the cold night air, holding a flashlight to illuminate the instrument bubble chamber and recording my observations on a clipboard. When the twins were born on March 5, 1968, she assumed full parental duties to allow me to continue to focus on my studies. Never once did she hound me to get up for a 2 A.M. feeding or wash the diapers or prepare formula. No new father of a single child, much less twins, had it as easy as I did.

I graduated first in my class and took an assignment to the backseat of RF-4C Phantoms, the reconnaissance version of that fabled fighter. I had never finished first in anything in my life and it wouldn't have happened without Donna.

Meanwhile, I continued to make calls to air force HQ begging for a pilot training position, but the requests were repeatedly denied. At my annual flight physical I hounded the surgeon about ways I might improve my eyesight. He said there were none. "Mike, your astigmatism is caused by a physical defect in the lens of your eye. There's nothing that will correct that defect." I refused to believe him and searched the library for a miracle . . . and thought I had found it in a book titled *Sight Without Glasses*. But after practicing the recommended eye exercises for months, my visual acuity did not change. I remained physically unqualified for pilot training. As I cursed my bad luck, Donna continued to preach patience. "God has a plan."

From Mather AFB we were transferred to Mt. Home AFB, Idaho, for my transition training to F-4s. It was here I had my first aviation near-death experience, not from an engine fire or hydraulic failure, but from airsickness. I was dying in the cockpit. I couldn't complete a flight without my head in a barf bag. I would come home and collapse in depression. The writing was on the wall: The squadron commander was going to eliminate me from training, if I didn't barf up my duodenum and die first. But Donna was there for me. Her shrine blazed like a sunspot. She circled her rosary like a Tibetan monk on a prayer wheel. But my situation was so perilous she wasn't going to leave it just to heaven to deliver a fix. Having suffered months of morning sickness while carrying the twins, she was an expert on puking and was convinced I could be cured with the right breakfast. The specifics of the meal she cooked for me have long left my memory, but it worked. No doubt it was just a placebo effect, but I didn't care. I got through a flight without seeing that breakfast again. And then another. And another. My self-confidence roared back. My flying career was saved by Donna.

From Mt. Home I was directed to Saigon, Republic of Vietnam. I would be flying with the 16th Tactical Reconnaissance Squadron from Tan Son Nhut Air Base. Donna and the babies would wait out my tour in a Kirtland AFB house in Albuquerque. During my Christmas leave before departure we were initiated into the reality of war. A sobbing Jackie Greenalch called to tell us her pilot husband, a close friend from Mt. Home, had been shot down and killed while flying his RF-4C. He had been in-country only a few weeks. With this grim news hanging over us Donna drove me to the airport. We stopped for coffee at a doughnut shop and she cried while a teenage server gawked. The good-bye was made even more painful by the strains of "I'm Leaving on a Jet Plane" coming from a back room radio. But there were no ultimatums or threats or pleading that I abandon the air force upon my return. She would give her all for me and my career, wherever it took us.

I returned from Vietnam in November 1969, and was transferred to RAF Alconbury in England, where I crewed RF-4Cs, now as part of the NATO forces confronting the Soviet Union. I continued to press air force HQ for entry into pilot training, but my requests were denied. Not only were my eyes a problem, but now I was too old. "Give it up, Lieutenant Mullane. Pilot training is out. It's not going to happen for you" had been the personnel officer's unvarnished assessment. Except for a handful of civilian scientists selected for the Apollo program, every NASA astronaut had been a test pilot. They had never selected a GIB—or guy in back—in fighter jets. At age twenty-six my dream of space-flight had ended. When I told Donna of this final rejection, she was unbowed in her faith. "It will all work out for the best."

England was a bittersweet four years for us. Every several months there were training accidents, some of which were deadly. We attended memorial services for Jim Humphrey and Tom Carr, killed in their take-off crash. Another crew disappeared over the sea on a night mission. A pilot died when his Phantom caught fire.

But there was also the fun of traveling through Europe on our vacations. We left the kids with a nanny and rented a sailboat with two other couples and sailed the Aegean Sea for two weeks. We sipped wine while watching the rays of a setting sun pierce colonnaded ruins, and we swam in water as clear as space. While anchored in deserted coves and cloaked in nights so black we couldn't see each other even while kissing, Donna and I made love as quietly as a prayer. We walked the streets of Rome and Edinburgh and Florence. We played in the snow in the Austrian Alps and watched London stage plays. On a visit to the coast of Spain, Donna became pregnant with our third child, daughter Laura. On this same trip I volunteered to take up a cape and fight a one-ton bull at an organized tourist function. Alcohol was involved in both moments.

Our European assignment was the first time Donna and I had enough long-term stability in our lives to really get to know each other. And two more different people have rarely been pledged in the banns of matrimony. Donna was a lady. She was polite, sober, and soft-spoken. I, on the other hand, was as coarse as a convict. The all-male experiences of West Point, the air force, and Vietnam made it impossible for me to form a sentence without the F word in it. I was loud, frequently obnoxious, and an out-of-control joker. On our sixth anniversary, at a squadron party, I presented Donna with a gift-wrapped painting. She was certain of the contents. For a year she had been hinting at how much she wanted an oil portrait done of her in her wedding dress. Not one to disappoint, I found an English artist, gave him a photo from our wedding, and asked him to capture Donna. But I also included a Polaroid photo of a topless Donna I had taken in our bedroom. I requested a watercolor of that shot too. It was the latter painting I first presented her at our anniversary party. She ripped into the wrapping, breathless to see herself in bridal splendor. When she peeled away the last paper and found two nipples staring her in the face, she nearly fainted. She clutched the painting to her body while the bewildered audience asked, "What did Mike give you?"

We had vastly different senses of humor and decorum. But these were merely the veneer of our personalities. At the most fundamental levels we were also light-years apart. Donna was rule-oriented, risk-adverse, inflexible, and easily stressed. When I tore the warning label from a new mattress, she was certain a SWAT team would come bursting through the door. Just missing an exit on a freeway would virtually paralyze her. Any time the gas gauge on the car dropped below one-half she became as nervous as a fighter pilot sweating out a midocean aerial refueling. She was obsessive-compulsive to an extreme even I couldn't touch. All in all she had a personality ill suited for a woman married into the nomadic and dangerous world of military aviation. It had to have been torture for her to kiss me off to work, particularly after the plane crash that claimed our neighbor, but I never heard a complaint. My career had become her career.

In 1974 we transferred to Wright-Patterson AFB, Ohio, where I entered the Air Force Institute of Technology in pursuit of a master's degree in aeronautical engineering. We had barely unpacked before we were once again on the move to Edwards AFB, California, where I entered the Flight Test Engineer School. In both places, Donna became sole parent as my studies consumed me.

In 1976, while I was finishing my Edwards assignment, NASA announced it would begin accepting applications for the first group of space shuttle astronauts. For the first time in the agency's history there would be an astronaut position, mission specialist, that did not require pilot wings. It was astounding news. I was now eligible to apply to be an astronaut. Not only was I eligible but my flying background, master's degree, and flight test engineering credentials made me a strong candidate. When I rushed home to tell Donna the news she smiled and said, "I told you. Everything works out for the best. God has His plan," and then she went to her shrine and lit another bonfire, this one of thanksgiving.

How much of my curriculum vitae did I owe to Donna? All of it. Every step in my career set me up to meet the challenges of the next step. If I had stumbled at any point, there would have been a hole in my life that would have put my astronaut application in the "nice try" pile. But I hadn't stumbled. Donna had provided me with the one thing I needed more than anything else . . . the opportunity to focus. Unlike many of my air force peers and nearly all of the TFNGs, I wasn't a gifted person.

I couldn't get ahead on innate brainpower. I was more like the Forrest Gump of MS astronaut applicants. For me to have passed through the wickets of navigator training, combat flying, graduate school, and flight test engineer training required the intense focus of a dung beetle. And it was Donna who provided me the freedom to focus—to pour my heart and soul into the task at hand, to volunteer for extra flights, to take on additional squadron duties, to stay late in the grad school labs. I was spared the major distractions of married-with-children life.

Countless twists and turns in my life had put me on a Florida beach on June 24, 1984, only hours from my first space mission—but none as significant as the night a teenage girl stepped out from a party to kiss me.

CHAPTER 19

Abort

Back at the crew quarters I changed into my athletic gear and headed for the gym. If I died on tomorrow's mission, I would die in perfect health. I weighed 145 pounds, ten pounds less than I had weighed twenty-one years earlier when I had graduated from high school. I doubt there was a pound of fat on my 5-foot-9½-inch frame. I could run five-and-a-half-minute miles . . . four of them back to back. I had a resting heartrate of 40. My ass was so tight I could have cracked walnuts between my butt cheeks.

As I pumped iron, I chuckled at the sight of the straw-filled archery bull's-eye in one corner. What candy-ass astronaut had requested that addition to the gym? Whoever it was, I hoped they could fly a shuttle better than they could shoot a bow. The plaster wall around the target had been shotgunned by errant arrows.

I left the gym for an outdoor run and found Judy stretching before her jog. We fell in together. It was early evening and KSC had emptied of workers. Our only company were mosquitoes, and they were a real incentive to keep the pace fast. Sweat came quickly, which was the whole purpose of the run. I wanted to dehydrate myself to minimize bladder discomfort in tomorrow's countdown. Other astronauts did the same. The few couch potatoes in our ranks tried to wring themselves dry by sitting in a whirlpool bath and drinking beer, counting on the diuretic effect of the alcohol and sweat to do the job.

Judy and I passed the black hulks of several alligators on the opposite banks of drainage ditches. I had once seen one of these creatures explode out of the water in a chase after an armadillo. Why they never chased jogging astronauts was a mystery to me. Even when we teased them, they did not react. I once watched Fred Gregory toss shells at a twelve-footer hoping to see it stir. As the missiles ricocheted from its scales, I warned Fred, "Those things can run twenty miles per hour when they're riled and that's a lot faster than you." But Fred continued his reptile target practice while answering, "Yeah . . . but that's on firm ground. If they're chasing me, they'll be slipping and sliding through shit and they can't run nearly as fast."

Judy and I discussed an issue we had heard about just before leaving the crew quarters. An engineer had found a potential flaw that could result in the failure of the burned-out SRBs to separate from the gas tank. Such a scenario would be fatal. The shuttle would never make it to orbit or achieve a successful abort dragging along nearly 300,000 pounds of useless steel. The good news was it would take several simultaneous failures in the circuitry for the SRB separation failure to occur. When the launch team asked Hank whether he was comfortable flying the mission with this failure mode in place, his answer was yes. That didn't surprise me. They might as well have asked a three-year-old if he wanted to eat his candy now or wait until tomorrow. If the engineers said, "We forgot to install the center engine. Do you still want to launch?" Hank probably would have said, "No problem. We'll just burn the two we have." Nothing was going to get in our way.

Judy and I continued into the KSC wilderness. The only sound was the buzz of cicadas. The dusk was deepening and an occasional firefly flickered over the ditches. Judy voiced concern that we might trip over an alligator. I told her the Fred Gregory story and she laughed. But in a rare moment of prudence, we decided to turn back.

We gradually slowed into a cool-down walk. I had come a long way . . . and I don't mean during the run. There was a fox of a woman at my side and I could actually think of her as a friend and equal. Those six years ago when we were standing together on the stage being introduced to the JSC workforce, I saw Judy with three strikes against her: She was civilian. She was a woman. She was beautiful. At the time I wondered how her beauty had played in her passage through the wickets of life to become an astronaut; wickets that, for the most part, had been male tended. Had she been waved through some of those gates because her smile had melted a professor or perhaps her dynamite body had influenced a male astronaut sitting on the selection committee? We males are suspicious of female beauty because we know ourselves too well.

But, over the years, Judy had proven she wasn't an astronaut because of her sex appeal or because of an abuse of the affirmative action program. She was an astronaut because she was qualified to be one. I had watched her fly formation from the backseats of T-38s and lead instrument approaches in bad weather and do it as well as me (and my backseat fighter and T-38 time had made me a damned fine instrument pilot). I had seen her expertly operate the robot arm. I had watched her rappel off the side of the orbiter mock-up in our emergency training, parasail into the water in our survival training, work 20 feet underwater in a 300-pound spacesuit. In simulation after simulation, she had instantly and correctly reacted to countless emergencies. I think the best testimonial for Judy's proficiency was the fact it was never a topic of astronaut scuttlebutt. In a strange way, that was the best compliment an astronaut could achieve, not being discussed behind his or her back. And I never heard Judy's name attached to a "Who let that bozo in the door?" comment. Over a beer or in a jog with a TFNG, I would hear comments about the misadventures of other astronauts. When one TFNG MS was removed from being a robot arm operator, it took about ten milliseconds before the reason was being shared in whispers. He maneuvered the arm like a fifteen-year-old kid learning to use a stick shift. There was locker-room gossip about an MS jeopardizing the deployment of a satellite because of a failure to follow the checklist. One TFNG accidentally engaged the shuttle backup flight system during a prelaunchpad test and caused a delay in the countdown. Judy's name was never in any of these conversations, the ultimate testimonial to her competency. She wasn't the smartest or quickest TFNG—Steve Hawley held that position. Judy was like me. We weren't stars. But we were solid, dependable. We could be counted on to get the job done.

Until my STS-41D association with Judy, I had believed it impossible for a man to be the close friend of an attractive woman. It was a fact of testosterone as irrefutable as gravity was a fact of nature. Men see attractive women as sex objects and that destroys any hope of close friendship. But I had discovered an exception to that rule. When a man and woman are thrown together for several years of training for a journey that has the potential to kill them, the man learns to see through the woman's youth and beauty and measure her proficiency. He learns to see her as somebody whose response in an emergency might mean *his* life or death. On that June evening, six years after we first met, I could now see and appreciate Judy's skills as an astronaut. I could trust her with my life. Tomorrow, I would do exactly that.

Several years later I would learn this friendship had placed my name on the office grapevine. After *Challenger,* I was in Hank Hartsfield's backseat on a T-38 flight. We had been sharing our thoughts on the disaster and the loss of so many friends when Hank had commented, "Judy's death must have been particularly hard on you." I was confused by the statement. The death of all the crew had been hard on me. I asked him, "Why do you say that?" To which Hank had replied, "Well . . . being that you two were sleeping together." I was stunned. I proclaimed innocence but I knew he didn't believe me. Maybe it is possible for a man to share a close relationship with a beautiful woman that does not include sex, but don't expect other men to believe it.

At the crew quarters I showered and returned to the main conference room. Hank was my only company. He was reading the newspaper, mumbling about the idiocy of liberals and their destruction of the country. "Goddammit, I wish Ted Kennedy would find another bridge . . . with deeper and wider water under it." I had long before learned not to respond. It would only elicit a filibuster on the topic. When Hank got wound up on politics, you could never escape.

Our satellite TV, for some fortuitous reason, received the Playboy Channel. I marveled at this fact as much as I marveled that alligators didn't chase astronauts. How did the Playboy Channel end up on the TV in the astronaut crew quarters? I suspect it was just one of those government snafus. There was a KSC bean counter somewhere who had contracted with a company for satellite TV and this was what we got. It would probably have taken multiple forms in triplicate and ten thousand taxpayer dollars to turn it off. I wondered if the signal was coming from a satellite a shuttle had previously placed in orbit. That would be a unique claim to fame: "I was the guy who put the Playboy Channel in space."

So, at T-12 hours and counting I was listening to Hank grumble, "Gloria Steinem should be in Ted Kennedy's car when he finds that bridge," while watching a topless model speaking about her turn-ons, "a six-pack belly and world peace," and turnoffs, "pollution and rude people."

I finally headed to my room for sleep. I knew that would be a struggle. I was bipolar with the frequency of a tuning fork, oscillating between fear and joy. The flight surgeon had given us sleeping pills but I had no intention of taking one. There was still a last physical exam ahead and I didn't want an adverse reaction to the pills prompting a medical question. There were plenty of MSes who would gleefully step into my shoes on a moment's notice. I wasn't about to give any of those vultures that opportunity.

I lay in bed and studied the room's only wall decoration, a framed photo of an exploding volcano. The photo was a time exposure so the glowing ejected lava was captured as arcing streaks against a black sky. Bloodred coils of molten rock snaked downward on the skirt of the mountain. I wondered what bureaucrat had been doing the interior decorating for the astronaut crew quarters and thought, *If it was my last night before a mission into space, what wall art would I like to reflect upon to calm my uneasy soul? I know . . . an exploding volcano with lots of fire and sparks!* It was like showing films of airplane crashes on an airliner as the in-flight movie. If you're going to hang a picture of something exploding, why not hang a photo of a NASA rocket exploding on the launchpad? That would be rich.

The only sound was a muffled, unintelligible voice coming through the steel wall next to my ear. Mike Coats was talking on the phone to Diane and his children. I had made my final call to Donna and the kids a couple hours earlier and had performed as poorly in that good-bye as I had in person on the beach. Even though I now had time to make another call, I did not. One more good-bye wasn't going to help me or Donna. Mike was a better man than I, God bless him.

At least I had done a good job of financially protecting my family in the event of my death. I had three insurance policies on my life. Months earlier I had written each insurance company explaining my pending shuttle launch and asking if there was any fine print on the policies that would negate the payout if I died on the shuttle. Each company had replied in writing that its policy would be unaffected by death by rocket. I had stapled each of those letters to the respective policies and put them with my will. Donna wouldn't have to deal with any surprises there.

What were the chances there was a Gideon Bible in the nightstand drawer? I wondered. There was not, thank God. It would have scared the shit out of me if NASA thought we needed one. "We're not sure this rocket will work, so here's our ultimate emergency backup, a Bible."

I didn't need a Bible to talk to God. I prayed for my family. I prayed for myself. I prayed I wouldn't blow up and then I prayed harder that I wouldn't screw up. Even my prayers reflected the astronaut credo, "Better dead than look bad."

At some point in the night, exhaustion overpowered fear and excitement, and I fell into a shallow sleep. The smell of cooking bacon woke me. The dieticians had arrived and were making breakfast. My stomach turned in disgust. The thought of food was nauseating.

I could hear the wake-up knocks on the doors of the other crewmembers and wondered how many of them were actually asleep. I could believe Hank had slept well. Anybody who could read a newspaper and deliver political commentary on the eve of a shuttle launch must have their shit together. But I imagined the rest of the crew had spent much of the night as I had, counting holes in the ceiling tile.

The knock came on my door and I opened it to Olan Bertrand's smiling face. Olan was one of the Vehicle Integration Test Team (VITT) members and would be a participant in our final prelaunch briefing. He was also a Louisiana Cajun with an accent as thick as a bowl of jambalaya. He mumbled something I interpreted as "The weather and the bird are looking good," but could have been "It's raining like hell and *Discovery* blew over." Only his smile told me it was the former and not the latter.

I showered and shaved, then trimmed my fingernails. Some of the early spacewalkers had painfully torn their nails on the inside of the suit gloves and had suggested contingency spacewalkers cut them short, too. I did so and filed them to snag-free crescents.

For breakfast I dressed in a mission golf shirt. I had no appetite, but it was a mandatory photo opportunity. A NASA cameraman entered to film us sitting around the table. I faked a carefree smile and waved. Most of us ate nothing or very lightly. I had a piece of toast. As a teenager I had always heard the "voice of NASA" say the astronauts were enjoying a breakfast of steak and eggs before launch. One bite of that fare and I would have vomited. Nobody drank coffee. That would have been bladder suicide.

After the cameraman was gone, I gave Judy my emery board. "You can do your nails during ascent." She laughed. It had been a running Zoo Crew joke that, as a Jewish American Princess (JAP), she would be giving herself a manicure during the countdown. With the nail file I included my latest JAP joke: "What does a JAP say when she inadvertently knocks over a priceless Ming dynasty vase, it shatters on the floor, and museum officials rush to the scene?"

Judy sighed in resignation. "What does she say, Tarzan?"

"She shouts, '*I'm okay! I'm okay!*' "

After the meal, we collected in the main briefing room for a teleconference to review the launch countdown status and the weather forecast. Everything looked good. The weather for Dakar, Senegal, Africa, was covered. It was our primary transatlantic abort site, just twenty-five harrowing minutes away from Florida via a wounded shuttle. I really didn't want to make my first visit to Africa in a space shuttle.

Next, we visited flight surgeons Jim Logan and Don Stewart in the gym for a cursory last exam. They checked our ears, throat, temperature, and blood pressure. I put myself in a happy place to ensure the last was within limits. Both doctors were good friends of the Zoo Crew, but if they had raised any medical issues at this moment, others would have later found their arrow-riddled bodies spread-eagled to the archery hay bale. We wouldn't have missed.

We then cycled through the bathroom for a next-to-last gravity-assisted waste collection. We'd have one more chance at the launchpad toilet. My self-imposed fast from liquids was working. I had no urges, but nevertheless I took advantage of the moment to squeeze out a few drops of urine.

I returned to my room and began to dress. While we had been at breakfast, the suit crew had arranged our wardrobe on the bed. The first item I donned was my urine collection device. I stepped through the leg openings and pulled the condom toward my penis. It looked incredibly small. Not the condom . . . my penis. I coached the recalcitrant appendage into the latex. It promptly slipped from my body. Apprehension had sucked every molecule of blood from my crotch. I doubted even a naked Bo Derek doing jumping jacks in front of me would have stirred life into this lizard.

I made a second attempt to get my sword into its sheath, this time taking the weight of the UCD bladder in my hand so I would stay attached. I Velcroed the device around my waist, accepting the results whatever they might be. I had no choice. There was a countdown clock ticking.

I finished dressing in my flight suit, then filled my pockets with spare prescription glasses, pencils, pressurized space pens, and barf bags . . . lots of barf bags. I put one in each of my chest pockets and a couple spares in other pockets. Would I be a victim of space sickness? I had been sick so many times in the backseats of various jets, I couldn't believe I would be spared in space. I toyed with the idea of taking one of NASA's antinausea pills, a mixture of scopolamine (a downer) and Dexedrine (an upper), but decided otherwise. I wanted to know my Space Adaptation Syndrome (SAS) susceptibility and drugs would camouflage it. Besides, I didn't think the pill would work. Months earlier I had been on a deep-sea-fishing trip with a group of astronauts, several of whom had taken the capsule. Some had still gotten seasick. Another guest on the same trip, who had also swallowed a ScopeDex, had been so suddenly struck with nausea he had vomited before getting to the rail. The memory of NASA's miracle SAS pill floating in a puddle of barf on the fantail of a fishing boat did nothing to inspire confidence that the pills would help me in space. I left them behind.

Fully dressed and with pockets loaded, I stepped from my room and joined the rest of the crew in a walk to the elevator. Judy was in front of me and I could hear the whooshing sound of her diaper plastic rubbing against her coveralls. I teased her, "You're getting a little broad in the beam, JR."

"Screw you, Tarzan."

How different from reality were all those science fiction movies of my youth. As Lloyd Bridges (Colonel Floyd Graham) and Osa Massen (Dr. Lisa Van Horn) boarded their *Rocketship X-M* in the 1950 Hollywood classic of the same title, I don't recall them commenting on the condoms and diapers they were wearing.

A group of NASA employees welcomed us with applause on our exit from the crew quarters. I wanted to embrace them and say, "Thank you for giving me this moment." They were the best in the world.

We stepped into the elevator and two heavyset men wearing tool belts followed us. I was shocked. We were on our way to fly the space shuttle and two blue-collar workers had decided to hitch a ride with us. When the elevator opened and the photographers got this picture, it was going to be a hoot. A bewildered Hank had to ask, "What are you guys doing?"

"We're elevator repairmen. There've been reports this elevator is giving you guys problems. We didn't want you to get stuck on your way to your rocket." We all laughed. NASA thinks of everything. A comforting thought at this moment.

But the workers were not needed. We creaked to ground level without a problem and exited the building to cheers and more applause from a larger group of the NASA team.

Outside, I immediately looked to the sky hoping to see stars, hoping for proof the weather was good. But the lights of the cameras filming our departure had ruined my night vision.

We climbed into the astro-van and began the drive to Pad 39A, the same pad from which Neil Armstrong had embarked on his historic journey to the moon fifteen years earlier. I wondered what his drive had been like. The van air-conditioning was making ours frigid. My skin was clammy and I was shivering. Nervous small talk occupied us. I hoped nobody could hear my heart. Each pulse seemed like a detonation.

We passed successive security checkpoints where the guards saluted or waved or flashed a thumbs-up. They had trucks parked nearby for their own evacuation to more remote points. Closer to the pad we passed several fire trucks and ambulances. Their crews were clad in silver firefighting suits and hovered near their vehicles. When the launchpad closeout crew departed, these men and women would remain in a nearby bunker, ready to race to our rescue if there was a problem. I couldn't imagine any problem involving 4 million pounds of propellant leaving anything to rescue. There were certainly six body bags in those ambulances.

I was as scared as I had ever been in my life. But at that moment, if God had appeared and told me there was a 90 percent probability I wasn't going to return from this mission alive and had given me an opportunity to jump from that crew van, I would have shouted, "No!" For this rookie flight, I would take a one in ten chance. I had dreamed of this moment since childhood. I had to go. Even if God had given me a vision of what the other nine chances meant, a vision of my charred remains being zipped into one of those body bags, I still would have declined His offer to exit the van. I had to make this flight.

I would later look back on my desperate need for this first mission and think how perverted it was. What type of a person puts their wife, their children, *their own life* second behind a need to ride a rocket? I believed that surely I was unique in this sick prioritization. But I discovered otherwise. In the weeks after STS-41D, Hank Hartsfield described to me his feelings before his first mission (STS-4). I was stunned to hear his admission of the exact feelings I was now experiencing. He recounted how he would rather have died on his first mission than never to have flown in space. We were like the Mount Everest climbers stepping over frozen corpses from prior climbing disasters in our quest for the summit. Like those climbers, we were motivated by a fear far greater than death—the fear of not reaching the top.

What a fraud astronauts practice on our fellow citizens. Most Americans see us as selfless heroes, laying our lives on the line for our country, the advancement of mankind, and other lofty ideals. In reality no astronaut has ever screamed, "For God and Country!" when the hold-down bolts blew . . . at least not on their rookie mission. We were all stepping into harm's way because we knew otherwise we would die as incomplete humans. There was room in our souls for noble motivations only after our pins were gold.

As *Discovery* came into view we leaned into the aisle to watch. A crisscross of xenon lights bathed her. Against the backdrop of early morning blackness she appeared as a newly risen morning star. If my heart had been in overdrive before, it now accelerated to warp speed.

At the pad we stepped from the van and looked up at our ship. In spite of my faith in physics, it didn't seem possible anything so gargantuan could rise from the Earth, much less achieve a 17,300-miles-per-hour speed at 200 miles altitude. The stack towered 200 feet above the Mobile Launch Platform (MLP), which, itself, loomed several stories above us. The 4½-million-pound mass was held in place by eight hold-down bolts, four at each SRB skirt. The SRBs were separated by nearly 30 feet to accommodate the blimpish diameter of the ET. The gray acreage of the MLP's underside formed a steel overcast. Three cavernous openings were cut through it to allow the flames from the two SRBs and the SSMEs to descend into the flame bucket and be diverted outward. During engine ignition a nearby water tower would be emptied into that bucket to protect it from heat damage. Giant plastic sausages of water were also slung in the two SRB cavities. That water would attenuate the acoustic shock waves the boosters developed, which could reflect upward to damage cargo in the payload bay.

The pad was eerily deserted. A vapor of oxygen swirled around the SSME nozzles. A flag of more vapor whipped from the top of the ET beanie cap. Shadows played upon that fog, creating a scene right out of a creepy science fiction movie. Loudspeakers boomed the prelaunch checklist milestones, a noise that competed with the deafening hiss of the engine purge. The few remaining workers, appearing Lilliputian next to the machine they serviced, performed their duties with quiet urgency. In the shadows the glowing yellow safety light-sticks Velcroed to their arms and legs made them appear skeletal.

We climbed into the pad elevator and shot to the 195-foot level. Hank and Mike walked immediately to the white room, a boxlike anteroom up against *Discovery*'s side hatch, where technicians waited to help us into the cockpit. Hank and Mike would be first inside. I had time to kill and walked to an edge to get a better view of the vehicle. *Discovery*'s belly of black heat tiles gave her a scaly, reptilian look. They contrasted sharply with the white thermal blankets glued on her top and sides.

I looked out at the Launch Control Center (LCC), three miles away. Donna and the kids would be inside. At T-9 minutes the family escorts would lead them to the roof to watch the launch. I wondered how Donna was handling the stress. I knew the kids would be okay, but she would be at her emotional limits.

"Hey, Tarzan, don't fall." Judy came to my side. The wind had whipped her hair into a black aura. She had an ear-to-ear grin.

I made the observation that it was scary looking over the railing from two hundred feet up. "I've got a fear of heights, JR. I can't get any closer."

She laughed. "Well, Tarzan, you're screwed. We're headed to two hundred miles."

We continued with small talk, each of us trying to distract ourselves from our pounding hearts. Then the two-minute warning call came for my strap-in.

I embraced her. "Good luck, JR. I'll see you in space." Since she would be in a mid-deck seat, I wouldn't see her until after MECO. It was the first time I had ever held her and I was struck by how petite she was.

"Roger that, Tarzan." She returned my squeeze and we parted.

I detoured to the pad toilet for a last go at urinating. The bowl was a pond of unflushed filth and toilet paper. The plumbing had been turned off hours earlier as part of the checklist for launchpad closeout. The workers had no option but to use this facility. I added my urine to the mess, reattached my UCD, then walked to the white room.

The closeout crew quickly harnessed me. We shook hands and I dropped to my knees and crawled through the side hatch. The cockpit was as cold as a meat locker. It occurred to me the chill was going to shrink a critical part of my body even further. If my UCD condom stayed attached, it would be a miracle.

I stood on the temporary panels covering the back instrument panel and struggled to put myself in the chair behind Mike Coats. Once in, Jeannie Alexander, another of the closeout crew, helped me with the five-point harness. As she worked at my crotch to make the buckle connections, I teased, "I'll give you all day to stop that." She had probably heard the same joke a hundred times. She connected my communication cord and emergency breathing pack, then clipped my checklist to a tether. Everything had to be secured. Anything that dropped during launch would be slammed into the back instrument panel by the G-forces, irretrievable until MECO. Finally she gave me a big smile and a pat on the shoulder and turned to help Steve Hawley.

I looked around the cockpit. Everything appeared as it had in the countless simulations except for the sparkling newness. *Discovery* even smelled new. Every piece of glass gleamed. There were no wear marks on the floors or on the most frequently used computer keys. There were no vacant panels or panels with somebody else's payload controls as we had frequently encountered in the JSC simulators. This was our bird. It was our mission software humming in her brain. We would be driving a brand-new vehicle from the showroom floor.

About ninety minutes to go. With each vanishing second my heart shifted into yet a higher gear. Thank God we weren't wired for biodata. That had ended back in the Apollo days. I would have been embarrassed for anybody to have seen my vital signs. I envisioned Dr. Jim Logan looking at them and saying, "It must be a bad sensor. Nobody's heart can achieve those rates without exploding."

Jeannie finished with Hawley's strap-in. Judy and Charlie Walker were belted in downstairs. The closeout crew wished us good luck, unplugged from the intercom, and was gone. We heard the hatch close. A moment later our ears popped as the cockpit was pressurized. The wait began.

It quickly became an agony, physical and mental. I wiggled under my harness to restore some circulation to various pressure points. In spite of my dehydration efforts and earlier toilet visits my bladder quickly neared the rupture point. What were the chances my UCD condom was still attached? It had been on too long for my body to still feel it and I was convinced all the crawling and wiggling I had done, not to mention the effects of fear and cold, had caused my penis to disengage. If so, I would be urinating into my flight suit. And I was certain there would be a lot of urine. I could imagine it soaking my coveralls, dripping from the seat onto the back instruments, and shorting out an electrical circuit. My "accident" would be a gossip topic for decades. "Remember that Mullane guy? He pissed his pants on the launchpad. They had to delay the launch to dry out the instruments." God, I'd rather blow up. I tried to hold on, but soon realized that would be impossible. Praying for a miracle that I was still safely ensconced in latex, I decided to give it a shot. But I quickly discovered it was impossible to urinate on my back. Even though the urge was overwhelming, painful, even, I strained but nothing happened. There are some things even the world's best training program can't prepare you for. In desperation I loosened my harness and struggled to roll slightly to my side. In that new position I was finally able to open the floodgates. After a moment I tried to put on the brakes to determine if I was leaking, but I would have had better luck damming the Atlantic. Urine poured from me like water into the flame bucket. I felt no spreading wetness so my miracle had been granted. The condom was still attached. I collapsed in glorious relief. You would have thought I had already reached MECO.

There was little to do in the cockpit. After some radio checks with the Launch Control Center, they moved on with their prelaunch activities. We were left alone. Others complained about the state of their bladders. Judy and Charlie joined in from downstairs. I didn't envy them their position. They had no instrument displays or windows. They would be riding an elevator with no idea of what floor they were passing. Judy reminded us she did not want to hear any sentences ending in the word *that,* as in, "Did you see *that*!" or "What was *that*!" We all laughed. When you are blind to the *that* being referenced, it would be very disconcerting to hear any such exclamations.

We fell silent and just listened to the LCC dialogue. When the Range Safety Officer's (RSO) call sign was heard there were some joking comments on the intercom to cover the fear his grim function generated. The RSO would blow *Discovery* from the sky if she strayed off course. If the RSO ever transmitted the Flight Termination System ARM command, a red light on Hank's instrument panel would illuminate as a warning. I wondered what sick engineer had thought that would be helpful.

With each passing minute the mood in the cockpit grew more intense. Then we heard the dreaded word *problem.* At T-32 minutes a problem was noted with the Backup Flight System (BFS) computer. The launch director informed us he would stop the countdown at the planned T-20 minute hold point while the experts sorted it out. There was a communal groan on the intercom. The flight rules would never let us launch without the backup computer working properly. After all the emotional capital we had invested to this point, the thought of getting out of the cockpit and repeating that investment tomorrow was enough to make us physically ill. We all prayed that the offending circuit would fix itself. But God didn't hear us. Following several minutes of troubleshooting, the LCC called, "*Discovery,* we're going to have to pull you out and try again tomorrow."

I was crushed, totally spent. We all were. Our nerves had been in constant tension for four hours and we had nothing to show for it. I looked forward to a repeat of this tomorrow like I looked forward to a root canal.

Within the hour we had been extracted from the cockpit and were on our way back to the crew quarters. The spouses were driven out for lunch. Donna put on a brave face but it couldn't mask her exhaustion. The other spouses looked similarly beaten.

Then, the script was replayed. The tearful good-byes. Another review of checklists. Hank's political commentary. A fitful sleep. The nauseating smell of cooking bacon. The wake-up knock on the door. Olan's mumbles.

Once again we entered the elevator to be joined by the same two maintenance men. Another couple of launch scrubs and we'd be old friends. We exited the building into the same camera lights, heard the same enthusiastic applause from our friends, and boarded the same chilly astro-van. Even the extreme fear of death that had accompanied me yesterday was back again. The first launch attempt had done nothing to mitigate it. And so was the greater fear . . . that I would never make this flight, that at the last second something would happen to steal my chance. I would be forever damned as an astronaut in name only. My astronaut pin would remain silver.

As Jeannie Alexander worked at my crotch to fasten the seat harness I joked, "I'm getting tired of this foreplay." She didn't have time to do more than just smile.

The hatch was closed and we were back into the wait. After yesterday's urinary challenges, I had been even more aggressive at dehydrating myself. But it didn't help. With my legs elevated there was a whole lake of fluid heading downhill into my bladder. Within an hour I felt as if I were going to burst.

The intercom fell silent earlier than it had yesterday. We were all too exhausted to continue our lame jokes. The whooshing of the cabin fan was the only sound. I watched the sky grow lighter and seagulls soar past the windows. I could tell by the way Hank's head lolled to the side that he had fallen asleep. How some astronauts could do that amazed me. I could no more have nodded off than could a man strapped to an electric chair. I was scared. But at that moment there was nothing in the world, including celebrity, wealth, power, and sex, that could have motivated me to give up that seat. Sitting in it, being an hour from orbit, I was the richest man on earth.

T-32 minutes came and went. Yesterday's comment about a glitch in the BFS computer was not repeated.

Hank woke up. "Did I miss anything?"

I thought of telling him he had slept for four years and Ted Kennedy was now president, but decided otherwise. If he had a stroke, it would surely delay the flight.

We entered the T-20 minute hold. This was as far as we had gotten yesterday. *Please, dear God, let the count continue.* Each of us had our ears hypertuned to the LCC dialogue, praying we would hear nothing about an off-nominal condition. We didn't. Right on time, we came out of the hold.

At T-9 minutes we entered our last planned hold. Again, there were no discrepancies and the LCC released the clock. I thought of Donna and the kids. They would now be walking the stairs to the roof of the LCC. *God help them,* was my prayer.

T-5 minutes. "Go for APU start." Mike acknowledged LCC's call and flipped switches to start *Discovery*'s three hydraulic pumps. The meters showed good pressure. *Discovery* now had muscle. The pumps tickled our backs with their slight vibrations. The movement was the first indication the vehicle was anything but a fixed monument.

The computers ordered a test of the flight control system and *Discovery* shuddered as her SSME nozzles and elevons were moved through their limits.

T-2 minutes. We closed our helmet visors. Hank reached across the cockpit to shake Mike Coats's hand. "Good luck, everybody. This is it. Let's do it like we've trained. Eyes on the instruments."

T-1 minute. *Please, God, if something bad is to happen on this flight, let it be above fifty miles.* My prayer was specific for a reason. By NASA's definition you had to fly higher than fifty miles altitude to be awarded a gold astronaut pin. If I died below that altitude, Donna and the kids would only have my silver pin for a memorial shadowbox.

T-31 seconds. "Go for auto-sequence start." *Discovery*'s computers assumed control of the countdown from LCC's computers. She now commanded herself. The cockpit was a scene of intense, silent focus. Again, I was glad my vital signs weren't on public display. My heart was now a low hum.

T-10 seconds. "Go for main engine start." The engine manifold pressure gauges shot up as valves opened and fuel and oxidizer flooded into the pipes. The turbo-pumps came to life and began to ram 1,000 pounds of propellant per second into each of the three combustion chambers.

At T-6 seconds the cockpit shook violently. Engine start. *This is it,* I thought. In spite of my fear, I smiled. I was headed into space. It was really going to happen.

5 . . . 4 . . . The vibrations intensified as the SSMEs sequentially came on line.

Then, the warble of the master caution system grabbed us. "We've had engine shutdown." I don't know who said it, but they were stating the obvious. The vibrations were gone. The cockpit was as quiet as a crypt. Shadows waved across our seats as *Discovery* rocked back and forth on her hold-down bolts.

We were all seized with a *What the hell?* wonderment. Something was seriously wrong. Hank punched off the master caution light and tone. The left and right SSME shutdown lights stabbed us with their red glare, meaning they were off. But the light for the center SSME remained dark. Surely it couldn't still be running? There was no noise or vibration. But if it was still running, we wanted it off. Whatever was happening, we wanted everything off. Mike repeatedly jabbed his finger onto the shutdown button. But there was no change in the light status.

Paramount on everybody's mind was the status of the SRBs. We had gotten to within a few seconds of their start. If they ignited now, we were dead. Generating more than 6 million pounds of thrust, they certainly had the muscle to rip out the hold-down bolts and destroy the vehicle in the process.

The diagnosis quickly came from LCC. "We've had an RSLS abort." *Discovery*'s computers had detected something wrong and stopped the launch—a Redundant Set Launch Sequencer (RSLS) abort. But what had gone wrong? Had a turbo-pump disintegrated? Had an engine exploded? Had hot shrapnel been hurled around our engine compartment? We were strapped to 4 million pounds of explosives and didn't have a clue what was happening a hundred feet below us.

And neither did the families. I would later learn how the abort had played out on the LCC roof. A thick summer haze had obscured the launchpad. When the engines had ignited, a bright flash had momentarily penetrated that haze, strongly suggesting an explosion. As that fear had been rising in the minds of the families, the engine start sound had finally hit . . . a brief roar. It had echoed off the sides of the Vertical Assembly Building (VAB) and then . . . silence. Donna had been convinced she was seeing and hearing an explosion. Fortunately the astronaut escorts had been there to ease her fear with an explanation of a shuttle-pad abort. No doubt they had done so with some private reservations. There had never been a shuttle engine-start abort.

Donna had crumpled into a chair and cried. Amy, our oldest daughter, had followed suit. They were crushed by the thought that it would all have to be repeated another day. Amy snapped, "Why don't they just put more gas in it and launch it now!" The thought of having to climb the LCC roof and endure another countdown torture made her wild with anger.

Back in the cockpit, things took a turn for the worse. Launch Control reported a fire on the launchpad and activated the fire suppression system. Water began to spray across the cockpit windows. What was going on down there?

In the midst of these terrifying moments I looked at Steve Hawley. He stared at me with eyes as big as plates. I knew that was my face—I was staring into a mirror. Then he commented, "I thought we'd be higher when the engines quit." I wanted to hit the SOB. I wanted to scream, "This isn't funny, Hawley!" And to think, six years earlier I had harbored doubts about the post-docs' mettle. Some of them, Hawley included, had steel balls.

Hank ordered all of us to unstrap from our seats and prepare for an emergency egress from the cockpit. If we chose to leave we'd have to run across the access arm to the other side of the gantry and jump in escape baskets. In just thirty seconds those would slide us a quarter mile away. We'd be able to wait out the problem in an underground bunker, assuming we could get there before the rocket exploded.

Judy crawled to the side hatch window and reported the access arm had been swung back into place and the fire suppression system was spraying water over it. She didn't see any fire. "Henry, do you want me to open the hatch?"

Judy's question elicited several exchanges of do we or don't we make a run for it. How bad was the fire? LCC's conversations seemed unpanicked and that gave us some reassurance that everything was under control. But LCC and MCC always seemed in control. That was their job, to calmly look at their computer screens of data and make robotic, emotionless decisions. Could they remain calm even as their creation was de-creating? I had no doubt. "We've had an RSLS abort" could well be engineer-speak for "Holy shit! Run for your lives! She's gonna blow!" No, I was not comforted by the calm of the LCC.

"Negative on opening the hatch, Judy." Hank decided we'd sit tight. It was a decision that might have saved our lives. The post-abort analysis determined the fire had been caused by some residual hydrogen escaping from the engines and igniting combustible material on the MLP. The gas flame may have been as high as the cockpit but since hydrogen burns clean we would not have seen it. We could have thrown open the hatch and run into a fire.

As time passed it became more evident we were in no immediate danger. We waited for the closeout crew to open the hatch. My thoughts turned as dark as space. My worst nightmare had been realized. I was still an astronaut in name only. For how long? It was obvious there would be no launch attempt any time soon. In my quiet hell I catastrophized. I built a scenario in which the engine problem was severe. The vehicle would be grounded for many months, maybe years, while the engines were redesigned and tested. Launch schedules would change. Our crew would be pushed backward in line, maybe even disbanded. I had gotten to within three seconds of a lifelong dream only to have it snatched away. For how long? Weeks? Months? Forever?

We exited the vehicle into the residual rain of the fire suppression system. It had soaked the gantry and now water was dripping from every platform, pipe, and cross brace. We were quickly drenched. Judy's hair took a big hit. She looked like a sodden cat.

In the astro-van we sat in our soaked flight suits shivering from the chill of the air-conditioning. That system seemed to have only two positions, cold and freakin' cold. Our physical misery was a perfect fit to the cloud of depression enveloping us. I wasn't the only one doing mental gymnastics and wondering how badly screwed we were.

The wives and kids were waiting for us at the crew quarters and there were a lot of tearful hugs. "Dad, we thought you had blown up!" Pat was quick to fill me in on the momentary horror they had lived on the LCC roof.

At a press conference we all lied about the tension in the cockpit following the abort and fire. Hank took most of the questions and did the Right Stuff routine of, "Aaawh shucks, ma'am. T'weren't nothing." He explained how we train for these things, how confident we had been in the LCC's reactions to the abort, how we had never doubted our safety. Meanwhile, I was wondering if I had shit in my flight suit.

We were released from our health quarantine to join our families. As expected, there would be no more launch attempts until the engine problem was identified and fixed. The shuttle program had just come to a screeching stop for an indefinite period.

Donna, the kids, and I returned to their condo to a boisterous party. While my aunts and uncles and cousins were bummed out the launch had been aborted, they were still having a great time. The sun was shining. The booze was flowing. It was a Florida family reunion. They were all on vacation and having a blast. And now my abort had made the reunion complete. If I had launched, the family would never have seen me. So they overwhelmed me with questions and requests for photos and autographs. Their enthusiasm was understandable. Most of them had not seen me since I had been selected as an astronaut.

"Mike, let me get a photo of you standing next to your little cousins."

"Mike, why don't you sit here with Grandma and tell her what it's like to be an astronaut."

"Mike, can I get twenty autographed photos for my neighbors back home?"

I wanted to crawl into a hole and die. After a couple hours I finally escaped to the beach and collapsed. I had been so close, three freakin' seconds, and now I might be at the back of a long, long line. I couldn't get that dismal thought out of my head. I closed my eyes and prayed for blissful unconsciousness. It was a prayer immediately answered. The exhaustion of the past few days had finally caught up with me and I fell into a deep sleep. Minutes later I was awakened by a gentle kick in my side. I squinted upward to see my eighty-seven-year-old grandmother. "Mike, get inside. Bobby wants to take some more pictures. You'll get sunburned out here."

God, take me. My new prayer was that a meteor would hit and put me out of my misery.

CHAPTER 20

MECO

A few days later our crew was back in Houston and facing the grim possibility our mission was going to be canceled. Payloads were stacking up. Every day a communication satellite wasn't in space meant the loss of millions of dollars of revenue to its operators. NASA HQ searched for a way to minimize the impact of *Discovery*'s delay to their downstream customers. They focused on combining the payloads of two missions and deleting one from the schedule. Everyone in the astronaut office knew a deleted mission meant a deleted crew.

It was a miserable two weeks as HQ debated the best adjustment to the flight manifest. Every imaginable rumor twittered up and down the astronaut grapevine. Who was going to get screwed? There was a general feeling among the other mission-assigned crews that we had had our chance. It was just our tough luck *Discovery* had misfired. It should be our crew who suffered the consequences. I couldn't blame them. I would have felt the same way in their shoes.

To make matters worse, there were no firsts associated with our crew to give us some HQ PR cover. White males on a space shuttle were as newsworthy as white males on a hockey team, and there were five of us palefaces on the crew. Judy was merely Sally Ride's runner-up. We had nothing in the way of a celebrity first to protect us from the ax, whereas other downstream missions included the first spacewalk by a woman and the first satellite retrieval mission. HQ was going to ensure those high-visibility missions flew as planned. I told Hank he should get a sex-change operation so we'd have the first transgender astronaut aboard to protect us. He declined.

Astronaut office politics were also a significant cause for worry. There wasn't an air force astronaut who didn't think George Abbey was navy-biased in his flight assignments. Nine of the first eleven shuttle missions had been commanded by active duty or retired navy astronauts. Hank Hartsfield was only the second astronaut with air force roots to command a mission. And a glance downstream showed crews commanded by Crippen, Hauck, and Mattingly—all navy. If anybody was going to get screwed because of *Discovery*'s delay, it wasn't going to be them. They had Godfather Abbey's protection. I felt as if the entire STS-41D crew were walking around with a sign pinned to our backs: "Cancel This Mission."

But it didn't happen. Instead, the ax fell on STS-41F, Bo Bobko's mission. His payload would be added to ours, while he and his crew would be cut adrift to find something else downstream. Bo was retired air force. The USAF astronaut contingent cursed Abbey . . . again. I felt bad for Bo and company, but not for long. We were Lazarus back from the dead or, in this case, back in the front of the line.

Our major payloads would now include three communication satellites and Judy's solar panel experiment. We would also have a new center engine. Ground tests had been unable to duplicate the problem with the original SSME. Engineers could only assume there had been some minute contamination in the hydraulic system, which had caused a fuel value to malfunction. As a precaution, the engine was replaced.

August 29, 1984, found us once again in the cockpit of *Discovery.* By now we were old hands at strapping in and waiting. But it didn't get any easier. My bladder continued to torture me. My heart continued to race away in fear. And it didn't get any easier for *Discovery.* At T-9 minutes our third launch attempt was scrubbed for a Master Events Controller (MEC) malfunction. I was going to vomit before I got into space.

August 30. Another day. Another launch attempt. While awaiting our turns to enter the cockpit Judy and I slapped mosquitoes off each other's back. The little bastards could drill right through our flight suits. Judy observed that the insects seemed worse than yesterday. I told her it was because our launch scrubs had trained them. "They knew we'd be standing here this morning. And they know we'll be standing here at this exact time tomorrow, the next day, and every day afterward. We're on their menu every day."

Judy brushed away my pessimism. "We'll do it today, Tarzan. I've got good vibes."

I didn't share her enthusiasm. I was emotionally exhausted. Clinical depression was on the horizon, suicide to follow.

A call came from the white room that it was my turn to be harnessed up. For the fourth time I embraced Judy. "This is the only thing fun about these scrubs. I get to hug you every morning."

Judy smiled. "That's sexual harassment, Tarzan."

"I hope so."

I wished her good luck and walked toward the cockpit.

She called after me. "See you in space, Tarzan."

As the hours ticked by, I began to believe Judy's vibes were right. The count proceeded smoothly. Milestone after milestone came and went without a negative word being spoken. The weather was great in Florida and at our abort sites. A little sunshine began to melt my black pessimism.

Then it happened again. We were notified the T-9 minute hold would be extended. This time the problem was with the Ground Launch Sequencer (GLS). I guessed we'd never get off the ground until everything broke at least one time. I ached for Donna and the kids, back on the LCC roof. It had to be killing them.

Just as the discussions on the GLS began to sound promising we were slapped with another problem. Some bozo in a light airplane had entered the closed airspace around the pad. We would have to hold until that plane was out of the area. The intercom seethed in our rage. We all simultaneously developed Tourette's syndrome. Even Judy swore like a convict. *Shoot the fucker down,* was the general consensus. Previous shuttles had been delayed for the same reason, as well as for pleasure boats violating the offshore danger areas. Every astronaut thought these violators should be shot from the sky and sunk in the sea. Even astronauts enjoying a smooth countdown had no tolerance for idiots getting in the way of their launch, much less a crew as abused as ours.

As we waited, the LCC cleared the GLS problem. Now it was a matter of waiting until the light airplane exited the area. After nearly a seven-minute delay, its pilot pulled his head out of his ass and flew off. We all wished him engine failure. The count resumed.

Mike started the APUs at T-5 minutes. They all looked good. The sweep of the flight control system followed. It was also error-free.

At T-2 minutes we closed our helmet visors. Bob Sieck, the launch director, wished us good luck. Hank acknowledged, thanking him and his team for their efforts. I was glad I didn't have to say anything at this point. My mouth was a desert.

T-1 minute. Hank reminded us, "Eyes on the instruments."

T-31 seconds. "Go for auto-sequence start." I made one last prayer for Donna and the kids . . . and again to God, "If you're going to kill me, please do it above fifty miles altitude."

T-10 seconds. "Go for main engine start." The engine manifold pressures shot up.

T-6 seconds. For the second time in my life I felt the violence of SSME start. Two months earlier I had thought these vibrations were a guarantee for liftoff. No longer. Until there were goose eggs on the clock I would remain skeptical.

5 . . . 4 . . . 3 . . . We finally entered new countdown territory.

2 . . . 1 . . . At zero there was no doubt we had finally slipped the surly bonds of earth. As the hold-down bolts were blown, we were slapped with a combined thrust of more than 7 million pounds. A new wave of intense vibrations roared over us.

"Houston, *Discovery* is in the roll."

"Roger, roll, *Discovery.*" *Discovery*'s autopilot was in control. Hank and Mike reached to their Attitude Director Indicator (ADI) switches and flipped them to change the mode of the ball. I watched Hank's ADI reflect *Discovery*'s tilt toward the risen sun. If our ascent was nominal, the ADI switch would be the only switch touched until MECO . . . 8½ minutes, 4 million pounds of propellant, and 17,300 miles per hour away. *Please, God, that it be so.* Having to use other switches could only mean one thing: something wasn't nominal. My eyes fell on the contingency abort cue card Velcroed to Hank's window frame. It detailed procedures for ditching the shuttle, which all of us knew would be death. NASA called all the other abort modes "intact aborts"—the orbiter and crew would be recovered "intact" either in the United States, Europe, or Africa. But they couldn't bring themselves to call a ditching abort a "not intact abort." Like sailors of old painting the decks red so the blood of battle wouldn't shock a crew, NASA camouflaged the ditching procedures with the title "contingency abort." One of the card's helpful suggestions was to ditch parallel to the waves. Astronauts joked that the contingency abort procedures were just something to read while we were dying. For some reason the joke seemed funnier while standing at the office coffee bar.

Except for the noise, vibrations, and G-forces, the ride was just like the simulator, which is akin to the circus Human Cannonball saying, "Except for the earsplitting explosion, the G-forces, and the wind up your nose, it's just like sitting on a case of unlit dynamite." NASA would never duplicate this ride in any ground simulator.

"Throttle down." We were forty seconds up and the vibrations intensified as the vehicle punched through the sound barrier. Everything was shocking the air . . . the giant, bulbous nose of the ET, the pointed cones of the SRBs, the orbiter's nose, wings, and tail, the struts holding everything together. The interplaying shock waves were an aerodynamic cacophony and the engines throttled back to keep the vehicle from tearing itself apart.

Our seats wiggled and groaned under the stress. I was amazed by the flexibility of the machine. It reminded me of times in my childhood when I would slide down a bumpy, snow-filled arroyo in a cardboard box. Now, as then, I wondered how my cockpit could stay together through all the bouncing and shaking.

"Throttle up." The air was thinning and the aerodynamic pressure decreasing. The three Rocketdyne beauties at our backs were once again spiraling to full power. What a rush it was to feel the buildup of thrust, just like jamming the throttles of a fighter into the afterburner detents. I suspect every shuttle pilot would have loved to snatch the controls from the autopilot and manually throttle the engines to full power. How many times in your life would you have 1.5 million pounds of thrust wrapped around your fingers?

The prayers flying from the souls of everybody in the cockpit were identical, that God would continue to smile upon the SSMEs. We most feared these engines, and for good cause. There had been many SSME ground-test explosions and premature shutdowns. We were also strapped to two SRBs, each burning nearly 5 tons of propellant per second, but nobody gave a second thought to them. No engineer had ever come to a Monday morning meeting to explain away a SRB ground-test failure. The SRBs had always worked. But even as we scorched the prayer line with our pleas for flawless SSME function, both SRBs were betraying us. A primary O-ring at different joints in each tube had failed to seal as the motors had ignited. Tentacles of flame from the combustion area had wiggled between the segment facings. Like something alive and trapped, the gas had been wild to escape. It had reached the leak points and started to consume the O-ring rubber. The leak on the left-side SRB was bad enough for hot gas to actually get past the primary O-ring. Though we wouldn't know it until after the *Challenger* disaster, we had just experienced the first case of what the Thiokol engineers would later define as "blow-by." Hot gas had penetrated into the space between the primary and backup O-rings. Had our leak continued moments longer, the primary and backup O-rings would have been consumed and history would have recorded the *Discovery* disaster instead of *Challenger.* It would have been Zoo Crew's names etched in an Arlington Cemetery monument. But the leak hadn't continued. Inexplicably the primary O-rings had resealed.

The clock was approaching T+2 minutes and the Gs rose to 2.5. An invisible hand pushed me deeper into the seat. I reached forward, drew my hand back, then reached forward again. The veterans had warned it was tough to maneuver an arm under G-loads and it was a good idea to practice in case an ascent emergency later required a reach for a switch.

"You see that?" At Hank's question I was reminded of Judy's warning about not ending any sentence with the word *that*. Needless to say, my ears perked up.

Mike replied, "Yeah, it looks like foam from the tank is flaking off."

Mike and Hank continued a brief discussion about the particles that were racing past the windows, sometimes striking them. There was no concern in their voices and I quickly dismissed their comments. The ET insulation foam was so light I couldn't imagine it would damage any part of the vehicle. Nineteen years later a briefcase-size chunk of foam ripping from the gas tank would doom *Columbia*.

"P-C less than fifty." Hank relayed the message on his computer screen that the chamber pressure inside the SRBs had fallen to less than 50 pounds per square inch. A loud metallic bang shook the cockpit and a flash of fire whipped the windows as the boosters separated from the ET. Both SRBs tumbled away to parachute into the ocean.

The sudden loss of 6 million pounds of thrust accompanied by dead silence caught me by surprise. Had all three of the SSMEs also shut down? I leaned to my left and stared at the engine status lights and for several heartbeats I expected to see them illuminate in a deadly red glow. But the lights remained off, the radios quiet. I swallowed back my heart. Apparently I had been asleep in training when somebody had described SRB separation and the quiet, velvet smoothness that followed. There was nothing wrong with the vehicle. *Discovery* had put most of the atmosphere behind her. There was no air to grip the machine or rattle us with shock waves. And the SSMEs were as finely tuned as a Rolex. They continued to deliver nearly 1.5 million pounds of thrust 100 feet behind our backs without a whisper of noise or ripple of vibration. The ride became as smooth as a politician's lie.

Hank's altitude and velocity tapes scrolled upward as the sky faded to abysmal black. Sunlight was streaming through the windows and yet the sky was utterly dark. It was my first real space experience, something I had never and could never experience as an earthling . . . simultaneous night and day, simultaneous high noon and deep midnight.

Our various abort windows began to open and close. "*Discovery,* you're two-engine TAL." We had acquired enough altitude and speed to fly across the Atlantic and land in Senegal, Africa, if one engine failed, a maneuver known as a Transatlantic Landing abort (TAL). NASA had positioned an astronaut at the Dakar international airport to help air traffic control personnel if we declared an abort. He also had our passports and visas. I had a vision of standing in the customs line at the Dakar airport in our shuttle flight suits with our helmets in the crook of our arms while a fez-headed, accented bureaucrat asked, "Anything to declare?" It was something I hoped never to experience.

"*Discovery,* you're negative return." The Return to Launch Site (RTLS) abort window closed. We were now too far from Florida and headed too fast to the east to be able to return to a landing at KSC. If an engine failed, we were committed to a "straight ahead" abort. That was okay by all of us. Nobody wanted to do a turnaround RTLS abort. It was an unnatural act of physics. If selected, it would pitch the shuttle around in an outside loop to point us toward Florida. But it would take minutes to cancel our several-thousand-miles-per-hour eastward velocity, so we would actually be traveling backward over the Atlantic. Ultimately we would end up as a million-pound helicopter, fifty miles high, with zero forward speed. Then, we would begin the slow acceleration toward our objective, Florida, only two hundred miles away. The experts swore it would work and Mike and Hank had practiced RTLS aborts in the sim about a thousand times, but nobody wanted to be the first to field-test the procedure.

Discovery continued a nominal ascent. Passing about thirty miles altitude, it occurred to me I could die without ever having seen the Earth from space. The shuttle's nose was so high and I was sitting so far aft in the cockpit I couldn't see anything of the planet. But there was a window above and just slightly behind my head and since the shuttle flies to orbit upside down, that window did provide a view of the Earth. It was a mighty temptation.

I looked furtively at Steve Hawley. His head was making small jerking motions like Data from *Star Trek* as he moved his eyes to every display. There wasn't an electron running in *Discovery*'s body that Hawley's brain wasn't also processing. With him at my side, I rationalized, I wouldn't be missed for a moment of sightseeing. Under the mounting G-forces I craned my neck upward and backward until I thought it would break. The contortion worked. I could see the Earth receding below us. Scattered cumulus clouds had been reduced to points of white. The variations in sea depth were evident in different shades of blue. It wasn't much of a view and I was condemning myself to one hell of a neck ache to capture it, but it was enough for the moment. If God took me now, at least I would have a story to tell while waiting in line for the down escalator to Bible/feminist/post-doc hell.

We were approaching fifty miles, the magic line that would officially make us astronauts. I had always thought this altitude requirement was bean-counter bullshit. Effectively it was a statement that riding a rocket didn't really get dangerous until you hit fifty miles. In reality if you didn't make it to fifty miles on the shuttle, it probably meant the machine had killed you, as was later to be the case with the *Challenger* crew. Mike Smith was a rookie killed on that mission and by the official definition he didn't die as an astronaut since he only made it to ten miles altitude. (Note to NASA: When the hold-down bolts blow, you've earned your gold.)

Hank gave us a countdown. "Here it comes . . . forty-eight . . . forty-nine . . . fifty miles. Congratulations, rookies. You're officially astronauts." We cheered. I suspect Judy, Steve, Mike, and Charlie were relishing the moment as I was. I experienced a momentary calm not unlike what I expect someone summiting Mount Everest experiences. There were still a few thousand things that could kill me, but their threat couldn't tug me away from the moment. I stared into the black and watched images of my childhood play in my mind's eye. I saw my homemade rockets streaking upward from the Albuquerque deserts, my dad on his crutches cheering. I saw my mom helping me bake my fuels in her oven and cleaning out coffee cans for my capsules. I saw myself lying in the desert watching Sputnik and Echo streak across the twilight sky. I had achieved a dream of ten thousand nights. I was an astronaut.

My distraction was only a few heartbeats in duration, but it seemed an age since the cheers had ceased. I looked at Steve. He was still mind-melding with *Discovery,* sucking in every byte. I got back on the instruments and listened for more of MCC's abort boundary calls.

"*Discovery,* you're single-engine TAL."

"*Discovery,* you're two-engine ATO."

"*Discovery,* you're press to MECO." This was the sweetest call of all. It meant we could still make it to orbit even if one SSME failed. As an astronaut had once joked, "Surely God couldn't be so mad at us that He would fail *two* engines."

At about eight minutes the G-forces hit three and the main engines throttled back to maintain that acceleration. This was necessary to prevent *Discovery* from rupturing herself. With a nearly empty gas tank the engines now had the muscle to overstress the machine. The reduction in power prevented that.

Hank's velocity tape raced upward . . . 20,000 feet per second . . . 21,000 . . . 22,000. Every 15 seconds *Discovery* was adding another 1,000 miles per hour to her speed. We were giddy with excitement, our laughs distorted by the G-loads.

"Houston, MECO. Right on the money." At Hank's call another cheer swept the cockpit. *Discovery* had given us a perfect ride.

CHAPTER 21

Orbit

MECO was silent. The Gs just stopped. I had no sense of being hurled forward as some space movies depict. There was no *thud, thunk, bang,* or any other noise to indicate the end of powered flight. MECO could only be noted as the termination of acceleration. In a blink we went from a silent 3-Gs to a silent 0-G.

At this point *Discovery* was headed for an impact into the Pacific Ocean. We still were not in orbit. The ascent was intentionally designed so as not to drag the 50,000-pound gas tank into orbit, where it would become a threat to populations below. Better to keep it on a suborbital trajectory, where its impact could be predicted. There was a heavy *thunk* in the cockpit as the ET was exploded away to continue toward a Pacific grave. Hank moved his translational hand controller to the up position, and the thrusters in the nose and tail fired to clear us of the tumbling mass. The nose jets, merely a few yards forward of the windows, hammered the cockpit as if howitzers were firing next to us. Checklists strained at their Velcro anchors.

Now clear of the ET, *Discovery*'s computers fired her Orbital Maneuvering System (OMS) engines, twin 6,000-pound trust rockets mounted at the tail. Compared with the SSMEs, these were mere popguns, giving us only a ¼-G acceleration. The engines burned for two minutes to finish the orbit insertion. Then our silent free-fall began. We were in orbit 200 miles above the planet traveling at a speed of nearly 5 miles per second. The entire ascent had taken just ten minutes. In all likelihood Donna and the kids had not even had time to walk from the LCC roof.

I watched Mike activate the switches to close the ET doors. These covered two large openings on *Discovery*'s belly through which passed seventeen-inch-diameter fuel and oxidizer feed pipes from the gas tank. These pipes had been disconnected during the jettison of the ET. Now the doors had to close over the openings to complete the belly heat shield. If they failed to close, we were dead . . . but endowed with the power to choose the manner of our deaths: slow suffocation in orbit as our oxygen was depleted or incineration on deorbit. The open cavities would be pathways for frictional heat to melt the guts out of *Discovery*'s belly on reentry. I didn't lift my eyes from the ET door indicators until I saw them flip to CLOSED.

I was still strapped to my seat and didn't yet feel weightless but the cockpit scene made it obvious we were. My checklist hovered in midair. A handful of small washers, screws, and nuts floated by our faces. An X-Acto blade tumbled by my right ear. *Discovery* had been ten years in the factory. During that time hundreds of workers had done some type of wrench-bending in her cockpit. While NASA employed strict procedures to keep debris from being lost in the vehicle, it was impossible to prevent some dropped items. Now weightlessness resurrected those from various nooks and crannies. A live mosquito also flew into view. It had entered through the side hatch during the many hours of prelaunch operations and had hitched a ride into space. I slapped it dead between my hands.

While I had trained for thousands of hours to immediately dive into the postinsertion checklist, I couldn't overcome the temptation to look at our planet, now filling the forward windows. Blue, white, and black were the only colors. Swirls of lacey clouds patterned an otherwise limitless expanse of deep blue Atlantic Ocean. All of this was framed in a pre-Genesis black. There was no blackness on Earth to compare . . . not the blackest night, the blackest cave, or the abysmal depths of any sea. To say the view was overwhelmingly beautiful would be an insult to God. There are no human words to capture the magnificence of the Earth seen from orbit. And we astronauts, cursed with our dominant left brains, are woefully incapable of putting in words what the eyes see. But still we try.

I forced myself back to the checklist as we configured *Discovery* for orbit. Steve Hawley and I disassembled our seats, and he floated them downstairs for stowage. During the Houston simulations we had nearly popped hernias while moving these 100-pound monsters. Now we pushed them with our fingers.

We loaded the orbit software into *Discovery*'s brain; her decade-old IBM computers didn't have the memory capacity to hold ascent, orbit, and entry software simultaneously. Next we opened the payload bay doors. The inside of those doors contained radiators used to dump the heat generated by our electronics into space. If they failed to open, we'd have only a couple hours to get *Discovery* back on Earth before she fried her brains. But both doors swung open as planned, another milestone passed.

As I worked, I wondered if I would get sick. I questioned every gurgle, every swallow. Is that bile I taste? The rational part of my brain said I was okay, but my paranoia twisted every gastrointestinal sensation into something ominous. I checked and double-checked and then triple-checked that my numerous barf bags were ready for a quick draw. The veterans had warned us the sickness could come on very suddenly. They were right. The curse hit. Not me, but Mike Coats. He retched violently into his emesis bag. I felt something warm touch my cheek and reached up to wipe away yellow bile. Other tiny bits of the fluid floated in the cockpit. Mike was learning what we would all soon learn—it is impossible to completely contain fluids in weightlessness. Though he had his bag at the ready, some barf had escaped. The odor permeated the tight cockpit. Mike sealed his bag but, with work to do, he couldn't leave his seat to stow it downstairs. I took it and floated to the wet-trash container. With somebody else's emesis smearing my cheek, the smell in my nostrils, and a warm bag of the mess in my hand, I had every trigger in place to get sick myself but still I felt fine. I began to think maybe I had dodged the SAS bullet.

Downstairs I got my first view of Judy, who was busy activating the toilet. Everybody was anxious for that to be declared operational. The weightlessness had liberated her black tresses to coil about her head like Medusa's snakes. She would have made a great cannon cleaner. I lifted the floor-mounted trapdoor to access the wet-trash container and shoved Mike's barf bag through the rubber grommet. I mimicked the garbage pit scene from *Star Wars* and pretended my hand had been grabbed by an alien creature living inside. I made a few jerking motions and screamed for Judy to help me. She grabbed my arm, pretending to assist my escape. We tumbled together like fifth-graders on a playground, laughing all the while. With the terror of launch behind us and the intoxicant of being *real* astronauts, we had been transformed into kids.

During a break in the work I went to my locker to change out of my coveralls and take off my UCD. I had often wondered how the privacy issue would play out when we finally got to orbit. In our training Judy had certainly seemed unflappable. She had not fled from the EVA simulation when Hawley and I had been standing naked in front of her rolling on condoms. Still, I wondered how the tight living conditions would affect her behavior. I waited until she had some upstairs duties to attend to and then stripped from my clothes. A few moments later, while I was completely nude and extracting underwear from my locker, Judy returned. She looked at me and said, "Nice butt, Tarzan," then went back to her work. For once, I was speechless.

This wasn't the only time that day Judy showed how comfortable she felt around us men. While she was searching for something in her own locker she pulled out a chain of tampons. Like a magician pulling out a seemingly endless rope of scarves from a hat, she kept pulling and pulling. Each of the products was shrink-wrapped in plastic, each precisely separated from the other. The floating belt had all the appearance of a fully loaded bandolier of cotton bullets. Judy smiled. "I can tell you that a man packed this locker." I laughed at the image of a crusty old NASA engineer addressing the issue of how many feminine hygiene products should be loaded. He probably got a number from his wife and then applied a NASA safety factor and then added a few contingency days on top of that number. And then, incanting Gene Kranz's famous *Apollo 13* challenge, "Failure is not an option," he added some more.

As she wrestled the belt back into its tray, Judy commented, "If a woman had to use all of these, she would be dead from blood loss."

Our first day continued with payload preparations. We rolled out the robot arm and closed the sunshades on our trio of satellites. Charlie Walker began work on his experiment. Mike loaded the IMAX camera. Throughout the mission he and Hank would be filming some space scenes for the Walter Cronkite–narrated IMAX movie, *The Dream Is Alive*.

At one point I was alone in the upstairs cockpit when Hank called to me from the toilet, "Mike, let me know when we're passing over Cuba." I grabbed a camera, assuming he wanted me to photograph the island as part of our Earth observation experiment.

"We're about five minutes out."

"Give me a countdown to Havana."

Shit, I didn't know where Havana was. I scrambled to find it in our booklet of maps. "Ten seconds, Hank. It's coming up quick. I'll get the photo for you." I aimed the Hasselblad and began to click away.

From below I heard Hank in his own countdown, "Three . . . two . . . one," followed by a cheer.

A moment later Hank's head popped above the cockpit floor. He wore an expansive grin. "I just squeezed out a muffin on that fucker Castro. I've always wanted to shit on that commie."

Every military astronaut was a Red-hater. We'd been shot at by communist bullets in Vietnam. Many had experienced long separations from families while deployed to remote Cold War outposts. Hank had claimed his small revenge by giving birth to an all-American turd two hundred miles above that commie clown. Hank wistfully continued, "Damn, I wish our orbit took us over Ted Kennedy." Mr. Kennedy was spared Castro's fate by our orbit path. The only parts of the continental United States we flew over were the extreme southern portions of Texas and Florida. Our orbit inclination (tilt to the equator) fixed the traces of our orbits between 28 degrees north latitude and 28 degrees south latitude.

For several hours we were immersed in our checklists to deploy our first communication satellite and its booster rocket. It, like the others, was destined for an orbit 22,300 miles above the Earth's equator. At that extreme altitude the orbit speed of the satellite matched the turn of the Earth so, to Earth observers, it would appear parked in the sky. At the contractor's ground receiving stations, satellite dishes could be pointed at the satellite and the Earth's rotation would do the tracking.

Hawley monitored the satellite-deployment computer displays from the front cockpit while Judy and I worked the release controls in the back. We opened the baby buggy–like sunshield, spun up the payload to 40 revolutions per minute (for stabilization during its uphill rocket burn), then activated the switches to pop it free of *Discovery.* The orbiter shivered as the 9,000-pound mass was shed.

The successful payload deployment refreshed our euphoria. We knew there were a hundred sets of very critical eyes watching our performance—all belonging to fellow astronauts. Any screwup would be our legacy. Astronauts have elephantine memories when it comes to crews who make mistakes.

Afterward we relaxed around our supper of dehydrated shrimp cocktail, beef patties, and vegetables. The food was packaged in plastic dishes and rehydrated with water from our fuel cells. After nearly burning a hole in my esophagus by swallowing a blob of inadequately hydrated horseradish powder, I learned to mix the food and water a bit longer. Fuel cell water was also used for drinking. It was dispensed into plastic containers, some of which contained various flavored powders (yes, including Tang). Because nothing can be poured in weightlessness, the drinks had to be sucked from straws. I quickly learned never to drink plain water. Iodine was used as a disinfectant and the water was tinged yellow with it and tasted of the chemical. While we ate much better than the early astronauts, who had to squeeze their food out of tubes, I still longed for the day the NASA food engineers would come up with dehydrated beer and pizza.

After cleaning up and cycling through our toilet we prepared for sleep. This was not a "shift" mission so we all slept at the same time. We would depend upon *Discovery*'s caution-and-warning system to alert us if something bad happened. Each of us had a sleep restraint, a cloth bag we pinned to the walls and zipped into. There was no privacy. Like bats in a cave, we bunked cheek to jowl in the lower cockpit. We slept downstairs because the lack of windows made it darker and cooler than the upstairs cockpit.

As I floated inside my restraint I joined in the chorus of complaints about a fierce backache. In weightlessness the vertebrae of the spine spread apart, resulting in a height increase of an inch or two. The strain on the lower back muscles is significant and painful. All of us but Judy were bothered by it. Why she was immune I had no idea, but she grew weary of our complaints and exploited her advantage: "I'm probably the first woman in history to go to bed with five men and all of them have backaches."

I couldn't sleep . . . and it wasn't because of any backache. I didn't *want* to sleep. I wanted to celebrate. From MECO to this moment, I had been too busy with checklists to really consider the life-changing experience of the past twelve hours. I had done it! I was an astronaut in the cockpit of a spaceship orbiting the Earth. I was living what Willy Ley had written about in *The Conquest of Space*. I wanted to scream and shout and punch my fists in the air. Fortunately for the rest of the crew, I didn't do any of those things. Instead, I floated my sleep restraint upstairs. *Floated!* God, I still couldn't get my mind around the reality of it. I tied the bag under the overhead windows and slipped inside. I would celebrate by sightseeing. Since the autopilot was holding the shuttle with its top to the Earth, I now had the planet in my face.

Other than the breath of the cabin fans and the white-noise hiss of the UHF radio, the cockpit was midnight still. In the silence I felt as if we had stopped dead in space. In all my other life experiences speed meant noise . . . the howl of wind gripping a cockpit, the roar of an engine. Now I was traveling at nearly 5 miles per second and there was only silence. It was as if I were hovering in a balloon, and the Earth was silently turning beneath me.

I was also gripped with a powerful sense of detachment from the rest of humanity. There was nothing at the windows to suggest any other life in the universe. I was looking to a horizon more than a thousand miles distant and could see only the unrelieved blue of the Pacific. In each passing second that horizon was being pushed another five miles to the east but still nothing changed. There was no vapor trail of a jetliner, no wake of a ship, no cities, no glint of Sun from a piece of glass or metal. There was no signature of life on Earth. And the view into space was even more lonely. The brilliance of the Sun had overwhelmed the faint light of the stars and planets. Space was as featureless black as the ocean was blue.

The Sun was intense and the cockpit grew uncomfortably hot. I pushed from my sleep restraint and hovered in my underwear a few inches from the glass. In my relaxed state my arms and legs folded inward as if trying to return to their fetal position. I had become a hairy *2001: A Space Odyssey* embryo.

The forty-five minutes of my orbit "day" drew to an end and I was treated to another space sight of such breathtaking beauty it would challenge the most gifted poet. As *Discovery* raced eastward, behind her the Sun plunged toward the western horizon. Beneath me, the terminator, that hazy shadow that separates brilliant daylight from the deep black of night, began to dim the crenellated ocean blue. High clouds over this terminator glowed tangerine and pink in the final rays of the Sun. *Discovery* entered this shadow world and I turned my head to the back windows to watch the Sun dip below the horizon. Its light, which to this moment had been as pure white as a baby's soul, was now being split by the atmosphere. An intense color spectrum, a hundred times more brilliant than any rainbow seen on Earth, formed in an arc to separate the black of earth night from the perennial black of space. Where it touched the Earth, the color bow was as red as royal velvet and faded upward through multiple shades of orange and blue and purple until it dissipated into black. As *Discovery* sped farther from it, the bow slowly shrank along the Earth's limb toward the point of sunset, diminishing in reach and thickness and intensity, as if the colors were a liquid being drained from the sky. Finally, only an eyelash-thin arc of indigo remained. Then it winked out and *Discovery* was fully immersed in the oblivion of an orbit night.

Suddenly the uniform black of daytime space was transformed into the stuff of dreams. The Milky Way arced across the sky like glowing smoke. Other stars pierced the black in whites, blues, yellows, and reds. Jupiter rose in the sky like a coachman's lantern. For planet and stars alike, there was no twinkle. In the purity of space they were fixed points of color.

I stared down into the dark of the Earth. Lightning flashed in faraway Central American thunderstorms. Shooting stars streaked to their deaths in multihued flashes. To the northeast I could see the sodium glow of an unknown city. At the horizon the atmosphere had a faint glow caused by sunlight scattering completely around the Earth. In this glow the air was visible as several distinct layers of gray.

I watched a satellite twinkle through the western sky. Though *Discovery* was in darkness, the other machine was far enough to the west to still reflect sunlight.

With the instrument lights off and the Sun gone, the cockpit chilled and I floated back into my restraint to attempt sleep. I had just nodded off when a streak of light flashed in my brain and startled me awake. Veteran astronauts had warned of this phenomenon. The flash was the result of a cosmic ray hitting my optic nerve. The electrical pulse generated by that impact caused my brain to "see" a streak of light even though my eyes were closed. I wondered what those cosmic rays were doing to the rest of my brain. *Oops, there goes second grade.*

I slept fitfully through the night, waking with each sunrise and whispering, "Wow!" At one point I floated into the lower cockpit to retrieve a drink container and entered a scene straight out of a science fiction movie. A light had been left on in the toilet and it dimly illuminated *Discovery*'s sleeping crew. They were in their restraints, some pinned to the forward wall, others stretched horizontally across the mid-deck. In the relaxation of sleep their arms floated chest high in front of them. It appeared as if they were in suspended animation. I was tempted to join them in the cool darkness, but the pull of the windows was too great. I floated back upstairs.

Reveille came in the form of rock music. It was traditional for the CAPCOM to provide music for MCC to send up as a wake-up call. But the tune was unrecognizable. Apparently NASA's budget was running low when it came time to procure speakers. Pop music from these Radio Shack rejects sounded like fingernails being drawn across a chalkboard.

To my surprise I did not wake up alone. My closest friend was alert and waiting. I had an erection so intense it was painful. I could have drilled through kryptonite. I would ultimately count fifteen space wake-ups in my three shuttle missions, and on most of these and many times during the sleep periods my wooden puppet friend would be there to greet me. Flight surgeons have attributed this phenomenon to the fluid shift that occurs in weightlessness. On the Earth, gravity holds more blood in our lower legs. In orbit that blood is equally distributed throughout our bodies. For men the result is a Viagra effect. There are beneficial effects for the female anatomy, too. The same fluid shift makes for skinnier calves and thighs and larger, nonsag breasts. If NASA wants to secure its financial future, it would be smart to advertise the rejuvenating effects of weightlessness. Taxpayers would demand that Congress quadruple NASA's budget to finance the construction of orbiting spas where visitors from Earth could turn back time.

Fortunately for me, my brain was quickly flooded with thoughts of the workday and my body melted in response.

On day two we successfully launched our second satellite, Syncom, but not without mishap. As Hank was filming its release with the huge and unwieldy IMAX camera, a shank of Judy's frizzed-out hair was snatched into the machine by the belt drive of the film magazine. It was as if her hair had been caught up in the fan belt of an automobile. She screamed and I grabbed at her tresses to prevent them from being ripped out of her scalp, but, with nothing to hold me in place, I tumbled out of control. Judy did the same. Through her increasingly urgent screams, I heard the camera labor to a grinding stop. The hair had clogged the motor, finally stalling it and popping a cockpit circuit breaker.

We cut Judy free with scissors. Strands of loose hair floated everywhere. They were in our eyes and mouths. Mike Coats, who was the principal operator of the IMAX, took the machine to the mid-deck and began work at restoring its operation. The hair was so thoroughly jammed into the motor gears we doubted the machine would ever pull another frame of film. IMAX was going to be severely disappointed. They had spent millions to fly their camera in space and we had only recorded a fraction of our film targets. Even if the camera could be cleaned of hair and made to work again, a quick glance at the flight plan showed the next several film opportunities were certainly going to be missed. IMAX would have to do some replanning. We all knew this was the type of trivial screwup that would become the focus of an otherwise successful mission. The press wouldn't talk about how our crew had successfully taken *Discovery* on its maiden flight or how we had successfully released thirty thousand pounds of satellites. Instead, they would zero in on our hair incident. But we had no alternative other than to come clean with MCC. The flight planners needed to assume the camera could be repaired and get started on rescheduling our targets.

But we males had been missing the *real* issue. As Hank picked up the microphone to call MCC, Judy lashed out at him with something along the lines of, "If you so much as breathe a word to MCC about my hair jamming the camera, I'll cut your heart out with a spoon." Or perhaps she threatened a more vital area of his anatomy. There was a brief moment as we struggled to understand Judy's rage. Then it dawned on us. She was only the second American woman to fly in space. The press had her under a magnifying glass, looking for the slightest flaw in her performance. The hair jam incident was just that: a mistake with her name on it. Not only that, it contained the worst possible sin against feminism. Judy had demonstrated, however innocently and however insignificantly, that women were indeed different from men.

Hank Hartsfield, a grizzled air force fighter pilot who had stared death in the eye on many a mission, now faced a man's worst nightmare—a *really* pissed-off woman. No communist gunner had ever appeared as deadly as did Judy at that moment. Under her searing glare Hank did what we all would have done. He wanted to return with all of his appendages, so he called MCC and told them the IMAX had a film jam and Mike was working to clear it. He made no mention of the cause of the jam. Eventually Mike was able to breathe life back into the camera. With rescheduled targets, he and Hank continued their filming while Judy stayed far, far away.

Nature finally caught up with me and I floated into the shuttle toilet to face what was truly the most difficult part of any spaceflight—a bowel movement. The toilet provided little privacy. It was situated in the rear corner of the mid-deck on the port side. There was no door, only a folding curtain that could be Velcroed across the mid-deck–facing entry. Another curtain was Velcroed to form a ceiling and isolate the toilet from the upstairs cockpit. The lack of privacy was intimidating. I felt like I was back on my honeymoon, preparing for my first married-life BM. We've all been there.

After I was inside the curtained box, I took the advice of shuttle veteran Bob Crippen and stripped naked. "It's a lot easier to wipe feces off your skin than it is to get it off your clothes" had been one of his STS-1 mission debriefing comments.

I located my personal urine funnel and twisted it on the end of the urinal hose, then loaded a disposable vacuum cleaner–like bag in a can on the left side of the toilet. Used tissue had to be placed in this bag. It could not be put in the toilet since that would require the ass to be lifted, which, in turn, could result in feces being released into the cabin. Suction at the bottom of the can would hold used tissue inside the bag.

I floated over the throne, lifted up on the thigh restraints, and twisted them inward to clamp my body to the plastic seat. Recalling my boresight alignment from the camera view in the toilet trainer, I wiggled my body until some freckles on my thighs were properly positioned in relation to the toilet landmarks. I switched on the toilet fan and welcomed the noise it generated. At least some of my BM noises would be camouflaged. Finally I pushed my penis into proper aim at the urinal funnel, reached for the solid waste collector lever, and pulled it back. Directly beneath me the waste opening was uncovered and the feces-steering airflow was activated. Suddenly a very sensitive part of my body was hit with a blast of chilled air. Few things are less conducive to promoting a BM than having cold air jetting around the principal performer in that act. The natural tendency is to clamp shut. But I convinced the orifice in question to ignore the cold gale and let fly. Simultaneously, I held the urine funnel at my front to collect my liquid waste. The vacuum flow into the urinal hose was very effective at sucking away the fluid until my bladder pressure fell. Then the urine refused to separate from my skin and a ball of it grew on the end of my penis. NASA's engineers had anticipated this aspect of fluid dynamics and had provided a "last drop" feature. By squeezing buttons at the sides of the hose, the suction was boosted and I was able to wet-vac myself of most of the fluid. As the slurping sounds of this operation came through the curtains, Hank hollered, "More than five seconds and you're playing with it, Mullane!"

The toilet was a bountiful source of male juvenile humor. By far the best toilet joke was pulled by Bill Shepherd (class of 1984). On one of his missions he carried a piece of sausage from his breakfast into the toilet. After finishing a bowel movement, he set the sausage free to float upstairs. As panicked crewmembers ricocheted from wall to wall in a mad retreat from the offending planetoid, Bill chased after it with a piece of toilet tissue. He finally grabbed it and then, to the horror of all, he ate it.

For me, cleanup was next. I used a tissue to blot the remaining dampness from my penis. Wiping after urinating was such a feminine act I almost felt compelled to hack up a luggie to reestablish my sexual identity. I pulled the solid waste collector lever closed, lifted my thigh restraints, and floated from the seat. I was now free to wipe myself, putting the used tissue in the disposable vacuum bag.

Besides cleaning myself, I also had to clean the toilet. It would be a serious violation to leave any fecal smears on or around the slide cover for the next user to confront. And there were always some smears. Even after the camera-aim practice in the toilet trainer in Houston, it was difficult to get a direct hit during a BM. The feces almost invariably made some contact with the inner sides of the collector hole. As one astronaut had once lamented, "Turds come out curved. If only they were straight, we might have better luck in cleanly using the toilet." I used a NASA-provided disinfectant to wipe away my smears and put the soiled tissue in the vacuum bag. I then sealed that bag and stowed it in a container at the back of the toilet. While the retention of solid waste and BM tissues suggested a bad odor problem could develop, the toilet designers had done an excellent job of routing airflow through the toilet and filtering it with activated charcoal filters. There were never any toilet smells in the cockpit.

Finally, I dressed. From start to finish, a task that might have taken me five minutes on Earth had consumed nearly thirty minutes in space (and covered about eight thousand miles). There are times in an astronaut's life he or she would pay dearly to have a gravity vector. Using the toilet is one of those times.

Our third and last communication satellite was successfully deployed on flight day three. Compared with the missions of the early space program this was blue-collar work, completely devoid of glory. We weren't beating the Russians to anything. We weren't planting an American flag in alien soil. On Earth, there was no Walter Cronkite removing his dorky glasses, wiping his forehead, and shaking his head in relief while telling a waiting, breathless world, "They've done it! The *Discovery* crew has just released another communication satellite!" The space program had become a freight service, justifiably ignored by the press and public. But not one of us in that cockpit was complaining. Even Hank would have hauled Lenin's taxidermied body into orbit and sung the "*Internationale*" for all the communists in the world, if that's what it took to put him in space.

On flight day four Judy activated the controls of our final major payload, a solar energy panel. A collapsible motor-driven truss unfurled the 110-foot-long-by-10-foot-wide Mylar sail out of its payload bay container. There were no active solar cells on the sail. The experiment was only to gather data on the dynamics of the deployment and retraction system. When the panel was completely up and in tension, she radioed MCC, "Houston, it's up and it's big." In numerous simulations Judy had joked that she was going to make that call. We had teased her about the obvious sexual innuendo. She made the call nevertheless. That was Judy. She could be extremely defensive of her status as a feminist standard

I was a child of the space race—twelve years old at the time of Sputnik I's launch, October 4, 1957. I wanted to be an astronaut from the moment I heard the word.

The Hugh Mullane family in Albuquerque, New Mexico, circa 1960. I'm second from the left. My dad was rendered a paraplegic at age thirty-three by polio. His leg braces are visible at his atrophied ankles.

At about age sixteen, I'm posed with one of my homemade rockets. My "capsule" was a coffee can; the nose cone was a rolled-up sheet of plastic. My mom and dad were huge supporters of my interest in space. After taking this photo, my dad drove me into the desert for the launch.

Donna and I walk out of the Kirtland AFB chapel, June 14, 1967. We both married for the wrong reasons—me for sex, she for escape from her parents. Somewhere in our marriage (thirty-eight years and counting), we fell in love. We would have three children, who would give us six grandchildren.

In 1969, I flew 134 combat missions in Vietnam in the backseat of the RF-4C, the reconnaissance version of the F-4 Phantom. My flawed eyesight prevented me from being a fighter pilot.

George Abbey, a midlevel bureaucrat, was God to the astronaut corps. He had supreme authority over shuttle mission assignments. Morale suffered significantly under his despotic and secretive leadership style, and many astronauts came to loathe him.

udy Resnik, the second American woman in space, helps me prepare for a pacewalk simulation in early 1984. In our year of training for our rookie nission, STS-41D, we became close friends. Judy opened my male, sexist-pig yes to the reality that women could do the astronaut job as well as any man. he would die aboard *Challenger* while flying her second space mission.

The heartrending final astronaut-spouse good-byes occur approximately twenty-four hours prior to launch at the astronaut beach house. For the *Challenger* and *Columbia* spouses, the good-byes were forever. The beach house sits on sacred ground.

Donna and I sought the privacy of the beach house sands for our rewells. Before all of my missions, told her, "If I die tomorrow, I died doing what I loved."

Donna slumps in emotional and physical exhaustion after one of my many mission scrubs. At T–9 minutes in the countdown, spouses and children are escorted to the roof of the Launch Control Center to watch the liftoff in the company of an astronaut family escort, aka "an escort into widowhood."

The STS-41D in-orbit crew self-portrait. I'm floating at the left (legs extended). At the bottom, from right to left, are Pilot Mike Coats and Commander Hank Hartsfield. At Judy Resnik's left side is Mission Specialist Steve Hawley. Payload Specialist Charlie Walker floats behind me. Judy received hate mail from a handful of feminists for the cheerleader effect the pose suggested.

A lifetime dream comes true—ating in *Discovery*'s pper cockpit on my rookie mission, TS-41D, August 30–September 5, 1984.

Donna greets me after landing at Edwards AFB from my first mission, September 5, 1984. Mission Commander Hank Hartsfield and his wife, Fran, are in the background.

Located a few blocks from the Johnson Space Center main gate, the decrepit Outpost Tavern is a popular astronaut hangout.

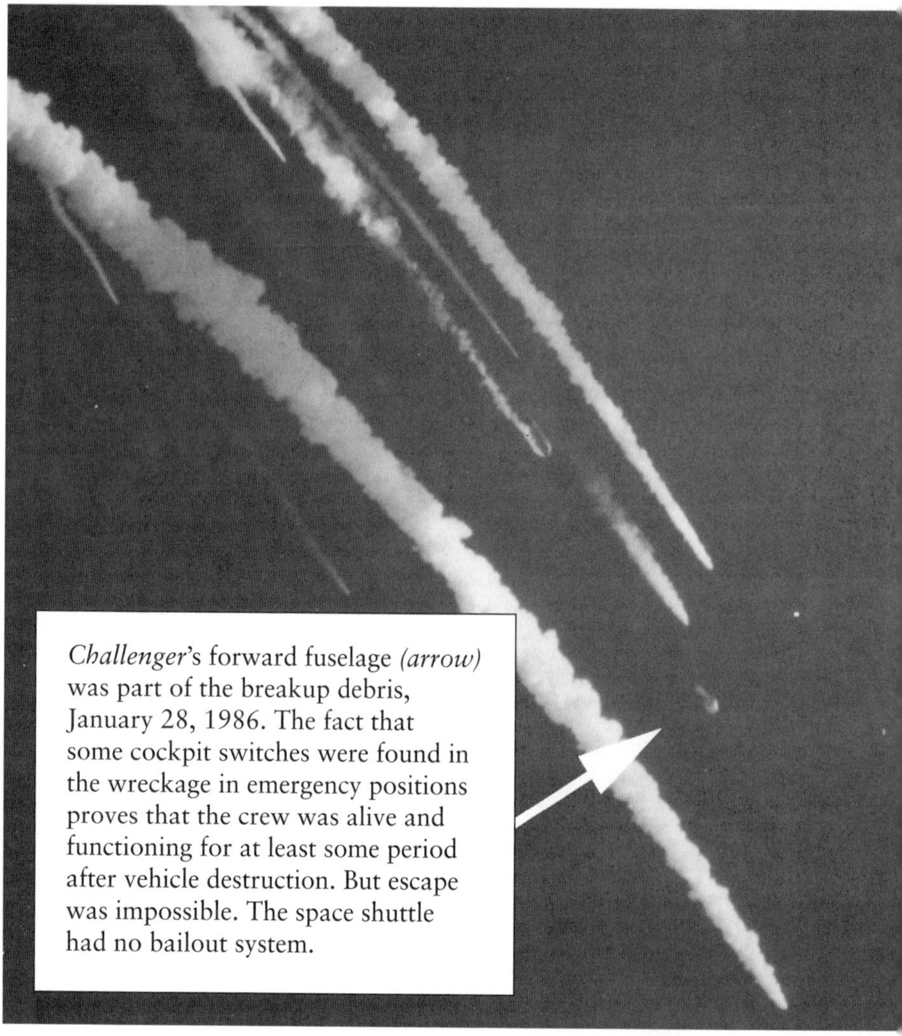

Challenger's forward fuselage *(arrow)* was part of the breakup debris, January 28, 1986. The fact that some cockpit switches were found in the wreckage in emergency positions proves that the crew was alive and functioning for at least some period after vehicle destruction. But escape was impossible. The space shuttle had no bailout system.

The STS-27 Swine Flight crew after arriving at the Kennedy Space Center for our December 2–6, 1988, mission. *From left to right:* Mission Specialist Jerry Ross and Pilot Guy Gardner. I'm standing in the middle. To my immediate left is Mission Specialist Bill Shepherd. Robert "Hoot" Gibson, the commander, is at the microphone.

Viewing the severe heat-shield damage sustained during our STS-27 launch. The tip of the right side SRB broke off during ascent and damaged seven hundred belly heat tiles, by far the worst shuttle heat-shield damage sustained prior to the *Columbia* tragedy. I'm leaning around Pilot Guy Gardner.

Christie Brinkley is all smiles while standing next to me at a Super Bowl XXIII halftime photo-op, January 22, 1989. Bill Shepherd is at my right. Guy Gardner stands on Christie's left side. No doubt it was meeting me that doomed her marriage to Billy Joel . . . or so I tell everybody.

My family at the astronaut beach house in late February 990, prior to my last mission, STS-36. Donna and I stand in the middle. My mom and our youngest daughter, Laura, stand at my right. Our twins, Patrick and Amy, stand at Donna's left. My dad passed away in 1988.

On the drive back from an STS-36 launch scrub, I hold my pressure-suit neck ring open for a flow of cooling air to escape. I forced the smile. Mission scrubs were always a crushing disappointment . . . and I had six of them.

The STS-36 (February 28–March 4, 1990) in-space crew photo. *From left to right:* Commander John "J.O." Creighton and Mission Specialist Dave Hilmers. I'm floating in the middle Next are Mission Speciali Pierre "Pepe" Thuot and Pilot John Casper.

Donna and I meet the Bushes in the Oval Office, May 1990. While leading us on a tour of the White House, Mrs. Bush displayed a hilarious one-of-the-guys sense of humor.

bearer but could then turn around and yuck it up with us guys about a solar panel erection. I often wondered if that was the reason she was flying as the *second* American woman in space—NASA management knew she wasn't a pure enough feminist to satisfy the NOW crowd.

After our major payload work, we gathered on the flight deck to accept a congratulatory call from President Reagan. Each of us was tense and nervous as we handed around the microphone to answer his questions. Thank God we were in space while a Republican president was in office. I shudder to think how Hank would have handled a call from a Democrat. He probably would have asked the president's latitude and longitude coordinates in anticipation of his next BM. When it was his turn, Mike Coats was able to deliver the pro-navy observation that most of what he saw from the windows was water. *That's why the navy is so important, Mr. President* was his implication. Hank Hartsfield defended the air force: "ALL of the Earth is covered by air, Mr. President." There are no circumstances under which astronauts will not compete. Even having the president of the United States in the conversation wasn't an inhibition.

While in the midst of this White House call, a cockpit alarm tone sounded. It was a "systems alert," an indication of a minor malfunction. Still, we needed to respond. In a grand display of the thoroughness of NASA's training we worked the malfunction while continuing to humor Mr. Reagan. Steve Hawley grabbed the massive shuttle malfunction book and began to move through the fault tree, pantomiming to Mike which computer displays to call up. When Hawley had the correct response identified, he passed the book to Judy, who was nearest the appropriate switch panel. She flipped a switch to activate a backup heater, the specified response to the alert. Meanwhile, the rest of us continued, "Thank you, Mr. President. Everything is just fine, Mr. President."

After our payload activities were finished, we posed for our weightless crew photo. It was a tradition for each crew to take a self-portrait in orbit. We dressed in golf shirts and shorts, set up a camera on the middeck, and activated the self-timer. To squeeze everybody into the frame, we posed in three tiers with Hank and Mike lowest, then Steve, Charlie, and me floating above them. Judy floated highest. While we didn't intend it, the pose suggested a cheerleader's pyramid. Adding to the effect were Judy's legs. They dominated the photo . . . tan, perfectly proportioned, beautiful. Judy would later receive hate mail from feminist activists who thought her pose was disgusting and degrading to women. Breaking barriers was a task fraught with all manner of perils.

Around this time in the mission, MCC became suspicious of a temperature indication in our urinal plumbing. Urine is collected in a tank that is periodically emptied via an opening on the port side of the cockpit. Heaters on the exit nozzle are supposed to ensure the fluid separates cleanly from the vehicle and does not freeze to it. But MCC noticed that the temperature at the nozzle was anomalous and suspected some ice might have formed on it during our last urine dump. No windows provided a direct view of the nozzle, so Hank Hartsfield was instructed to use the camera on the end of the robot arm to take a look. We had TVs in the cockpit to monitor the camera view. When Hank positioned the arm, we saw we had grown a urine-sicle.

The image suddenly explained a mystery from STS-41B. After that mission landed, engineers were puzzled to find damage to several heat tiles on the port-side OMS pod at the rear of the fuselage. The damage had certainly occurred during reentry because the same heat tiles had been visible from the back windows during the orbit phase of the mission and the crew didn't see any damage. A similar urine-sicle must have formed during the waste-water dumps on the STS-41B mission. During reentry, the ice had broken off and flown backward, hitting and damaging the OMS pod tile. MCC was now concerned *Discovery* could suffer the same or worse damage. Theoretically it was possible the heat tiles could be so damaged by the ice, *Discovery*'s tail could burn off. I had imagined many scenarios in which my life could be threatened as an astronaut—engine failures, turbo-pump explosions, decompression—but I had never imagined a threat from a frozen block of urine. I had an image of Peter Jennings reporting, "The astronauts were killed by their own urine." It wasn't a heroic-sounding epitaph.

Letting the Sun melt the ice wasn't an option. In the vacuum of space, water doesn't exist in liquid form. It goes from ice to vapor in a very slow process called sublimation. We wouldn't be able to stay in space long enough for sublimation to get rid of our hitchhiker. So MCC directed Hank to tap the ice away using the robot arm.

Then came the bad news. We were told we could not use the urinal for the rest of the mission for fear another ice ball could jeopardize us. We would have to urinate in "Apollo bags." These bags had been *the* toilet for the Apollo astronauts and were stowed aboard the shuttles for just this type of contingency. NASA wasn't going to prematurely end a billion-dollar space mission because of a failed toilet. To our great relief we would still be able to use the solid waste collection feature of our commode. We wouldn't have to use the bags for our BMs as did the Apollo astronauts (they were *real* men).

I looked at Judy. "I sure bet you have penis envy now."

She tersely replied, "I'll manage."

The CAPCOM went on to explain there was enough remaining volume in the waste-water tank for about three man-days of urine. It was obvious to us what they were thinking: Judy would be able to use the urinal for the rest of the mission. We men could get by with the bags. All of us thought this was fair enough, but Judy saw a feminist trap. If she used the urinal while we men were stuck with the bags, word would eventually get around. It would be another damning sin against the feminist cause. In fact it would be a far more egregious sin than the hair-jam incident. Her use of the urinal would be a shout from the rooftops that a penis *was* necessary to deal with certain shuttle emergencies. Judy wasn't going to fall into the trap. She elected to use the Apollo bags like the rest of us.

I have no idea how Judy managed with the bags but I'm sure she paid a messy price for her feminist stand. It was a mess even for us males. On my first attempt, I just held the bag around myself and let fly. Bad idea. The urine splashed into the bottom of the bag and bounced right back, soaking my crotch. Not only that, some fluid escaped and I became the proverbial one-armed paperhanger, trying to hold the bag at my crotch and blot the little yellow planets out of the sky with a tissue in my free hand. Others made similar rookie mistakes. But we quickly came up with a solution. We stuffed washcloths in the bottom of the bags. In weightlessness the "wicking" action of cloth was still effective. We could aim our stream onto the cloth and the fluid would be wicked away instead of splashing around. There was just one catch: If we urinated too fast, the wicking action couldn't keep up with the stream and splashing would result. If we slowed our stream too much, the fluid wouldn't separate from us and a large ball of urine would grow on our penises. We learned it was necessary to very precisely regulate our urine flow to achieve a stream of perfect balance. Even then, there would always be a significant "last drop" that had to be wiped away with a tissue.

Our greatest challenge occurred when we had bowel movements. It was virtually impossible to regulate urine flow while bearing down for a BM. On the second day of our toilet purgatory, I heard another Hank Hartsfield cheer rise from the toilet. "I did it! I did it!"

Since Cuba wasn't at our nadir, I couldn't imagine the source of Hank's glee. "What did you do, Hank?"

"I took a shit without pissing!"

From the look on Hank's face you would have thought his earlier turd had reentered the atmosphere and nailed Castro right between the eyes. But I could appreciate his joy. Turning off one's urine while having a BM was a real trick. The things they didn't teach us at astronaut school.

As our washcloths were consumed we turned to using our socks. When I had exhausted my extras, I began to use my towels. On one occasion with my bladder near rupture, I threw a covetous glance at the clean socks Judy was wearing. I flew straight at her and began to rip them from her feet. She knew exactly what I was doing and jokingly screamed, "Help! I'm being socked!"

By the final day of the mission our wet-trash container was becoming seriously overburdened. Under us floated a volume of vomit, urine, and decomposing food containers. My earlier *Star Wars* prank about alien creatures living in the trash container didn't seem so funny now. Nobody wanted to put their hands in the mess. We would jam our urine bags past the grommet, jerk away, and quickly rip into an alcohol hand wipe.

As we configured *Discovery* for our last sleep period, I repeated my day-one routine. I moved my sleep restraint upstairs and tied it beneath the overhead windows. I intended to stay awake as long as possible to stuff my brain with space memories. While I had every intention of making this trip again, I couldn't be sure there would ever be a second opportunity. *Discovery*'s engine problem had delayed the program by two months. What other problems were lurking? Could one result in a program delay of years, or even total program cancellation? Even if the shuttles continued to fly on schedule, office politics could end my career. It was impossible to know where you stood with Abbey. He might never assign me to another mission. I was going to assume these would be my last hours in space and I wasn't going to waste them sleeping.

So I watched the Kalahari Desert pass beneath me and rammed its beauty into my overflowing memory banks. The Atlantic blue contrasted sharply with the ocher colors of Saharan Africa. Enormous sand dunes shouldered the beach and rippled inland like tan water. I watched clouds of every imaginable shape and texture: circular swirls of low pressure areas, wispy mare's tails, cumulonimbus monsters with anvil heads stretching across the sky like the headdresses of Indian chiefs. In sunset and sunrise terminators, thunderstorms cast hundreds of mile-long shadows. Fair-weather cumulus clouds floated over oceans like popcorn scattered on blue carpet. Unseen jet streams rippled solid blankets of white like a stone dropped into heavy cream.

I crossed Africa in minutes and raced over Madagascar in seconds.

The Indian Ocean was another vast, empty blue. The 3,000-mile brown continent of Australia came and went in ten minutes. Then *Discovery* was once again over Pacific skies. The view of that ocean always intimidated me—its blue seemed as infinite as space. How much greater its immensity must have seemed from a Polynesian outrigger or from the decks of Magellan's ships. We astronauts are frequently characterized as heroes and heroines for sailing into a great unknown. In reality no astronaut has ever sailed into an unknown. We send robots and monkeys ahead to verify our safety. Magellan didn't put a monkey on a ship and wait for it to safely return before going himself. He and those Polynesians set sail without maps, without weather prediction, without a mission control, without any idea of the immense emptiness that lay beyond their meager three-mile horizons. It is laughable to compare astronauts with those explorers. The next humans who fly into a great unknown will be those souls who set sail to Mars and watch our planet dim to a blue-white morning star.

I watched as city lights took on the form of glowing spiderwebs with bright, sodium-yellow interiors and major roads radiating outward and ring roads completing the web effect. I watched lightning begin at one end of a weather front and ripple like a sputtering fuse for hundreds of miles to the other end and then start again. And every ninety minutes I would watch the incomparable beauty of an orbit sunrise. I would watch as a thin indigo arc would grow to separate the black of nighttime Earth from the black of space. Quickly, concentric arcs of purple and blue would rise to push the black higher and higher. Then bands of orange and red would blossom from the horizon to complete the spectrum. But only for a moment. The Sun would finally breach the Earth's limb and blast the colors away with its star-white brilliance. I wanted to scream to God to stop *Discovery,* to stop the Earth, to stop the Sun so I could more thoroughly enjoy the beauty of that color bow.

When sleep finally overtook me, I'm sure I slept with a grin.

CHAPTER 22

Coming to America

The next morning we prepared for reentry. We scrubbed *Discovery*'s walls and windows clean. An earlier crew had turned over a dirty vehicle to their ground team. Small bits of vomit, food, and drink had been found dried to the walls. This pigpen crew quickly became a joke on the astronaut grapevine. We weren't about to let that happen to us. After nearly six days with six people locked inside, *Discovery* was soiled with the same flotsam but we polished her to a shine.

We followed the flight surgeon's recommended protocol of consuming salt tablets and fluids. The excess liquid would increase our blood volume and help minimize the possibility of the reentry G-forces pulling blood from our brains and causing blackout. I also donned my anti-G suit as another defense against G-induced unconsciousness. The suit looked like cowboy chaps and was zipped over my legs and around my stomach. It contained air bladders that could be inflated to squeeze those body parts and restrict blood flow from the upper torso and head. I would later find out Judy did not put on her anti-G suit and suffered for the omission. After landing she was deathly pale, sweating profusely, and unable to stand from her seat for many minutes.

We closed our payload bay doors, strapped into our seats, and then flipped *Discovery* backward so the thrust from the firing of her OMS engines would slow us down. The deorbit burn only braked us by several thousand miles per hour but that was enough to dip the low point of our orbit into the atmosphere. After the burn was complete, Hank maneuvered *Discovery* into a nose-forward 40-degree upward tilt so that her belly heat shield was presented to the atmosphere.

Discovery was now a 100-ton glider. She had no engines for atmospheric flight. We began the long fall toward Edwards AFB, California, twelve thousand miles distant. We were coming to America. On the way, *Discovery* would be enveloped in a 3,000-degree fireball and at the end of the glide Hank would get only one chance at landing. In spite of these daunting realities I didn't fear reentry as I had feared ascent. There were no SSMEs or turbo-pumps to fail and endanger us, and reentry lacked the rock-and-roll violence of ascent. I shouldn't have been so confident. There were still plenty of ways to die on reentry and landing. The STS-9 crew had almost found one with their hydrazine fire. In 1971, three cosmonauts were killed on reentry when their capsule sprang a leak. They had not been protected with pressure suits and their blood boiled inside their bodies. We were not wearing pressure suits, so a pressure leak would kill us in the same manner. Of course, none of us could see the future, but on February 1, 2003, the STS-107 crew would find death on reentry due to damage sustained to *Columbia*'s left-wing heat shield. Foam had shed from the ET during launch and punched a hole in it. *Discovery* was flying with the identical heat shield and Mike and Hank had seen foam shedding from the tank during our ascent. We could have been falling into the atmosphere with a hole in our wing and been blissfully unaware. No, I shouldn't have been so confident in a safe trip home.

For the first half hour after the deorbit burn, there was no indication anything had changed. It felt as if we were still in orbit. We had fallen a hundred miles closer to the planet but the air was still so tenuous it had no observable effect. Then a lost M&M candy appeared from a corner and began a very slow fall. It was my first indication we were no longer weightless. The fringes of the atmosphere were finally slowing us.

At 400,000 feet above the Pacific Ocean, atmospheric friction began to heat the air. The glow in the cockpit windows changed from orange to red to white hot. I twisted my head to look upward through the ceiling windows. A vortex of white-hot air streamed away, flickering like a ribbon in the wind. I was seeing *Discovery*'s wake. The superheated air on the belly was wrapping around the vehicle and combining above her to form a wake of plasma. It streamed off into infinity. In spite of the incredible light show, the cockpit was quiet. There was no wind noise, no vibration.

Discovery completed several role-reversals to manage her energy. As thin as it was, there was still enough air to produce lift and the autopilot commanded the vehicle into alternating 75-degree banks to use this lift to pull her off centerline. She was standing on alternating wings, skidding into the Earth's atmosphere like a snowboarder braking to a stop. She was flying a giant *S* across the Earth, lengthening the distance to the runway, to give her more time to lose altitude. If she had attempted to dive straight ahead we would have been incinerated.

Our computer displays showed *Discovery* as a bug tracking down the centerline of a fan of green energy lines. She was flying like a dream. In spite of the fire outside the windows and *Discovery*'s bizarre maneuvers on our instruments, I felt completely secure. The cockpit was as comfortable as a womb.

The Reaction Control System (RCS) thruster lights flashed intermittently to indicate they were firing to hold our attitude. Just a fraction of a degree in error and we would be tumbled out of control. If it happened, the Pacific would swallow our ashes. NASA wouldn't find a trace of us.

Deeper into the atmosphere the G-forces increased to the maximum of 2. In any other circumstance this would have been a trivial force. A modern fighter can subject its pilot to 9 Gs. But for an astronaut returning from days of weightlessness, the feel of the G-forces was significantly amplified. It seemed as if an elephant were on my shoulders. I was being crushed into my seat. The weight of the helmet made it difficult for me to hold up my head. My vision began to tunnel, as if I were looking through a straw. I knew from my fighter jet experiences tunnel vision was an indication of approaching blackout. The vision area of my brain wasn't getting enough oxygenated blood. I inflated my anti-G suit to the maximum setting and the air bladders squeezed my belly button nearly to my spine. I simultaneously bore down with my gut muscles, all in an effort to tourniquet my waist. It worked. My vision cleared.

Passing 200,000 feet we began to hear the faint rush of wind around the cockpit. *Discovery* was transforming herself from a spacecraft to an aircraft. Mike deployed the air data probes to give us better airspeed and altitude information. We flew into sunlight. It was still twilight below us but the Sun had dawned at 100,000 feet. As *Discovery*'s velocity fell below the speed of sound, her shock waves, which had been trailing her, now zoomed ahead. A buzzing vibration shook the vehicle at their passage.

At seventy thousand feet the steering rockets on *Discovery*'s tail stopped controlling her attitude. She was now fully an aircraft, a creature of the air. Hank took control from the autopilot. While he could have taken control at any point during the reentry, there had been no reason to do so. The runway wasn't visible until the final ten minutes of flight.

The dry lakebed of Edwards AFB, which had welcomed countless machines from the edge of space, now welcomed *Discovery.* Hank guided her over the runway and then banked into a wide, sweeping left turn toward final approach. With her short, stubby wings she was a poor glider and he kept her in a kamikaze-like dive at nearly 350 miles per hour. From the cockpit it appeared as if we were diving straight into Earth. At 1,800 feet above the ground he started his flare. At 300 feet Mike lowered the landing gear. *Discovery* touched the sand in a perfect landing, just as the dawn was breaking. Hollywood couldn't have written a better ending.

"Houston, wheel stop." Hank made the call.

"Roger, *Discovery.* Welcome home."

Our cheers had hardly died before all of us were wondering, *When will I be able to do this again?*

CHAPTER 23

Astronaut Wings

I was drunk on joy and beer. We were headed back to Houston on the NASA Gulfstream jet with our wives. A cooler of beer had been placed aboard and I was doing my best to ensure it was empty by the time we got to Ellington Field. I couldn't sit down. I couldn't stop talking. I was giddy and silly and, oh, so happy. I was the bride on her wedding day, the child on Christmas morning. Periodically I would sit with Donna and try to describe the things I had seen, but as soon as I would get started on one memory, another would pop up and I'd be off on its telling. I never finished a sentence. I would leap to my feet and pace the aisle. I was incoherent with joy. I was now a *real* astronaut. I was a *live* astronaut. The latter fact was something I had never really expected. Subconsciously, I don't think I ever believed I would survive this mission and now that I had, I was wild to celebrate life. I was the soldier back from combat. I had walked the narrow precipice of death and had not fallen.

The others stared at me like I was nuts, which of course I was. In one insane moment I bet everybody I could drop a can of beer and catch it before it hit the floor. I was past the bulletproof stage of intoxication and had entered the weightless stage. The results were predictable. I ended on my hands and knees chasing the foaming, rolling can while the others laughed at my floor show. I didn't care that I was making a fool of myself. There was nothing anybody could have done or said to diminish my celebration.

A crowd of family, friends, and NASA employees greeted us at Ellington Field. Some of the family members and office secretaries had fashioned welcome-home signs. I saw my three children in the front row wearing huge smiles of pride and relief. We wouldn't get any NYC ticker-tape parades but this was better. The people behind the ropes were NASA family. They had put me into space. I loved them all and given a chance I would have kissed each and every one.

A microphone was provided so we could say a few words of thanks. Stepping forward for my turn, I tripped on my own feet. It wasn't because of my intoxication . . . or at least, not entirely. My sense of equilibrium had been affected by weightlessness. It was a common, short-term aftereffect of spaceflight. I have no idea what I said, but it didn't generate any groans of embarrassment from my compatriots behind me, so I guess I did okay.

The ceremony ended and we walked into the crowd. Someone shoved another beer into my hand. I hugged my kids. Amy, my sensitive child, was full of tears. Pat and Laura were smiling. Only my death would have pulled tears from them. I worried the press might find me. I could see their vans and knew they were somewhere in the crowd. The last thing I wanted was to have a camera in my face. That would have been sure to dampen my fun. But I need not have worried—nobody was interested in male astronauts when Judy was around. She looked stunning. We had all showered at Edwards and donned fresh flight suits, and Judy had applied lipstick and a little makeup. She was holding the spray of roses given to her at Edwards. Unlike her predecessor, she had graciously accepted them. At that moment she was everything to everybody, the feminine feminist. The press was all over her. Fortunately, her hair was so big the shank she had lost in orbit wasn't noticeable enough to generate questions.

Gradually the celebration dissolved and Donna and I drove home with the kids to get on with the rest of our lives. That night, as we lay in bed, I joked with Donna about the flight surgeon's warning to purge my sperm.

She laughed. "That's so romantic, Mike."

"But the doc says it's mutant, radioactive!" The doctors were serious about such purges for men still in the procreating mode, the fear being that some of our swimmers could have been damaged by space radiation. In one of the Monday meetings, after hearing the warning repeated, one TFNG had shouted, "Give me a break. I'm purging as fast as I can!" Our baby-making days were over, so Donna knew the flight surgeon's comment didn't apply to me. Nevertheless, we followed the doctor's orders, celebrating as lovers do.

Afterward, we held each other and I was finally calm enough to describe some of the things I had seen. I told her of sunrises and sunsets that would make every future Earth rainbow I ever witnessed a disappointment. I told her of oceans that seemed infinite, of lightning and shooting stars, of a blue-and-white planet set in abysmal black. And I told her I wanted to do it all over again. There were other TFNGs already in line for a second flight. Some were doing spacewalks. Some were operating the robot arm. Some were flying high-inclination orbits where they would get to see all of the United States and most of the inhabited earth. Vandenberg AFB in California was being modified to launch shuttles into polar orbits. Some lucky TFNGs would be on those flights. My just completed mission of whirring around the Earth in a near equatorial orbit and throwing a few toggle switches to release a couple communication satellites seemed ho-hum compared with what was on the horizon. I was discovering what every other TFNG was learning: There were gradations in the title "astronaut" and we all wanted to be on top of the scale. As neophytes we had seen a flight into space, *any* flight into space, as total fulfillment of our life quests. But as we moved into the ranks of veterans, our hypercompetitive personalities created a TFNG hierarchy. For pilots, the command of a rendezvous mission was the most desired prize. For MSes, the A-list astronauts were those who flew the Manned Maneuvering Unit (MMU) on tetherless spacewalks. Very close behind were MSes who did traditional tethered spacewalks. The next tier down were MSes who used the robot arm to grab free-flying satellites. At the bottom of the pile were those sorry souls doing actual science in the bowels of a Spacelab. While many of the scientist MSes really enjoyed Spacelab, most of the military MSes wanted nothing to do with it. Piloting an MMU or operating a robot arm had a lot more sex appeal and generated a lot more personal fulfillment than watching a volt meter on some university professor's experiment. The Untouchables of our strange caste system were those MSes engaged in the Spacelab missions dedicated to life sciences. They collected blood and urine and butchered mice and changed shit filters for primates (and I don't mean the marines). I lit candles at Donna's home shrine to carry my prayer to heaven that I would never be assigned to a Spacelab mission.

As I recounted for Donna the incredible experience of spaceflight and expressed my intention to do it all over again, I was sure she was disappointed. I was sure she would have much preferred a "rest of our lives" scenario that had me returning to the air force for a staff position somewhere in space command that might lead to a star on my shoulders. I was sure her preferred scenario did not include the selection of another potential escort into widowhood, another good-bye walk on the beach house sands, and another T-9 minute vigil on the roof of the LCC. But she would have died before she would have ever put her feelings first—it was the Catholic in her. In her high school marriage course lesson plan is this statement: "The happiness of the woman is found in dependence on her husband. She's happiest when she is making others happy. Selfishness is the greatest curse to a woman's nature." Through her childhood the nuns had browbeaten her to believe that her feelings didn't count, that her lot was to mourn and weep in the "vale of tears" called life. Personal happiness? Fuhgeddaboutit. Her existence was to be one of sacrifice; sacrifice for her husband and sacrifice for her children. Her reward would come in the next life. No, Donna would never ask me to leave NASA. Her Catholicism had given me a free pass to pursue my own fulfillment.

Before we had even gotten our Earth legs back the Zoo Crew was in the JSC photo lab editing our mission videos and compiling a movie to take on the road to show the world. No longer would I have to interpret what others had done before me. I had loathed that ritual—going to the luncheons of America and prefacing my comments with, "I haven't flown in space yet, but those who have say . . ." It was like a minor league baseball player getting onstage and saying, "I've never played in the *Show,* but those who have say . . . ." Now I had my own space story and movie to support it.

In the rented ballroom of a local country club, a few weeks after our landing, Mike, Judy, Steve, and I climbed onstage to accept the coin of the realm from Hank—a gold astronaut pin. None of us cared that we had previously handed over a check for $400 to pay for the pins.

We began our postflight appearances. One of my first was to Albuquerque, where I was given the keys to the city. My flight had not conferred any celebrity. There were no cheering throngs at city hall, just the mayor's staff, my mom and dad, and a handful of other family and friends. But at least I was introduced by my own name. I had attended an earlier event where Steve Hawley had been introduced by the NASA administrator as "Sally Ride's husband." Talk about living in a cold shadow. Hawley's marriage to Sally had put him on the far side of the moon. When officials in his hometown of Salina, Kansas, told Steve they were putting a sign on a nearby highway proclaiming his astronaut status, Steve had joked with us, "It'll probably say, 'Hometown-in-law of Sally Ride.'" Judy, too, had to walk in Sally's shadow. At several appearances she was referred to as Sally Ride. So I was happy to be "Mike Mullane" at my Albuquerque homecoming instead of "an astronaut who flew in space with the husband of Sally Ride."

It was a few hours after the Albuquerque ceremony that my grandmother, who had flown from her Texas home to participate in the festivities, died quietly while napping at my parents' house. Margaret Pettigrew had been born on a Minnesota farm in 1897, six years before the Wright brothers had flown an airplane. In her early childhood, the horse had been the primary mode of transportation. At age eighty-seven she had stood on a Florida beach and watched her grandson ride a rocket into space. Incredible.

Accompanied by our wives, NASA flew us to Washington, D.C., for a glad-handing event with members of Congress. Our crew assembled in a reception line while congressmen and senators passed by, shaking our hands and offering their congratulations. I wondered if Ted Kennedy would appear. If he did, I was certain Hank would blow a brain vein, but the senator from Massachusetts was a no-show.

The reception provided another opportunity for me to observe the pull of Judy's flight-suited beauty. It was Jovian. During a break in the greetings Steve whispered in my ear, "Watch their eyes as they shake my hand." I was confused by his comment, but only until the next senator passed. As the politician pumped Steve's hand, his head was turned and he was smiling directly at Judy. Steve was invisible. I watched several times and every man did the same thing; focused on Judy while handshaking with Steve. Steve could have greeted each of them with "Kiss my ass, Senator," and they would not have heard. They had come into the gravitational pull of Judy's beauty and were deaf and blind to the males next to her. Judy handled it with her usual aplomb, being equally gracious to the old lechers as well as the young ones. Of course, every politician wanted a photo next to her. I watched as one deftly folded his cigarette and highball behind his palms and out of view while the photographer snapped a shot. As quick as the flash faded, the cigarette and drink reappeared. He could have gotten a job as a Vegas illusionist.

On this same trip, Mike and Diane Coats and Donna and I, along with Admiral Dick Truly and a handful of other senior NASA officials, traveled to the Pentagon for Mike's and my astronaut wings ceremonies. The gold astronaut pin was a NASA tradition. The military recognized their astronauts in a separate ceremony with the pinning of aviator wings bearing the astronaut shooting star on the center shield. Every military astronaut considered the award of astronaut wings to be the highlight of their careers. Mike and I were no exception. We had dreamed of this day as ensign and second lieutenant. For me, the ceremony would hold even greater significance. I would become the rarest of USAF weapon systems operators (guys in back of fighters). I would be the first WSO astronaut. It was a very small first, to be sure, but I was looking forward to hearing the USAF acknowledge it.

Diane and Donna were just as thrilled as Mike and I. It was their payday for a lifetime of sacrifice for their man's career, for the terror of the T-9 minute walk to the LCC roof. They would be recognized and toasted for their contributions and bravery. They wore new dresses and shoes and had perfect hair and makeup. I hadn't seen Donna look more radiant and more expectant since she had walked down the aisle in her wedding gown. She loved the pomp and circumstance of formal military events, as our astronaut wing pinnings promised to be.

Our first visit was to the chief of naval operations for Mike's ceremony. As we approached the CNO's office we were greeted by the CNO himself, Admiral James D. Watkins, beaming with almost fatherly pride and stepping forward to heartily shake Mike's hand and hug Diane. Behind him waited a gauntlet of lesser admirals. They were dressed in whites, their epaulets dripping in gold braid, their chests festooned with ribbons and gold wings. It was as if every flag officer in the U.S. Navy had come to congratulate Mike. Each of them smiled broadly and rendered Mike and Diane a deference becoming royalty. Donna and I and the rest of the NASA entourage greeted the CNO and then melted to the sides of the room to let the spotlight focus on Mike and Diane. We'd have our fifteen minutes of fame in a moment.

While waiting I looked at the CNO's wall art. There were gold-gilded paintings of Old Ironsides firing a broadside into an enemy ship, of dogfighting Japanese Zeros and Corsairs, of battleships pounding an enemy atoll. The art complemented the statement the CNO was making with the party, that the U.S. Navy was a service of unmatched history and glory, and new astronaut Mike Coats was the latest addition to that history.

White-gloved stewards orbited the gathering and served finger sandwiches and pastries from silver platters. I looked at Donna. She was in heaven. This was pomp and circumstance beyond anything she had expected and she knew she was up next.

The CNO began the pinning ceremony with comments on the importance of space to U.S. Navy operations. He highlighted the fact that one of our STS-41D communication satellites was a navy fleet UHF relay. He thanked Mike for laying his life on the line for the navy and thanked Diane for her years of wifely support. He then invited Diane to do the astronaut wing-pinning. In word and deed he made her feel that Mike's new wings were as much her award as they were his. The CNO then led his throng of admirals in loud applause. The entire program had been first-class from start to finish.

We all bid our thanks and departed for the office of the vice chief of staff of the air force, General Larry D. Welch. For some reason the chief himself was unavailable but we didn't care; we were certain the number-two man in the U.S. Air Force would take good care of us. Donna was biting at the bit to get there. The gleam in her eye said it all. She was anticipating the identical "Queen for a Day" treatment she had just seen rendered to Diane. So were we all.

Our first indication that things were to be a little different on the air force side of the Pentagon occurred as we neared the office. No generals awaited us. Instead, a lowly captain rose from his desk, welcomed us, and then said, "Please wait here. I'll see if the general can see you." I felt Donna tense at my side. If there was any pomp and circumstance around, it was well hidden. I whispered, "Maybe the party is set up in a different room."

She replied tersely, "I hope so." I was beginning to have a very bad feeling.

The captain emerged from the vice chief's office. "The general is now ready to see you."

Jesus, I thought, *this has more the air of a court-martial than an awards celebration.* I could hear Donna's molars grinding in her rising anger. The rest of our entourage exchanged wondering looks. The contrast to the manner of welcome given Mike and Diane at the CNO's office could not have been greater.

Our group entered the vice chief's office and my worst fears were realized. It was just him, General Larry Welch, and his aide. There was no celebratory cake—no celebratory anything. Even the room seemed cheap compared with the CNO's office.

I presented the general with a framed photo of the launch of STS-41D and tried to inject some levity into what was unfolding as a severe embarrassment for our group. I joked, "General, the only way the space shuttle could look better was if it had *USAF* emblazoned on the wings."

The general didn't find the comment the least bit amusing. Instead, he launched into a discussion on the air force budget and how important it was for money to be spent on the development of a new cargo airlifter, not on a new air force–manned space program. I wanted to scream, "It was a joke, general!"

The rest of the ceremony—if it could be called that—was quick. The general pinned the wings on my uniform, shook my hand, and posed for a photo. He made no comment about the fact I was the first nonpilot air force officer ever to fly in space. Then the aide hustled us out of the office so the general could get back to work on those airlifters. I had never been more embarrassed for my service. USN Admiral Truly had seen the debacle. Mike and Diane had seen it all. The NASA officials with us had seen it. The navy treated theirs like royalty; the air force treated Donna and me like an interruption. I wanted to crawl under a rock.

I held Donna's arm as we walked from the office, and I could feel her trembling in rage. She had received no acknowledgment from General Welch. This was supposed to have been the highlight of my career, and, by proxy, her life. She had put me in that rocket. To do it, she had buried friends, and consoled widows, and kissed her husband off to war, and endured four shuttle countdowns including one engine-start abort. As we exited the office, Donna cursed under her breath. It was a mark of her extreme outrage: I was the foul mouth of the family—Donna never swore. The aide was close enough to hear the word, but I doubted he had any idea as to the reason for the outburst. I knew the general was clueless about how close he had come to feeling the wrath of a woman scorned. His obliviousness reminded me of something an air force pilot had once said in Vietnam, "We've all seen tracers coming at us and think that's the closest we've come to death. In reality, some gomer in a rice paddy has probably fired an old single-shot rifle at us and the bullet passed within a foot of our heads and we never knew." As a combat veteran, I'm sure General Welch had his "I was this close to death" stories, but in reality the closest he ever came to death was by the hands of my wife in his Pentagon office, not in the skies of Vietnam.

The manner in which I had been treated cleared up one source of wonder for me. Over the years, I could not understand why the air force hadn't done something about Abbey's preferential treatment of the navy astronauts. The next three missions to follow STS-41D were all to be commanded by navy pilots. On one of those, navy captain Bob Crippen would be flying his third mission as a shuttle commander before his peer, USAF colonel Karol Bobko, would fly his first. Why didn't the USAF see this as I and the other air force astronauts did—a slap in the face of the air force? Now I had my answer. The shuttle program and its air force astronaut corps were invisible to the top leadership in the U.S. Air Force. We were interruptions to more pressing business. When I got back to JSC, I spread the depressing word to others within the air force community. "Don't expect help from the Pentagon. We're on our own," was my message. Abbey could do whatever he wanted with us and there would be no outrage from our leaders. We were a forgotten squadron.

CHAPTER 24

Part-time Astronauts

The shuttle program introduced several new crewmember positions, besides mission specialists, to the business of spaceflight. There were payload specialists (PSes) like Charlie Walker, who operated his McDonnell Douglas Corporation experiment on STS-41D. There were also military space engineers (MSEs), officers the Department of Defense wanted to fly on some secret missions. There were European scientists assigned by the European Space Agency (ESA) to fly as PSes on Spacelab missions. The Canadian Space Agency was supplying the shuttle robot arm, so some of that agency's astronauts were put aboard the shuttle. NASA was also promising seats to other nations as a marketing tool. *Launch your satellite on the shuttle and we'll throw in a ride for one of your citizens.* An example of this program was when Prince Sultan Salman Al-Saud of Saudi Arabia flew as a passenger on a mission carrying a Saudi communication satellite. Another category included U.S. passengers, for example schoolteacher Christa McAuliffe. And, finally, there were a handful of politicians who used their lawmaking positions to assign themselves to shuttle missions. What all of these people had in common was that they were not career NASA astronauts and usually flew only a single mission. They were part-time astronauts.*

Training for part-timers was limited to their experiments, shuttle emergency escape procedures, and habitability practices: how to eat, sleep, and use the toilet. Mission commanders provided their own additional training in the form of the admonishment "Don't touch any shuttle switches!"

Another thing these people had in common was that, to a large degree, they were not welcomed by NASA astronauts, particularly by mission specialists. Before *Challenger,* twenty-two out of a total of seventy-five MS-available seats were filled by personnel who were not

*There were exceptions. Charlie Walker flew three missions and many of the Spacelab PSes flew multiple times too.

career NASA astronauts. That hurt. The line into space was long and these part-timers made it longer. No, they were not welcome.*

While it would be easy to discount MS complaints about stolen seats as nothing more than union-esque protectionism, there was a legitimate reason for us to want the part-timer programs to be canceled. There was potential for these astronauts to imperil us. Imagine being in an airliner and hearing this comment over the intercom: "Ladies and gentlemen, this is the captain speaking. We are cruising at 35,000 feet so sit back and enjoy the flight. Oh, by the way, we have Mr. Jones up here in the cockpit. He doesn't have a clue what all these switches are for but I've told him not to touch any. I assume he won't. And I don't really know how this guy would respond in stressful situations since I've only known him for a short time. But my airline headquarters says he'll be fine. Of course, they know him even less than I do, but what the heck, he seems like a nice guy. So don't worry when you see me step out of the cockpit to use the toilet. Mr. Jones should be fine sitting up here by himself."

Beginning in 1984, NASA HQ began putting a lot of Mr. Joneses in the shuttle in the form of part-time astronauts and we didn't really know who they were. I mean *really* know. I doubt many of them really knew themselves, at least in the sense of how they might react in stressful, even life-threatening, situations.

Military aviation, the background of many astronauts, is a dangerous and stress-filled occupation, frequently complicated by long separations from spouse and family. It is quick to eliminate the slow and the weak, either through an early death or administrative action. It is for this reason most aviators have an intrinsic trust of other aviators who have survived this winnowing process and a deep suspicion of passengers who, for whatever reason, are given cockpit access. This was the reason most of the military TFNGs had harbored doubts about the post-docs and other civilians when we had first come together in 1978. Who were these people? What stress-filtering processes had they been through? How were they going to react in dangerous situations? They had a lot to prove, and they did. NASA's astronaut training program made sure they had continuing chances to prove themselves in environments where mistakes could kill. They regularly flew in the backseat of T-38 jet trainers. They experienced sphincter contractions like the rest

*Again, there were exceptions. Most astronauts felt the European Spacelab and Canadian astronauts, as well as McDonnell Douglas's Charlie Walker and a handful of other part-timers, were valuable additions to crews.

of us during various in-flight emergencies and bad weather instrument approaches. They went through sea-survival training. They dressed in spacesuits and trained in vacuum chambers where one mistake would give them a few seconds to feel their blood boil inside their body before death came. After several years of this stress exposure, the military TFNGs had come to trust our civilian counterparts. They had *earned* that trust. But the part-time-astronaut training program was measured in months and didn't provide the sustained and comprehensive stress-testing needed to truly evaluate a person's mettle. Part-timers got a ride or two in the Vomit Comet, a couple rides in the backseat of the T-38, and some sea-survival training. These were helpful evaluation venues, but hardly sufficient. So, it didn't surprise any TFNG when disturbing stories about the behavior of some of these part-timers began to make their way to the Monday meetings.

One shuttle commander told of being very concerned about his part-timer's interest in the side hatch opening mechanism. The shuttle side hatch is very easy to open, intentionally designed so because of the *Apollo I* tragedy. The initial Apollo capsules had a complex opening mechanism that is believed to have hindered that crew's escape from their burning cockpit. Determined not to repeat that mistake with the shuttle, engineers designed its hatch to open with just one turn of a handle. And the hatch opens *outward.* Since the shuttle flies in the vacuum of space with the cockpit pressurized at 14.7 pounds per square inch, there are thousands of pounds of force acting to push the hatch open. If the handle was ever turned to the open position in space, the hatch would explode outward, immediately decompressing the cockpit and killing everybody aboard. Knowing this, how would you feel if a person *you really didn't know* took an unusual interest in the hatch opening system? I daresay you would feel as that commander had . . . very concerned. It was after this mission that a padlock arrangement was placed on the hatch handle and only commanders were given the key.

Another part-timer story involved a PS on a mission from hell. First, he fell victim to space sickness. Then, his experiment failed. After years of peer reviews and shuttle delays, he was finally getting his one and only chance to operate the device in space. Its failure severely depressed him and he surrendered to episodes of crying. But this was just the beginning of his torture. He turned out to be a cleanliness freak. What he imagined life would be like aboard the space shuttle for two weeks with possible vomiting, no running water, and few changes of clothes was anybody's guess. Living aboard the shuttle doesn't leave its occupants feeling springtime fresh. If the toilet had functioned normally, the part-timer in question might have had a chance. But as luck would have it, the commode suffered a low-airflow malfunction. In his debriefing the commander had explained the situation: "We had to use our glove-wrapped fingers to separate the feces from our bodies." The already stressed-out PS now faced another significant challenge. His solution was to refuse to allow himself a BM. Over several days he miserably constipated himself, which aggravated his depression. A doctor aboard eventually convinced him to take a laxative, but afterward he refused to eat any solid foods to avoid more BMs. This lack of nutrition further compromised his mental and physical health. In debriefing, the mission CDR summarized the situation he had faced: "I had a depressed, crying, constipated PS on my hands. I thought I was going to have to place him under a suicide watch." It was only by the grace of God that some of these part-timers didn't cause problems that would have jeopardized mission success or worse.

STS-51G, which included the Saudi prince and a Frenchman, provided another part-timer story. (Among TFNGs, STS-51G was known as the "Frog and the Prince" mission.) Prince Al-Saud brought a handful of experiments from Saudi universities to fly into space along with a request to observe the new moon, which would be visible at the end of the flight. NASA approved this request and included it in the Crew Activity Plan (CAP), giving it the label LCO, or Lunar Crescent Observation. The LCO was actually religious in nature. The mission was going to occur in the ninth month of the Muslim calendar, the fast of Ramadan. This period of fasting and spiritual contemplation ended at the sighting of the new crescent moon. Prince Al-Saud just wanted to be a space observer to the end of the fast of Ramadan. The new moon–observation request was apparently approved by HQ without knowledge of its religious importance. When the mission commander, Dan Brandenstein, later learned of its significance, he was concerned the prince might be planning to use the shuttle as a 200-mile-high minaret to make a religious announcement over the radio. If that happened, the American press would fillet NASA for allowing a U.S. spacecraft to be employed by a foreign national for religious purposes. Knowing he would be at the center of that shit-storm, Brandenstein confronted the prince and made him agree on the exact wording he would use if he discussed the moon observation on the air-to-ground link, wording devoid of anything religious. While this issue was merely a distraction for Brandenstein, it was one he didn't need. Shuttle commanders had enough on their plate getting ready for a flight. They didn't need to be worrying about what passengers might say over the radios.

It wasn't just mission commanders who were bothered with part-timer issues. While I was a CAPCOM for the Frog and the Prince mission, Shannon Lucid's bare legs were an issue that came to my desk. (Shannon was an MS on the crew.) Like most crews, the STS-51G astronauts had changed into shorts and golf shirts for their orbit operations. There had been several TV downlinks in which Shannon had been seen working in her shorts. Prior to the orbit news conference, the public affairs officer sent the flight director a note requesting that the crew "dress in pants for the press conference." When the note came to me I understood its intent. Public affairs was concerned the Arab world might find it offensive for one of their princes to be seen hovering in midair with a woman's naked legs prominently displayed next to him. I tossed the note in the garbage. HQ could fire me but I wasn't going to tell an American woman to modify her dress to accommodate the values of a medieval, repressive society where women couldn't drive cars, much less fly space shuttles. I wanted to call Shannon and tell her to wear a thong for the press conference. The irony wasn't lost on me. I was taking a stand for women's rights! Feminist America owes Mike Mullane one. As it was, the framing of the camera for the press conference only captured the crew's upper bodies. Shannon's legs, covered or not, were not visible.

Other cultural issues with foreign nationals surfaced. One guest crewmember told his CAPCOM he wanted the national anthem of his country played every morning as the wake-up music . . . and he wasn't joking. The request was denied. Another foreigner provided the name of the immediate family member he wanted on the LCC roof to watch his launch. NASA assumed the woman was his wife but it turned out to be his mistress. He had left his wife at home.

Another example of the negative impact of the part-timer program on mission operations occurred with the fatal *Challenger* flight. The primary objective of that mission was actually to launch a several-hundred-million-dollar communication satellite that was critical to NASA's and the U.S. Air Force's space operations. But an outsider would never have known that from the way HQ acted. In their eyes the mission was to put a teacher in space. If the satellite was deployed, well, that would be nice, too. But as long as Christa McAuliffe's space lesson got beamed into every elementary school in America, the mission would be successful. Unfortunately, as the mission moved toward launch, a weather delay pushed the flight twenty-four hours to the right—and Christa's space lesson to a Saturday. For NASA's PR team this was a disaster. The space lesson would not be live. It would have to be recorded and rebroadcast. To the surprise of no astronauts, NASA went to work to revise the flight plan and move the space lesson to a school day. All who were aware of what was going on were outraged. Astronauts and MCC live by the motto "Plan the flight and fly the plan." Tens of millions of dollars are spent in simulations to prepare crews for their missions. The Crew Activity Plan, the mission bible, is fixed early in the training flow for the very purpose of ensuring the crew and MCC are thoroughly prepared. Major CAP modifications, even months from a launch, are rarely done and then only when essential to the success of the primary mission objective. Significant flight plan changes close to a launch merely to accommodate a secondary mission objective were unheard of. For any other mission, if someone suggested a flight plan rewrite *twenty-four hours prior to launch* to facilitate a secondary objective they would have been staked out in the launchpad flame bucket. But STS-51L wasn't any other mission: It was the "Teacher-in-Space" mission. The flight plan was rewritten.

After *Challenger*'s loss, Commander Dick Scobee's effects were cleaned from his desk. Among those was a list of notes he had been keeping for his postmission debriefing. One of those notes was critical of the impact a secondary mission objective—Christa's space lesson—was having on his primary mission, the satellite deployment. Of course as a commander, he could have refused to allow the flight plan change, just as Brandenstein could have demanded the Ramadan lunar crescent observation be removed from his mission. But neither man made such demands, no doubt because they worried about the effect on their careers. Telling HQ no in any organization isn't usually a good career move.

The part-timer program that many TFNGs found particularly offensive was the "Politician in Space" program. Even though astronaut-senator Jake Garn (R-Utah) and astronaut-congressman Bill Nelson (D-Florida) were huge NASA supporters, professed the political ideals of many astronauts (I would vote for them), and were very likeable men, they committed the grievous sin of using their lawmaking clout to jump to the front of our line. Garn and Nelson both tried to excuse their actions with the claim that a flight into space would give them a better understanding of NASA's operations and make them more effective supporters of the agency, but many of us found that rationale seriously deficient. If I walked into Congress an hour before a critical vote and assumed Garn's or Nelson's seat to cast their ballot, would I then understand the intricacies of congressional lawmaking? Not in the least. To do that I would have to spend months, if not years, observing behind-the-scenes lobbying, the committee meetings, and political maneuvering leading to the vote. So it was with NASA. Anybody wishing to understand its operations needed to go behind the scenes: to KSC to understand the flow of hardware, to JSC to watch Mission Control in action, to MSFC to understand the difficulties associated with developing propulsion systems, to every NASA center director's office to understand the conflicting pressures of budget, schedule, and safety they labored under. Riding a space shuttle was no more a window into NASA's operations than casting a vote in Congress was a window into congressional operations. But riding a shuttle, like casting a critical senatorial vote, is a lot more glamorous.

In early 1985, NASA HQ announced Senator Jake Garn would fly on STS-51D. The astronaut grapevine said Garn didn't so much as request a flight, as specify to NASA which flight he would take. Supposedly he required a flight in early 1985 to ensure minimum conflict with his senatorial duties and his reelection campaign. We also heard that four other politicians, hearing of Garn's assignment, immediately asked NASA for their own flights, and NASA HQ had requested JSC to start looking at reducing the number of MSes on missions to accommodate them and the growing list of other passengers. It was a kick in the balls and ovaries to astronaut morale. A disgusted Steve Hawley suggested that all of us should walk out on a strike and refuse to fly any missions until HQ desisted in their efforts to give MS seats to part-timers. What an image that comment conjured—astronauts walking a picket line in front of the JSC gate chanting, "Hell no, we won't go!"

Garn was a rarity in Congress—he had actually done something in his life besides lawyering. In that, he should be cheered. He was a former navy pilot and brigadier general in the Utah Air Guard. When he reported to JSC for his eat/sleep/toilet training, he came across as easygoing and approachable. With his military aviation background he had no trouble fitting in. Nobody feared he would have a mental breakdown in space or do something dumb in the cockpit that might threaten a crew or the mission. He had a lot to recommend him to our ranks, except that he hadn't paid the dues to get there—a lifetime of brutal work and fierce competition. Of course we treated him with respect, but our displeasure was evident in subtle rebellions. Before he arrived at JSC a sign-up sheet briefly appeared on the astronaut office bulletin board for people who wanted to take an eight-week course to become a senator. When his mission was delayed for several weeks, the office jokers spun this sarcastic entertainment:

Question from the press for Senator Garn: "Senator, how do you feel about your mission being delayed?"

Senator Garn: "I'm terribly disappointed since I've trained for *hours* for the flight."

During his mission Garn suffered one of the more legendary cases of space sickness. There were whispers he was virtually incapacitated for several days. (A flight surgeon would later tell me they jokingly adopted the "Garn Unit" as a measure of quantifying nausea among astronauts.) But his illness pointed to another danger of flying non–mission essential passengers of any ilk aboard the shuttle: If they had a serious health problem, the mission might have to be terminated early. It could happen. While NASA's prelaunch physicals were thorough, they could easily miss a ballooning aneurysm or a plaqued-up artery or a kidney stone. If a mission ended early due to a serious medical problem, it would mean the enormous risk the crew took to get in space, not to mention the hundreds of millions of dollars of launch costs, would be for naught. Another crew might have to risk their lives to repeat the mission and NASA might have to burn another pile of money. While mission termination for health reasons was a possibility with any crewmember, it was a *necessary* risk for all mission-essential crewmembers. Not so with a passenger.

In the fall of 1985 it was announced that Congressman Bill Nelson would also fly a shuttle mission. Another groan arose from the astronaut office. No doubt the biggest groan came from another part-timer, Greg Jarvis. Greg was an employee of Hughes Space and Communications Company, a major supplier of communication satellites. He was flying in space to observe the deployment of one of his company's products and to perform some in-cockpit experiments on the physics of deployments. Garn's flight assignment had already pushed him to the right on the schedule and he had finally ended up on STS-61C. It was while he was on a trip to JSC to pose for an official crew photo that HQ announced Nelson would replace him. The justification was that the Hughes satellite, which had originally been scheduled to fly on STS-61C, was having technical problems and was going to have to be deleted from the cargo manifest. Since one of the major purposes of Jarvis's shuttle mission was to observe a Hughes satellite deployment, it made sense, HQ intimated, to move him and give his seat to Nelson. This sounded reasonable—except for the fact NASA moved Jarvis to a mission that did not have a Hughes payload. That made it clear to TFNGs he was being removed for one reason only—to make room for Nelson. Now it was apparent to every astronaut that our management was useless when it came to confronting politicians. Anybody could be bumped off any flight at any time to accommodate the whims of a congressman or senator. While it was just part-timer Jarvis getting the giant screw now, no TFNG MS felt immune. Next time it might be one of us airbrushed out of a crew photo like some disgraced Politburo member so a politician could be painted in. It was just one more threat to our place in line and we knew we could forget about protection from our JSC management. They were facilitators. The politicians could have their way with us.

NASA bumped the oft-abused Jarvis one mission to the right. The next time he would pose for a crew photo would be for STS-51L, the mission that would kill him. He would die on a mission that had no Hughes satellite to deploy, the singular event that had been the original justification for his assignment to a shuttle flight.

When Congressman Nelson arrived at JSC he was eager to secure a part to play on his mission. NASA obliged him by rolling out the old standby: photography. The congressman, like Garn, would be taking photos of various geologic, meteorologic, and oceanographic phenomena. But Nelson didn't want to be "Garn-ed." He wanted to be a contributing crewmember and do something really important. There was just one problem. None of the principal investigators of any of the experiments manifested on the mission wanted Nelson anywhere near their equipment. They were getting one chance to fly their experiments, had been working with the astronauts for months on how to best operate the equipment, and had no desire to have a nontechnical politician step in at the last moment and screw things up. Nelson continued to press the issue, but Hoot Gibson, the mission commander, remained firm . . . his mission specialists would do the major experiments. The jokers in the office quickly latched on to Nelson's enthusiasm to operate an "important experiment" and exaggerated it as his "quest to find the cure to cancer."

With the manifested experiments off limits, Nelson hit on the idea of taking photos of Ethiopia in the hopes they could help humanitarian agencies dealing with the drought that was ravishing the country. This well-meaning intention was exaggerated in office gossip as Nelson's second mission objective: "To end the famine in Ethiopia."

Finally, he threw out a real bomb. He wanted NASA to work with the Soviets and arrange an in-orbit gabfest between him and the cosmonauts aboard the Salyut space station. At this moment in history, the Cold War was still very frosty. The complications, both technical and political, to pull off this spacecraft-to-spacecraft link would be difficult and time consuming. The crew wanted nothing to do with it. The MCC flight directors wanted nothing to do with it. To the astonishment of all, even

Nelson's appeals to NASA HQ fell on deaf ears. Nobody wanted to touch this turd. The office gossips had a field day. They created a third Nelson mission objective: "To bring about world peace by talking to the Russian cosmonauts." The wits got more ammunition when the Salyut cosmonauts unexpectedly returned to earth, supposedly because one of them had become ill. Astronauts joked that the commies ended their mission as soon as they heard Nelson wanted to talk to them. Even they didn't want to be part of that bullshit.

These exaggerated Nelson mission objectives—cure cancer, end the famine in Ethiopia, and world peace—generated this joke among TFNGs:

Question: "Do you know how to ruin Nelson's entire mission?"
Answer: "On launch morning tell him they've found a cure to cancer, it's raining a flood in Ethiopia, and the Berlin Wall is coming down! He'll be crushed."

Neither Garn nor Nelson should feel abused at being the butt of an office joke. If you're going to get in the game, you can expect some hits. We've all been there.

The passenger program didn't end with Nelson's landing. Next in line was Christa McAuliffe's initiation of the teacher-in-space program. And it wasn't supposed to end with her. NASA HQ was dreaming of flying other passengers. There were rumors Walter Cronkite and John Denver were being considered for flights. TFNGs greeted these rumors with head-shaking despair. The part-timer program was not only taking seats from us and flying people who were scaring the dickens out of some crews, it was also an immoral program. Individuals who were clueless about the risks of spaceflight were being exploited for public relations purposes. The entire part-timer program was built on the lie that the shuttle was nothing more than an airliner, which just happened to fly higher and faster than a Boeing 747. The very act of assigning a schoolteacher and mother of two to a shuttle mission dramatically reinforced that lie. But every astronaut knew what the shuttle was—a very dangerous experimental rocket flying without a crew escape system. Christa McAuliffe's death on *Challenger* would finally open HQ's eyes to that fact and the agency ended the passenger program . . . with one notable exception—John Glenn.

I was a retired astronaut when I heard the news that seventy-seven-year-old Mr. Glenn had been assigned to fly on mission STS-95. Had NASA completely forgotten *Challenger*? Glenn may have been a former astronaut and he may have been a national hero (he had been *my* hero when I was a child) and he certainly understood the risks, but he would still be flying the shuttle as a non–mission essential passenger for PR purposes. Forget all that claptrap about his geriatric studies. That was another NASA fig leaf to cover a powerful politician. If geriatric research in space was so important, why was NASA pushing older astronauts *out* of the cockpit? Story Musgrave was a six-time shuttle veteran and a card-carrying AARPer who had been moved out to pasture. No . . . when Mr. Glenn lifted off, he was just another politician using his power for personal gratification. In Glenn's case he was also a part-timer whose advanced age added greater health risks to the mission than any part-timer before. It was insane. It was wrong. It was immoral. NASA Administrator Dan Goldin, who approved the mission, needed a time machine to go back and stand at Christa McAuliffe's graveside ceremony. Maybe seeing her weeping family would have opened his eyes to the possibility he might have to hand Mrs. Glenn a folded American flag during an Arlington ceremony while facing this thought, *I let this man die on a lark*.

When I heard that Administrator Goldin had suggested to the press other geriatrics would fly on the shuttle after Glenn, it was too much for me. I emailed an astronaut friend who was consulting for NASA and who had contacts among HQ managers. I asked him if NASA had lost its mind in putting Glenn aboard a shuttle, and if there was any truth to the press reports that other geriatrics would also fly. He replied that NASA had no intention of flying any more geriatrics and that "most NASA folks will tell you that the whole thing [flying Glenn] is a dumb idea, but not too dumb to actually do. In other words NASA believes chances are excellent it will turn out okay, and why not suck up some badly needed PR." I was astounded by his answer. NASA was pressing ahead with a "dumb idea" and relying on chance it wouldn't end badly. Apparently nothing had been learned from *Challenger*. Russian roulette with the O-rings had brought us to that tragedy and now NASA was back at the game with Glenn's mission.

I emailed my reply: " . . . you remember what *Challenger* was like. The team killed seven people. It wasn't an accident. Afterward, we could all see how dumb we had been. This situation with Glenn sure takes me back to pre-*Challenger* thought processes. . . . These 'little things' add up. They embolden people to try other things that might be just a little dumb. This Glenn thing isn't happening in a vacuum with no future ramifications."

I wrote an editorial for *Aviation Week & Space Technology*, a major aerospace publication, concerning Glenn's mission. The piece was published in the September 21, 1998, issue. I closed it with these comments: "It bodes very poorly for any team when management needlessly accepts risk and then silently hopes for the best. It's little things like this that ultimately pave the road to another *Challenger* . . ."

Five years later, in 2003, another commission would investigate the *Columbia* tragedy. Its conclusions would hauntingly mirror those of the *Challenger* Roger's Commission—cultural issues within NASA had led to *Columbia*'s loss. No one should have been surprised. The lessons of *Challenger* had been forgotten long before *Columbia* was dust falling through the Texas sky. Watching Mr. Glenn strap into the shuttle was proof of that.

CHAPTER 25

The Golden Age

If ever there was a Golden Age for the space shuttle program, that period was 1984 to *Challenger.* In those two years there were a total of fifteen successful shuttle missions, ten of those coming in the final twelve months. The shuttle would never again achieve that flight rate. In April 1985, *Discovery* and *Challenger* were launched only seventeen days apart, another STS record. (The seventeen-day record marks the interval between successful launches. *Challenger*'s final mission was launched only sixteen days after a *Columbia* mission.) The missions were coming so fast that shuttles were simultaneously being readied for launch on pads 39-A and -B. KSC was looking like a spaceport out of science fiction.

The history recorded in this Golden Age was remarkable. It included the world's first tetherless spacewalks by jet pack–wearing astronauts, the first on-orbit repair of a satellite by spacewalkers, and the first retrievals and return to earth of malfunctioning satellites. With its fifty-foot-long robot arm and spacewalking astronauts, the shuttle repeatedly demonstrated its unique ability to put man to work in space in ways never before possible. It was also during this period that the orbiters *Discovery* and *Atlantis* joined *Columbia* and *Challenger* to complete the four-shuttle fleet. And that fleet showed its muscle: Twenty-three satellites, totaling 142 tons of payload, were deployed from shuttle cargo bays. Just as NASA had promised, the shuttle was doing it all . . . launching commercial satellites, DOD satellites, and science satellites.

On the surface things looked glorious for NASA. But there was a problem: Getting to the twenty-plus missions per year that would give the shuttle a cost-competitive advantage over other launch systems was proving to be a much more formidable task than expected. The shuttle was a voracious consumer of man-hours. After every landing there were thousands of components that needed to be inspected, tested, drained, pressurized, or otherwise serviced. There were 28,000 heat tiles and thermal blankets on the vehicle. Each one had to be inspected. Mission-specific software had to be developed and validated. Payloads had to be installed and checked out. Severely hampering every turnaround was the lack of spare parts. Just-landed orbiters were being cannibalized of their main engines and other components to get the next shuttle ready. The necessary requirement to meticulously document all work was another drag on vehicle turnarounds: Just tightening a screw generated multiple pieces of paperwork. The joke within the astronaut corps was a space shuttle could not be launched until the stacked paper detailing the turnaround work equaled the height of the shuttle stack . . . two hundred feet.

At just ten missions per year the shuttle was driving the system to its knees. The message was the same everywhere: "I need more people. I need more equipment. I need more spare parts." But NASA didn't have the money to buy these things. While commercial customers offset a portion of the expense, the cash flow was nowhere close to making the shuttle the pay-as-you-go enterprise promised years earlier to Congress. Significant taxpayer money was needed to underwrite the program, and those funds were fixed in the budget. The launch rate had to be doubled with the funds available. The end result was that more was being demanded of the existing manpower and equipment to achieve a higher flight rate. Everybody had a story about how this was overwhelming the various NASA teams. I recall being with an MCC controller when his boss brought in more work for him. The controller objected, "I haven't had a day off in six weeks. My wife and kids don't know who I am." The supervisor was sympathetic but had no other option. "We're all in the same boat. I don't have anybody else. You've got to do it." I could see it in both of their faces. They were exhausted, totally burned out. And they weren't the exception. In many areas NASA only had a first string. There was no "bench" to call on for substitutes. One of our STS-41D prelaunch hangar tests of *Discovery* had been botched for that reason. The first string had been supporting the pad checkout of the shuttle being readied for the next launch, so the contractor had scraped together a team for us from God-only-knew-where. One of the technicians had apparently been called from home because he arrived in the cockpit with the smell of alcohol on his breath. It was an outrageous violation and Hank Hartsfield confronted the man's supervisor about it. He apologized for the intoxicated worker as well as for the entire test debacle, adding, "I don't have enough people to cover everything."

The story was no different for the engineers at the SRB Thiokol factory in Utah. The pressure to keep flying was hammering them even while they were struggling with a major anomaly. The O-ring problem first seen on STS-2 had not gone away. In fact, it had gotten worse. Beginning with STS-41B, launched in February 1984, and up to *Challenger,* only three missions did not have O-ring problems. The other fifteen flights of this period returned SRBs with eroded O-rings. Astonishingly, in nine of these fifteen flights, the engineers had recorded "blow-by," in which heat had not only eroded the primary O-rings but, for very brief moments, had gotten past those rings. On STS-51C, the blow-by had been exceptionally significant. That mission had launched in January 1985, after the stack had waited on the pad through a bitterly cold night. Engineers suspected that cold had reduced the flexibility of the rubberized O-rings, which, in turn, had allowed a more significant primary O-ring leak, resulting in a more significant blow-by. But in all cases none of the observed erosion equaled what had been recorded on STS-2's damaged O-ring, and that mission had been fine. In effect the STS-2 experience had become the yardstick against which all following O-ring damage was being measured. If the damage was less (and it always was), then it was okay to continue flights. In what would later be defined as "normalization of deviance" in *The Challenger Launch Decision* by Diane Vaughan, the NASA and contractor team responsible for the SRBs had gotten away with flying a flawed design for so long they had lost sight of its deadly significance. The O-ring deviance had been normalized into their judgment processes.

There were a handful of individuals who resisted this normalization of deviance phenomenon. Thiokol engineer Roger Boisjoly was one. In a July 31, 1985, memo to a company vice president, Boisjoly expressed his concern about continuing shuttle flights with the SRB O-ring anomaly. He concluded the memo with this prophetic sentence: "It is my honest and very real fear that if we do not take immediate action to dedicate a team to solve the problem with the field joint [a reference to the O-ring] having the number one priority, then we stand in jeopardy of losing a flight along with all the launch pad facilities." Boisjoly feared a catastrophic failure at booster ignition that would not only destroy the shuttle and kill her crew, but would also destroy the launchpad.

Another engineer, Arnold Thompson, wrote to a Thiokol project engineer on August 22, 1985: "The O-ring seal problem has lately become acute."

An October 1, 1985, interoffice Thiokol memo contained this plea: "HELP! The seal task force is constantly being delayed by every possible means." In his last paragraph, the memo's author, R. V. Ebeling, obliquely highlights the major problem of the operational STS . . . not enough people. "The allegiance to the O-ring investigation task force is very limited to a group of engineers numbering 8–10. Our assigned people in manufacturing and quality have the desire, but are encumbered with other significant work." He finished his memo with the warning, "This is a red flag."

Another indication of the crushing workload being borne by the Thiokol engineers is found in an October 4, 1985, activity report by Roger Boisjoly. "I for one resent working at full capacity all week long and then being required to support activity on the weekend . . ." The operational shuttle program was devouring people.

Astronauts remained ignorant of the O-ring bullet aimed at our hearts. It was never on the agenda of any Monday meeting. None of the memos being circulated at Thiokol made it to our desks. But there were other things happening in the Golden Age of which we were aware—terrifying near misses.

On April 19, 1985, as *Discovery* landed from STS-51D at KSC, the brake on the inboard right-side wheel locked on, resulting in severe brake damage and the blowout of the tire. Unlike large aircraft, which have engine trust-reversers to aid in stopping the machine, the shuttle is completely dependent on brakes . . . and it lands 100 miles per hour *faster* than airplanes of comparable size. (A deployable drag chute was added in 1992.) When a shuttle touches down, it is a hundred tons of rocket, including several tons of extremely dangerous hypergolic fuel, hurling down the runway at 225 miles per hour. While the shuttle runways at KSC and Edwards AFB, at 3 miles in length, are sufficiently long for stopping, they are only 300 feet wide. A perfectly landed shuttle is only 150 feet from an edge, an eye blink in time at those speeds. It was a minor miracle that *Discovery* didn't experience directional control problems as a result of the blown tire and careen off the runway.

STS-51F experienced the second engine-start pad abort of the shuttle program. While not really a near miss, pad aborts have the potential to become dangerous. Afterward, I watched that crew put on their Right Stuff, no-big-deal faces for the press, just as we had done following our 41D pad abort. Astronauts are great actors.

STS-51F also became the first shuttle mission to perform an ascent abort when *Challenger*'s center SSME shut down nearly three minutes early. It was later determined that the malfunction was due to two faulty engine temperature sensors. There had been nothing wrong with the engine. With only two SSMEs, the crew was forced into an Abort to Orbit (ATO). Fortunately, this was the safest of aborts. The shuttle had been high enough and fast enough at the time of the engine failure to limp into a safe orbit on its two remaining engines. Had the engine failure occurred earlier, the crew would have faced a much more risky 15,000-mile-per-hour, thirty-minute TAL to a landing at Zaragoza, Spain.

Having experienced both an engine-start abort and a powered-flight abort, the 51F crew had gone through ten lifetimes of heartbeats. After they returned, astronauts joked that a cocked, loaded gun pointed between the eyes of any of them would not have elicited the slightest fear response. The mission had desiccated their adrenal glands.

STS-61C (Congressman Nelson's flight), the last mission prior to the *Challenger* disaster, experienced a pair of bizarre and dangerous malfunctions even before it was launched. During a January 6, 1986, countdown attempt, a temperature probe inside one of *Columbia*'s propellant pipes broke off and was swept into a valve that controlled fluid flow to an SSME. Unknown to anybody, the valve was jammed in the prelaunch open position. Engineers in the LCC noted the temperature sensor was not responding, but erroneously assumed it was due to an electronic malfunction. It had not occurred to anybody that the probe might have actually broken free and was floating around in *Columbia*'s guts. The countdown continued using a backup temperature sensor. The mission was ultimately scrubbed for other reasons and the valve jam was discovered in the countdown reset. Had *Columbia* launched, there was a good chance the jammed valve could have caused a turbo-pump to overspeed and disintegrate during the engine shutdown sequence at MECO. The resulting shower of hot steel inside the engine compartment would probably have trashed the vehicle hydraulic system, dooming the crew on reentry.

During the same 61C countdown, a malfunction of a different valve (this time on the launchpad side of the plumbing) caused the drain back of a large amount of liquid oxygen from the gas tank. For a variety of technical reasons, the LCC had remained ignorant of the lost propellant. The shuttle very nearly lifted off without enough gas to reach its intended orbit. The crew's first indication of a problem would have come when all three SSMEs experienced a low propellant level shutdown somewhere over the Atlantic. How high and fast they were at that moment would have determined whether the crew lived (TAL, AOA, or ATO abort) or died (contingency abort). Again, the day was saved when the launch was scrubbed for unrelated reasons and the drain-back problem was discovered in the turnaround.

These near misses should have been warning flags to NASA management that the shuttle was far from being an operational system. They were indicative of the types of problems that occur in the early test phase of any complex aerospace machine. Every military TFNG had seen it happen in new aircraft systems they had flown. In fact, we were used to having urgent warnings appear on our ready-room B-boards concerning newly discovered failure modes on aircraft types that had been seasoned in decades of operations. It is the nature of high-performance flying. The machines are extremely complex and operate at the edge of their performance envelopes. And the space shuttle was about as high-performance as flying got. There were certainly more surprises awaiting us in its operations. In fact, if the shuttle program should survive for a thousand flights, I am certain engineers will still be having occasional moments of "Holy shit! I never expected to see *that* happen."

The shuttle was not operational and the close calls—STS-9's APU fire, STS-51D's brake problem, STS-51F's ascent abort, and STS-61C's valve problems (not even considering what was going on with the SRB O-rings)—were clear warnings to that effect. Yet, nothing changed. The shuttle continued to fly with passengers and without an in-flight escape system, the two most visible manifestations of the operational label. Senior management saw the dodged bullets as validation that shuttle redundancy would always save the day. Meanwhile, astronauts saw the near misses as indicative of the experimental nature of the craft. When backup systems saved the shuttle, we cheered the genius of the engineers just as management did. The gods of Apollo were damn good. But we also knew these incidents were just the tip of the iceberg. There were more unknowns lurking in the shuttle design, and when they finally reared their ugly heads, redundancy might not be enough to save us.

Astronaut concerns about the shuttle's operational label, the lack of an escape system, and the passenger program should have been heard by every key manager, from Abbey to the JSC center director to the NASA administrator. But they were not. We were terrified of saying anything that might jeopardize our place in line to space. We were not like normal men and women who worried about the financial aspects of losing a job, of not being able to make the mortgage payment or pay the kids' tuition. We feared losing a dream, of losing the very thing that made us *us*. When it came to our careers, we were risk averse in the extreme. Effective leaders would have done everything possible to eradicate that fear. George Abbey, the JSC director, and the NASA administrator all should have been frequent visitors to the astronaut office, actively polling our concerns, and each visit should have started with these or similarly empowering words: "There is nothing you can say to me that will jeopardize your place in the mission line. Nothing! If you think I'm doing something crazy, I want to hear it." I had experienced this form of leadership many times in my air force career. I saw it on an F-4 mission in which a general officer was serving as my pilot. I was a first lieutenant—and terrified. I had never flown with a flag officer before. But this man was a leader who understood how fear could jeopardize the team and did his best to eliminate it. As my foot touched the cockpit ladder, the general stopped me and said, "See these stars," and pointed to his shoulder. "If I make a mistake they won't save our lives. If you see anything that doesn't look right on this flight, tell me. There's no rank in this jet. Flying is dangerous enough as it is without having crewmembers afraid to speak up." It was an empowering moment. The astronaut office desperately needed the same empowering moments, but they never came. Fear ruled—a fear rooted in Abbey's continuing secrecy on all things associated with flight assignments. We kept our mouths shut.

It was in the Golden Age that Judy Resnik was assigned to her second mission, STS-51L. She would join TFNGs Dick Scobee, El Onizuka, and Ron McNair as well as pilot Mike Smith (class of 1980) for a flight aboard *Challenger.* Christa McAuliffe, a New Hampshire schoolteacher, would later join the crew. Her assignment to 51L was linked to Judy's. NASA logically wanted Christa to fly with a veteran female astronaut. Greg Jarvis, another part-timer, would ultimately draw a *Challenger* slot when Congressman Bill Nelson bumped him from STS-61C.

I don't blame Nelson or Abbey or anybody else for how the chips fell on the *Challenger* crew composition. Only God can explain the how and why of that. In fact, many months prior to *Challenger,* Mike Smith was named as a backup to a mission pilot who was suffering a potentially career-ending health problem. That pilot recovered and Smith wasn't needed. But had the sick pilot's convalescence taken just a few more weeks, Mike would have flown on the earlier mission and another pilot would have died on *Challenger.*

I congratulated Judy and the others at their Outpost celebration. With a gold pin in my bureau drawer it was easy to be sincere. No more fake smiles. Still, I felt a touch of envy. The 51L crew would be deploying an IUS fitted with a NASA communication satellite. The Boeing engineers had finally fixed that booster rocket so Judy had a proven payload. It was one less thing to get in the way of her launch date. She would have a second flight long before I would and that was something to envy.

In spite of the record number of missions in 1985 and flight opportunities for astronauts, morale continued to suffer under the leadership of John Young and George Abbey, particularly the morale of the USAF pilots. Air force pilot Fred Gregory filled my ear on a T-38 mission. "Of the twenty-eight CDR and PLT seats available on the first fourteen missions, only six have been filled with air force pilots. Fifteen went to navy pilots." Fred went on to complain that of the six CDR and PLT seats available on the first three Spacelab missions, four were being filled by air force pilots. (He was one of those four.) He didn't have to explain the meaning of the latter statistic: If any space missions could be considered routine, they were the Spacelab missions, and the USAF astronauts were getting more than their fair share of those. The navy pilots were getting the challenging and historic missions that included hands-on-the-stick rendezvous time and interviews on national TV. The most egregious example of an air force TFNG being screwed was when pilot Steve Nagel was assigned to fly his first mission—not as a PLT, but as a mission specialist! Even some navy astronauts were outraged by this travesty. Steve was known to be a far superior pilot and to have much better judgment than several of the USN pilots who had drawn front-seat assignments. And Abbey's preferential treatment of the navy didn't just stop with shuttle crew assignments. He also picked navy astronauts (Walker, Gibson, and Richards) to serve as directors of NASA's flying operations at Ellington Field, and navy pilot Don Williams was assigned a position in the JSC Shuttle Program Office.

Ironically, the flight assignment situation with the air force pilots turned in my favor. On February 6, 1985, Abbey phoned me (no office visit this time) to tell me I was being assigned to the first shuttle mission to fly from Vandenberg AFB in California. Abbey had finally drawn the air force's attention when he assigned Bob Crippen, a navy captain, to command the most "air force" of all missions—the first Vandenberg flight. The air force was the lead service in DOD military space operations, and it was a fact of orbital mechanics that many of their satellites had to be launched into polar orbits. For a spy satellite to see all of America's potential enemies, it has to have a view of all the Earth. A satellite orbiting around the Earth's poles gets such a view as the Earth spins underneath it. But it is impossible to launch polar orbiting satellites from the Kennedy Space Center, because a north- or south-directed launch from KSC would endanger populations below the rocket flight path. Polar orbiting satellites have to be carried into orbit by rockets launched from Vandenberg AFB, located near Point Conception, California. A rocket launched on a southern trajectory from this point will achieve polar orbit while flying safely over the ocean. The air force had spent a decade and several billion dollars building a shuttle launchpad at Vandenberg AFB. It was their launchpad and the first mission to be flown from it would carry an air force payload. The air force had wanted it commanded by an air force pilot, but Abbey had other ideas and assigned Bob Crippen. In the ensuing discussions between the USAF and NASA, the air force had accepted Crippen, but with the caveat that the majority of the rest of the crew would be air force. (Or so the rumor mill had it. As always, there was nothing but rumors on the subject of flight assignments.) In a strange twist, I became a beneficiary of Crippen's commandership of the first Vandenberg mission, a fact made clear to me when Crippen later commented, "You have the right color uniform for the flight."

I was deliriously happy about my good fortune. The Vandenberg mission was going to be a true first. It would carry me and the rest of the crew into polar orbit, something no human had ever done. The poor schmucks flying out of KSC on the commercial communication satellite deployment missions only got to see a narrow strip of the Earth between 28 degrees north and 28 degrees south latitude (as I had done on STS-41D). How boring. In a polar orbit we would see all of the Earth. We would fly *through* the northern and southern lights. We would fly over the Greenland ice cap and the mountain ranges of Antarctica. We would pass over all of the Soviet Union. It was a mission Hank Hartsfield would have loved—he could have made the Kremlin a target for one of his BMs. I was back in my pre–STS-41D frame of mind. I was mad to get into space on this mission. But the liftoff date—originally scheduled for spring 1986—was slipping to the right. The new Vandenberg launchpad and launch control center had to be finished and checked out. The State Department had to complete its negotiations to secure shuttle abort landing rights on Easter Island's runway, a task being made more difficult by a Soviet Union disinformation campaign that shuttle operations would destroy the island's stone figures. The Soviets understood that most of the payloads carried out of Vandenberg would be spying on them and were doing their best to lay down obstacles.

STS-62A's slippage provided time for me to pull other duties, including several missions as a CAPCOM. There were no *Apollo 13* dramatics on any of these flights but, like everything else in the astronaut business, even the mundane can be unique. One Saturday night I was on CAPCOM duty and nearly comatose in boredom. The orbiting crew was engrossed in their experiments and the shuttle was performing flawlessly. On rev after rev all I did was make Acquisition of Signal (AOS) and Loss of Signal (LOS) calls as the shuttle passed in and out of the coverage of various tracking stations. I tried to maintain an appearance of busy professionalism, knowing the public affairs wall-mounted cameras were focused on me. When no video was being streamed from the shuttle, the NASA PR officer would switch to these MCC cameras. Cable companies broadcast "NASA Select" video to their subscribers, including most astronaut households. My image was being dumped into living rooms throughout Clear Lake City and across America. Aware of this, I resisted the impulse to pick ear hairs and instead opened a shuttle malfunction checklist and pretended to study it. My eyes glazed over and my head nodded.

When my console phone rang I was instantly alert. The MCC phone numbers were unpublished. If a phone was ringing it was official business. I was glad for the interruption . . . *anything* to break the monotony. I snatched the receiver and answered in a crisp military manner, "CAPCOM, Mike Mullane speaking."

What came into my ear was a soft, feminine voice. "Raise your hand if you want a blow job."

I bolted upright. Was I hallucinating? "Pardon me" was the only rejoinder I could muster.

"Listen up, Mullane! I said, raise your hand if you want a blow job."

It is in the DNA of men to respond to such a proposition in the affirmative, so my hand shot up like the space shuttle. The flight director and a couple of nearby MCC controllers looked at me like I had just had a seizure. No telling what the space geeks around the country watching me on TV thought had happened.

My brain quickly replayed the conversation and I identified the voice, a TFNG wife. It was a Saturday night. Somewhere there was an astronaut party. Someone had turned on the TV to check on the progress of the shuttle flight and found me bobbing toward unconsciousness. A crowd had gathered at the TV while this woman was given the CAPCOM phone number and made her call. I could imagine the roar of laughter when the party audience had seen my hand jerk skyward.

Now it was my turn to shock the caller. "You know this phone call is being recorded." She just laughed me off. It was no more possible to embarrass this particular woman than it was to embarrass Madonna. But the call *had* been recorded. All MCC telephone conversations are recorded for accident investigation purposes. Somewhere in the National Archives are audiotapes with historic quotes from the space program, like Alan Shepard's "Let's light this candle," and Neil Armstrong's "Houston, the *Eagle* has landed," and Gene Kranz's "Failure is not an option," and a TFNG wife's "Raise your hand if you want a blow job."

There were Monday meeting discussions that proved almost as attention-grabbing as this proposition. We received a status report on the subject of herpes-infected monkeys. STS-51B, a Spacelab mission, was to carry several primates as part of their life-science research and it was feared the virus, which was common in monkeys, could infect the crew. Needless to say this was a briefing that brought out the best in the Planet AD crowd.

"If you don't screw the monkeys, you won't catch herpes" came one call from the cheap seats.

"Good luck restraining the marines" came another.

"The ugliest one will come back pregnant by one of you air force perverts."

As this inter-service banter continued, one of the post-docs was able to shoulder in a valid question. "Why don't they just fly clean monkeys?"

The presenter replied, "It's difficult and expensive to find herpes-free monkeys." Then he added, "The scientists believe the herpes risks to astronauts are acceptable. They think there's a greater chance of the shuttle exploding than the crew contracting herpes." The scientists were right. Nobody on the 51B crew would be worried about catching herpes from a monkey while sitting on 4 million pounds of propellant.

Weeks later, during a STS-51B simulation, the Sim Sup introduced a simulated monkey "malfunction." It wasn't a herpes outbreak, but a monkey death. This was to help prepare the MCC PR people to deal with nightmare antivivisectionists scenarios. (A group of these people were protesting NASA's Spacelab animal experiments.) Per Sim Sup's instruction, the crew reported the simulated monkey was sick and bloated. A short while later they made the "monkey has died" call. At about the same time in the simulation, Sim Sup also introduced a human medical problem for the MCC flight surgeon to work—pilot Fred Gregory was ill with a fever and a urinary tract infection. The nearly simultaneous monkey illness and Fred's simulated infection had Fred vigorously defending himself in the simulation debriefing: "I did not violate the monkey!"

The herpes-infected monkeys made the flight and, as far as anybody knew, none of the crew caught the virus, not even the marines. And neither did any of the monkeys later give birth to an air force pilot's simian bastard. Nor did Fred come back with a urinary tract infection.

At another meeting one of the female physician astronauts presented some life-science findings derived from Spacelab animal experiments. "Newly born mice appear healthy but, in weightlessness, they are unable to stay on their mother's teats to nurse."

The comment elicited a Beavis and Butt-Head reaction from the Planet AD crowd. "Dude, she said teats." A wave of giggles swept through our ranks. One USMC astronaut whispered, "Sucking tit in zero-G sounds like a job for a marine."

Another life-science experiment presented to astronauts involved the insertion of an instrumented hypodermic needle into an astronaut's body to measure zero-gravity veinous blood pressure. A Spanish Inquisitor would have blanched at the size of the experiment needle. I asked, "Where are you going to find a vein large enough to stick *that*?"

Physician (and former marine fighter pilot) Norm Thagard joked, "The dorsal vein of the penis will work." On Planet AD everybody was a comedian.

The briefer assured us the penis would not be a target, but wherever the needle was destined it wasn't going to be fun. Needle-oriented experiments always seemed to be part of Spacelab missions, a fact that generated this office joke.

Question: "Why do Spacelab missions require a crew of six MSes/PSes?"

Answer: "Five are needed to hold down the experiment victim."

At yet another Monday meeting the topic was the STS-51F space cola war between Coke and Pepsi. That mission carried experimental zero-G-functional cans of each soft drink. The crew was to evaluate them in the hope carbonated beverages could be added to the menu. Not surprisingly, both soft drink companies wanted theirs to be the first cola consumed in space and called for their political connections to make that happen. Astronauts would hear the issue had reached all the way to the

White House. A disgusted John Young returned from one management meeting and said the first-cola-consumed-in-space topic had occupied hours of the committee's time. That prompted a growl from the back ranks: "Sure hope they're spending as much time working on the things that can kill us."

As the Coca-Cola Company was the first to come to NASA with the suggestion of flying their product, they won the battle. The 51F crew was ordered to take photos of the consumption of the drinks with the date/time recording feature of the NASA cameras in the on position. That data conclusively established that Coke was the first cola consumed in space. But since shuttles have no refrigerators, the beverages had to be consumed at room temperature. That fact doomed the experiment to be a disappointment. STS-51F was the first and last cola flight.

On January 27, 1986, I jumped in a T-38 and, along with the rest of the STS-62A crew, flew to New Mexico for some payload training at Los Alamos National Laboratory. While the primary business of the lab was nuclear weaponry, it was also involved in passive military space experiments. Some of these were to be payloads on our Vandenberg flight.

We landed in Albuquerque and took a lab-chartered flight to the small Los Alamos airport. After checking into a motel, I called Judy at the KSC crew quarters to wish her good luck on tomorrow's mission. I also teased her about the black cloud of delay that seemed to follow her. Her mission had already recorded two launch scrubs, one on January 25 for bad weather and then the next day for a problem with the side hatch.

"So you're the bad-luck person who caused all our *Discovery* scrubs."

"I don't think so, Tarzan. It was Cheetah." She was right about Hawley. Steve now had the unenviable record of nine strap-ins for two flights. Judy was only working on her sixth strap-in.

I asked her how the launch looked for tomorrow. "Good, except it's supposed to be cold, down in the twenties. We're worried about ice in the sound suppression system."

"It's all these shuttle launches that are changing the weather."

She chuckled at my reply.

I kept the call brief knowing she probably had others to receive or make. "I just wanted to say good luck, JR. Please tell the others the same for me." These were the last words I would ever speak to her.

"Thanks, Tarzan. I'll see you back in Houston." These were the last words I would ever hear from her.

The last hope to save *Challenger* passed that night. When the Thiokol engineers learned of the extremely cold temperatures forecast at KSC, they convened a special teleconference with their NASA counterparts and argued that the mission should be delayed until the temperature warmed. Their justification was the fact that STS-51C, launched a year earlier with the coldest joint temperature yet—53 degrees—had experienced the worst primary O-ring blow-by of any launch. They suspected the cold temperatures had stiffened the rubberized O-rings and adversely affected their ability to seal. With an estimated joint temperature of about 30 degrees for *Challenger*, the same thing could happen tomorrow, they argued. They recommended the launch be delayed until the joint temperature was at least 53 degrees. The suggestion brought a fusillade of objection. One NASA official responded, "My God, Thiokol, when do you want me to launch, next April?" Another said he was "appalled" by the recommendation to postpone the launch. They correctly pointed out that there had been blow-by observed after launches in warm weather, a fact that suggested there was no correlation between temperature and the probability of O-ring failure. The arguments continued for several hours but, in the end, Thiokol management caved in to NASA's pressure and gave the SRBs a go for launch. The Golden Age had only hours remaining.

CHAPTER 26

Challenger

After waking on January 28, I flipped on the TV to see what was happening with *Challenger.* The STS-51L countdown was running two hours late. I had plenty of time for my morning run so I dressed in my sweats and stepped into the crystalline twilight.

Few cities in America are more beautifully sited than Los Alamos, New Mexico. Set on a shoulder of a dormant volcano at an elevation of 7,200 feet, it commands a godly view of the Rio Grande Valley and the Sangre de Cristo Mountains to the east. The city is built upon multiple mesas separated by dramatic mini-grand canyons. The soil is soft volcanic tuff and eons of erosion have sculpted the terrain into bizarre and breathtaking shapes.

While Los Alamos was a joy for the eye, it was a pain for the lungs. In its thin air I was unable to keep the pace I regularly ran at sea level and I throttled back to a more leisurely jog. The dawn was pinking the eastern sky while a nearly full moon graced the west. I steered myself on a path through a forest of ponderosa pine, the scent of their needles perfuming the air. A herd of white-tailed deer, long accustomed to humans, didn't bolt at my appearance.

I ran for half an hour and then dropped into a cool-down walk, enjoying a moment of total contentment as I did so. I was in top physical condition. I was a veteran astronaut. I was in line for a second spaceflight, a *fantastic* second flight. There were probably no more than six or seven missions between me and polar orbit. I could easily visualize *Discovery* on the Vandenberg pad, now that I had a photo on my office wall of *Enterprise* on the same pad.* Several months earlier NASA had airlifted that orbiter to Vandenberg for a pad fit-check and the photos taken had captured her as *Discovery* would soon be seen, standing vertical against a backdrop of California hills. It was an image that set my soul soaring.

After a shower and breakfast, I rendezvoused with the rest of the crew and drove to the lab to meet the principal investigators of our payloads. By now *Challenger*'s launch was only a few minutes away, so we delayed our training to watch it. We knew this launch, unlike other recent ones, would be covered on TV because of the public's interest in schoolteacher-astronaut Christa McAuliffe.

I couldn't sit down. As a rookie, I had been fearful while viewing shuttle launches. Now, I held a veteran's terror for what was at hand. I nervously paced behind the others. The TV talking heads focused on Christa, showing clips of her in training, then live shots of her students awaiting the blastoff. There was a carnival atmosphere among the children.

As the NASA PR voice gave the final ten seconds of the countdown, I was in prayer-overdrive, begging God for a successful launch. My motivation wasn't all selfless: There were still a thousand things that could come between me and my Vandenberg mission, and STS-51L was one of them. Another pad abort or, God forbid, an abort into Africa or Europe would have a serious impact on the launch schedule. The ripples of delay would push 62A even farther to the right.

At T-0 the SRBs blossomed fire and *Challenger* was on her way. The

***Enterprise,* the first orbiter, was never designed for spaceflight. It was used in pre–STS-1 glide tests off the back of NASA's 747 carrier aircraft.

TV only covered a moment of ascent and then cut to the trivia of the morning. Bob Crippen spun the dial to other stations hoping for more coverage but there was none. Even the novelty of a schoolteacher couldn't buy NASA more than a minute of airtime.

We turned off the TV and gave our attention to a principal investigator of an experiment that would be in our cargo bay. As we were about to follow him to the hardware, Jerry Ross decided to give the TV another shot, "Maybe they'll have an update on the launch." He turned it on. What we saw immediately shocked us to silence. *Challenger*'s destruction had already occurred. We were seeing a replay of the horror. We watched the vehicle disintegrate into an orange-and-white ball. The SRBs twisted erratically in the sky. Streamers of smoke arced toward the sea.

For several heartbeats there was not a sound in the room. Then the exclamations came. "God, no!" Guy Gardner bowed his head and cried visible tears. I just stared in a dazed silence. Most of the others did the same. A few of the lab personnel wondered aloud if the crew had bailed out. I answered their question. "There's no ejection system on the space shuttle. They're lost."

The TV focused on Christa McAuliffe's parents. They were in bleachers in the press area and appeared merely confused. I could read the question on their faces: *Are the smoke patterns in the sky part of a normal launch?* Their daughter was already dead and they didn't know. I silently cursed the press for continuing to focus on them. It was the ultimate obscenity of that terrible morning.

I phoned Donna. She was sobbing. Even though the NASA PR announcer was only saying it was a major malfunction, she was familiar enough with the shuttle design to know it had no escape system. I didn't have to tell her the crew was dead. I suggested she pick up the kids from school. The press was going to be everywhere and I didn't want them shoving a camera in their faces. "Just keep them at home." I told her to expect me that afternoon. I knew we wouldn't be staying in Los Alamos.

I next called my mom and dad in Albuquerque. Dad, the big-hearted, sensitive Irishman, was crying. As always my mom was unbendable iron. I knew she was dying inside, but there was no way she could verbalize those feelings.

As expected, Crippen wanted to get back to Houston as quickly as possible. We drove to the Los Alamos airport and took the lab charter flight to Albuquerque. Within an hour of our arrival there, we were in our '38s headed home. I was in Crippen's backseat, in the lead aircraft of the three-ship formation.

As we climbed to altitude, ATC cleared us direct to Ellington Field and added, "NASA flight, please accept our condolences." I was certain those same sympathies were being offered to NASA crews everywhere as they hurried home. The entire nation was grieving.

The rest of our flight continued in silence. At each ATC handover the new controller would offer a few words of comfort and then leave us alone. There was no chatter among our formation on our company frequency. Crippen was silent on the intercom. We were each cocooned in our cockpits, alone with our grief. I watched the contrails of the other '38s streaming away in billowing white and prayed for the *Challenger* crew and their families.

My thoughts returned to the last time I had seen the crew—before they entered health quarantine—more than two weeks ago. I passed them on their way to a simulation. Each wore the thousand-watt smiles of Prime Crew. I shook their hands and wished them good luck and added a hug for Judy. With my arms around her I whispered, "Watch out for hair-eating cameras." She laughed. It was the last I would see of her and the others. Now, their shredded bodies were somewhere on the floor of the Atlantic. Friends whose joyous faces I had watched only two weeks ago were now being discussed by the TV talking heads as "remains." I could feel Judy's arms on my back and her hair brushing my cheek in that last hug. Now those arms, that hair, her smile were gone. They were just . . . remains. Though I'd known it would happen to some of us one day, I still could not come to grips with the reality of it. They were gone. Forever.

My only comfort was in my belief that their deaths had been mercifully quick, the instantaneous death we all hope for when our time comes. In one heartbeat they had been feeling the rumble of max-q (maximum aerodynamic pressure) and watching the sky fade to black and anticipating the beauty of space and then . . . death. I was so certain of it. How could anybody have survived the ET explosion? The cockpit was only a few feet from it. There were more than a million pounds of propellant still remaining in the tank when it detonated. The explosion must have destroyed the cockpit and everything in it. The more I dwelled on it, the more certain I was. They died instantly. I would later learn how wrong I was.

My thoughts drifted to the cause of the disaster. The video replays on TV showed fire flickering near the base of the orbiter just before vehicle destruction. Had an SSME come apart as so many of us had feared would one day happen? I was certain the SRBs had nothing to do with the disaster. They were seen flying after the breakup. It was to be expected their flight would be unguided and erratic, but other than that they appeared fine. Again, I would be proven wrong on all counts.

I asked Crippen what he thought had caused the tragedy. "I don't know. But whatever it was, we've all ridden it."

He was right. Whatever it was, the same mechanism of death had been with all of us on every mission. How close had it come to killing me on STS-41D? A second? A millisecond? Had the *Challenger* SSME that failed (and I was *sure* one of them had exploded) been one of the engines powering *Discovery* on my first mission? It was entirely possible. The engines were frequently interchanged among orbiters.

As I fell deeper into melancholy another thought wiggled its way to the fore. I hated that I couldn't keep it at bay but, like smoke under a door, it crept in to choke off every other thought. *What was* Challenger*'s loss going to do to me?* To ask such a question at this moment defined me as one sick bastard but try as I might, I could not stop it. I suspected every other TFNG was similarly stricken. Would the shuttle ever fly again? I thought of my morning run and how perfect my future had seemed, with images of polar orbit spaceflight filling my brain. Now those images blurred like a mirage.

Our flight entered the Ellington landing pattern, each pilot following Crippen's peal-off "break" to circle for touchdown. As we were marshaled to a parking spot, I searched the guest waiting area, expecting to see someone from the press. I dreaded the thought of speaking to them. But the only person to greet us was Donna. Crying, she walked to the side of the jet and rushed into my arms.

At home my fourteen-year-old daughter, Laura, informed me that someone from the newspaper had called and when she told them I was out of town they interviewed *her*. I was outraged. They had taken advantage of her naïveté to ask questions about my STS-41D experiences with Judy. "Daddy, they asked me how you felt when you saw *Challenger* blow up." It was a good thing I hadn't fielded that question. I could imagine the answer that might have leaped from my mouth. I had already passed the denial phase of grieving and had entered the anger phase. I told the kids to let the answering machine take the calls. I didn't want to speak to anybody in the press.

That evening there were church services throughout Clear Lake City. Donna, the kids, and I went to our parish church, St. Bernadette's. It was packed. I wasn't the only astronaut parishioner. There were a few others. Our friends and neighbors came to us and sobbed their condolences. Complete strangers did the same. The grief was beyond anything I would have ever predicted.

At the request of some of the parish members, my son had put together a slide and music show to play as people entered the church. Pachelbel's Canon and Copland's "Fanfare for the Common Man" accompanied slides depicting shuttle launches, spacewalkers, and other space scenes. There was a slide from STS-41D showing a grinning Judy with her cannon-cleaner weightless hair. When it appeared on screen, people were overcome, laughing and sobbing at the same moment. *Watch out for hair-eating cameras.* The slide resurrected from my memory the words I had spoken to her two weeks earlier. I closed my eyes. I wanted to cry, like the others around me, but I couldn't. That gene just wasn't in me. I was my mother.

The next day Donna and I drove to visit the widows. We first went to June Scobee's home. The street in front of her house was a mob scene. A large crowd of the curious filled the neighbors' driveways and lawns. The elevated microwave poles of news vans provided a beacon that drew a slow current of cars through the neighborhood streets. Power cables crisscrossed sidewalks. Technicians shouldered cameras and framed their news reporters with the Scobee home in the background. It would have been chaos but for a contingent of local and NASA police who kept everybody from June's front door. Several NASA PR personnel were teamed with the police to recognize and allow astronauts and other NASA VIPs to enter the home. Donna and I were waved through the cordon.

The house was filled with family, friends, and several other astronauts and wives. June was the picture of exhaustion, her face puffy and tear-stained. She and Donna hugged for a long moment, each crying into the neck of the other. As they parted, I embraced June, clumsily mumbling my sympathies then fading out of the scene as another visitor came to her. I observed how much better the women were at handling the situation. They easily conversed with June. The men mimicked my awkward performance—a quick hug, a few words, and then escape to a corner where they fidgeted uncomfortably.

The rest of the day was a blur of grieving women and children as we made our rounds to the other widows. Lorna Onizuka was incapacitated by her loss. She refused to see anybody and rumors circulated that she had not given up hope that the crew would be found alive somewhere.

A few days after the tragedy, I flew to Akron, Ohio, for a memorial service for Judy. Most of the astronaut office made the trip. On the flight Mike Coats astounded us with news on the cause of the disaster. "It was a failure of the O-rings on the bottom joint on the right side SRB.

There's video of fire leaking from the booster." He had been appointed to the accident board and had seen the films at KSC. Just by happenstance the video had been recorded by a camera whose signal was not being fed to the networks. Nobody at the LCC or MCC had been aware of the leak. We were all stunned. The SRBs had never been a major concern to us. *So much for being* certain *an SSME had failed,* I thought.

Mike also recounted his disgust with how the families had been handled immediately after the disaster. He had encountered them in the crew quarters three hours after *Challenger*'s destruction. They were clamoring to return to Houston but NASA was holding them at KSC, supposedly to retrieve their luggage from the condos for the return flight. But Mike didn't believe it. "The women said they didn't care about the luggage. They wanted to leave immediately. They were being held so Vice President Bush could fly to the cape and offer the nation's condolences." He sarcastically added, "The wives had to cool their heels so the VP could feel better." I didn't blame Bush—his intentions had been noble. But the incident was just another example of how useless NASA HQ was when it came to standing up to politicians. They should have explained the situation to the White House and immediately flown the wives to Houston. The VP could have consoled them there.

Judy's hometown memorial service was held at Akron's Temple Israel. A photo of her replaced a casket. Death in the arena of high-performance flight frequently left only that, a memory. Judy and the others had been perpetually frozen in their vibrant youth.

Judith Arlene Resnik, dead at age thirty-six, was eulogized into a person I didn't recognize, as heroic as Joan of Arc and flawless as the Virgin Mary. In multiple Houston ceremonies I had heard the same glowing praise bestowed on the other crewmembers. I excused the excess. It was the perfection the living always demand of their fallen heroes and heroines.

As I listened to Hebrew prayers being said for my friends, guilt rose in my soul. Every astronaut shared in the blame for this tragedy. We had gone along with things we knew were wrong—flying without an escape system and carrying passengers. The fact that our silence had been motivated by fear for our careers now seemed a flimsy excuse. There were eleven children who would never again see a parent.

The news of NASA's and Thiokol's bungling of the O-ring problem quickly reached the astronaut office and had a predictable effect. We were bitterly angry and disgusted with our management. How could they have ignored the warnings? In our criticisms we conveniently forgot our own mad thirst for flight. If NASA management had reacted to the O-ring warnings as they should have and grounded the shuttle for thirty-two months to redesign and test the SRB (the time it took to return the shuttle program to flight after *Challenger*), some of the loudest complaints would have come from astronauts. We were as guilty of injecting into the system a sense of urgency to keep flying as the NASA manager who had answered the Thiokol engineers' worries with "My God, Thiokol, when do you want me to launch, next April?" Only janitors and cafeteria workers at NASA were blameless in the deaths of the *Challenger* Seven.

After the White House learned of the O-ring history, it concluded there was no way NASA could conduct an impartial investigation into itself. President Reagan ordered the formation of the Roger's Commission to take over the investigation. (The commission was named for its chairman, former attorney general William P. Rogers.) As I followed their reports, I learned that my rookie flight, STS-41D, had been one of the fourteen O-ring near misses. In fact it had been the first to record a heat blow-by past a primary O-ring. As Bob Crippen had said, "Whatever it was, we've all ridden it." I wondered why STS-51L had been lost and not 41D? Would the "bump" of just one more max-q shock wave on *Discovery*'s flight have opened the SRB joint seal enough for total O-ring failure and death? Only God knew that answer. But on August 30, 1984, the breeze from Death's scythe had fanned my cheek.

In the weeks after *Challenger* I went to work each morning wondering why. I had nothing to do. A handful of astronauts were appointed to support the Roger's Commission but I wasn't one of them. My phone rarely rang. There were no payload review meetings to attend, no simulations to fill the hours. In the astronaut office safe was a preliminary copy of my classified STS-62A payload operations checklist, something I had been devouring in my pre-*Challenger* life. But now it sat abandoned. I saw no reason to continue my training for the Vandenberg mission. It was obvious the shuttle would not fly for a very long time and when it did it wouldn't be from California. Rumor had it the USAF was going to bail out of the shuttle program altogether and go back to their expendable rockets. They had never been fans of launching their satellites on the shuttle in the first place. Congress had rammed that program down their throats. The air force had rightly argued that when expendable rockets blew up, they could be fixed and returned to flight status within months, whereas the human life issue of manned vehicles could delay their return to flight for years. During that lengthy delay national defense could be jeopardized. That was exactly where *Challenger* had put the air force. It was easy to believe the rumors that the air force was going to walk away from their investment in the Vandenberg shuttle pad.

There was also a very big technical reason Vandenberg was dead. Because rockets being launched into polar orbit lose the boost effect of the eastward spin of the Earth, they cannot carry as much payload as eastward-launched KSC rockets. To recover some of that payload penalty, NASA had developed lightweight, filament-wound SRBs for use on Vandenberg missions. If Thiokol had been unable to seal a *steel* booster, the thinking went, how much more difficult would it be to seal one made of spun filament and glue? No one expected the lightweight Vandenberg SRBs to be certified now. The space shuttle would never see polar orbit, and neither would I. I removed the Vandenberg photo from my wall and placed it in the bottom drawer of my desk. I didn't want to be reminded.

It was impossible to escape the torment that was *Challenger.* In a walk down the hall my eyes would catch the 51L office nameplates. On a visit to the mail room I encountered the staff moving the *Challenger* crew photos to the "Deceased Astronauts" cabinet. I wanted to cry. I wanted to stand there and just weep. But the Pettigrew in me denied that release.

The flight surgeon's office informed everybody that Dr. McGuire—one of the psychiatrists who had interviewed us during our TFNG medical screening, in what now seemed like a different life—would be available for counseling. Some of the wives sought his therapy, Donna included. Most people in my mental condition would have jumped at the opportunity for some help. But most people were not astronauts. I was dyed through and through with the military aviator's ethos that psychiatrists were for the weak. I was an astronaut. I was iron. So I held it all in. If I could hold an enema for fifteen minutes, I could hold all this in and deal with it myself. I would cure myself of depression or survivor's guilt or post-traumatic stress syndrome or whatever it was that ailed me . . . probably all of the above.

Six weeks after *Challenger,* NASA announced they had found the crew cockpit wreckage in eighty-five feet of water. It contained human remains. I had been hoping the wreckage would never be found, that the cockpit and crew had been atomized at water impact. If it had been me, that's how I would have wanted it. Let the Atlantic be my grave. But as shallow as the wreckage rested, NASA had no option but to pull it up. Otherwise it would eventually be snagged on a fishing net or discovered by a recreational diver.

Having been at an aircraft crash site, I suspected the condition of the remains was horrific. The cockpit had been sheared from the rest of *Challenger*, and after a 60,000-foot fall had impacted the water at its terminal velocity of nearly 250 miles per hour. At that speed the Atlantic would have been as unyielding as solid earth. I couldn't imagine the remains would allow the pathologists to learn anything. And I was equally certain nothing relevant to the tragedy would be discovered on the voice recorder, even if it was in good enough condition to be read. Dick Scobee's "Go at throttle up" was uttered only a couple seconds prior to breakup and there was nothing out of the ordinary in his call. Obviously he and the rest of the crew were unaware of their problem. And nothing could have been recorded after breakup because the recorder lost electrical power and stopped at that instant. Not that there would have been anything to record. I remained convinced the crew had been killed outright or rendered unconscious when *Challenger* fragmented.

After the remains were removed, TFNG Mike Coats and several other astronauts examined the wreckage. Mike returned to Houston with the comment, "The cockpit looks like aluminum foil that had been crushed into a ball." It was largely unrecognizable as a cockpit, a fact that didn't surprise me. He added, "I saw a few strands of Judy's hair in the wreckage . . . and I found her necklace." He didn't have to say any more. I knew the necklace. Judy always wore it . . . a gold chain with a charm displaying the two-finger-and-thumb sign language symbol for "I love you." She had a hearing-impaired family member and the necklace was a display of her support for those with similar handicaps. The image Mike's words conjured would not leave me. Like the flash of a camera, I continued to see it no matter where I looked—the crushed cockpit, Judy's hair, her necklace.

The remains were held at Cape Canaveral Air Force Station for pathologists to identify. A few weeks later I watched NASA's TV broadcast as a procession of hearses drove onto the KSC runway and unloaded seven flag-draped caskets. Each was accompanied by an astronaut. A military honor guard reverently carried the remains into the belly of an air force C-141 transport aircraft. There was no dialogue to accompany the TV footage. The silence made the images even more heartrending. The camera followed the plane as it rolled down the runway and receded to just a dot in the sky. The *Challenger* crew was finally returning to their families.

On May 19 a horse-drawn caisson slowly bore the remains of Dick Scobee toward his final resting place in Arlington National Cemetery.

The day was sultry and the air tinted with the odor of horse dung and freshly mowed grass. A military band, playing a medley of patriotic arrangements, led our procession. A formation of skin-headed GI pallbearers, dressed in mirror-polished livery, marched with them. Another group of buzz-cut soldiers bore the American flag and other standards streaming blue and red battle ribbons. Rivulets of sweat poured into their eyes from under their headgear, but they did not break the precision of their march to wipe it away. The astronaut corps and a handful of our spouses trailed the entourage. Between music selections the drummer maintained a solo staccato. The clop of hooves on the cobblestone mingled with the tapping of the women's heels to compete with the drummer's cadence. A symphony of other mournful sounds tugged at the heart: the choking sobs of women, the creak of the caisson, the groan of the leather tack, the jingle of a bridle.

The chaplain conducted a brief graveside service. Then an honor guard fired a rapid three-shot rifle salute, each shot punctuated by the metallic tinkle of the ejected brass. The young children and some of the adults startled visibly at the loudness of the firings. Other soldiers lifted the flag from the casket and folded it with machinelike precision. It was handed to George Abbey, who, in turn, presented it to June Scobee. A flight of four NASA T-38s zoomed into view in fingertip formation. Over the grave the number-two pilot jerked his plane upward and disappeared into the clouds leaving the missing man gap. Then the play of "Taps" drew out a new wave of sobs.

My grief wasn't refreshed in any way by the scene. It couldn't be. I had reached my limits of that emotion. But as the notes of "Taps" floated in the air I was stirred anew in my anger at NASA management. This should have never happened. It was completely preventable. There had been four years of warnings.

I wondered if any of them at that grave felt culpable. I suspected those who knew nothing of the O-ring problem, and most at JSC and HQ had not, felt they were off the responsibility hook. In my book they were not. It wasn't an O-ring failure that brought us to this Arlington service. That was merely a symptom. The real failure was in the leadership of NASA. Over many years it had allowed the agency to degenerate into a loose confederation of independent fiefdoms. As proof of that, the Roger's Commission was finding that many at MSFC had been aware of the O-ring issue, but the problem had not been communicated to the appropriate offices at HQ and JSC, including Young's and Abbey's offices. Neither did the Thiokol engineers' eleventh-hour worries about launching in cold temperature get to the launch director at KSC. And astronaut concerns about the lack of an escape system and the passenger program were unknown to NASA's senior management, of that I was certain. NASA was filled with incredibly talented people, some of the world's best. But the agency lacked the leadership necessary to bind everyone together into an effective and safe team. The NASA administrators were largely budget lobbyists beholden to the White House and Congress. They didn't lead NASA. They certainly didn't lead *me*. I couldn't recall any administrator ever visiting the astronaut office to solicit our opinions. I had heard one TFNG grumble, "We should fly every new NASA administrator on a shuttle mission. Maybe if they had the shit scared out of them they'd be more beholden to us." That was a part-timer program I would endorse.

"Who led NASA?" was the question. Nobody. That's why we were standing in Arlington listening to "Taps" for Dick Scobee. It was even a mystery to me who led my fiefdom. Who was in charge at JSC? George Abbey seemed to be absolute ruler of his own little duchy. Even now, a previously planned new astronaut selection was still rolling along. The shuttle wouldn't fly again for years. Why bring in more astronauts now? We couldn't understand why the JSC director or NASA HQ didn't order a stop to it. It was more proof to us that when it came to anything associated with astronauts, *everybody,* including the JSC director and NASA administrator, worked for Abbey. He didn't answer to anybody. How many other similarly independent fiefdoms existed within NASA? What were their kings like? What frustrations burdened their serfs? I could only speak of my own. Lack of leadership at JSC and in Washington, D.C., had allowed the astronaut office to become dominated by fear. Even outsiders had become aware of it. In a vitriolic March 12, 1986, memo addressed to John Young, Colonel Larry Griffin, commander of the air force detachment to NASA (he was not an astronaut), wrote, " . . . my personal experience in working with the astronaut office is that nearly everyone there is absolutely afraid to voice any opinion that does not agree with yours. You criticizing anyone for 'pressure' is ludicrous when the primary axiom in the astronaut office is, 'Don't cross John if you ever want to fly.' That's pressure!" Colonel Griffin had it slightly wrong. We were afraid to voice any opinion that did not agree with that of Young *or* Abbey. Did the JSC director or the NASA administrator have any idea how fearful we were of our management? If they had been involved in our lives, they would have known and could have fixed the problem. That's what good leaders do.

I couldn't point to any single individual and say, "He did it!" but, collectively, NASA management put Scobee and the other six in their graves. I wanted them all gone. So did most of the astronaut corps. But we were so jaded by our NASA experiences, we doubted it would happen. Already it was more than three months since *Challenger* and there had been no firings. I saw the future and it looked remarkably like the past. Of course there would be new "oversight committees" and a new "safety emphasis," but to a significant degree the same people would remain in leadership positions and that meant nothing would really change. I would later hear a TFNG describe it perfectly. "You can paint a different tail number on the squadron dog [referring to the most malfunction-prone jet] but it's still the same dog."

I walked away from the Arlington ceremony angry, bitter, depressed, and guilt-ridden . . . making a mental note to tell Donna that if I died on a shuttle mission I didn't want Abbey or Young or anybody from NASA HQ anywhere near my grave. I certainly didn't want any of them handing her the flag from my coffin. (Upon my return to Houston, I did make that request of Donna.)

The only positive thought I could muster was that at least there would be no more scab pulling. The crew was buried. Now the healing could begin.

But God granted us only the briefest of reprieves. A week after Scobee's funeral, astronaut Steve Thorne, class of 1985, died in an off-duty recreational plane crash. It was another body blow to the astronaut corps.

CHAPTER 27

Castle Intrigue

Several weeks after *Challenger* I was finally given a job: to review the design of the Range Safety System (RSS). NASA wasn't just focusing on the SRB O-ring design. It wanted to be certain there were no other deadly failure modes lurking in other shuttle components. Astronauts were assigned to work with experts from every subsystem to root out any safety issues. I was assigned the RSS, the system designed to terminate the flight of an errant shuttle. It would prove to be an assignment that would nearly terminate my career.

Most astronauts grudgingly accepted that the RSS was needed to protect civilian population centers. But there was no denying we hated it because it directly threatened our lives. Over several months I traveled to Cape Canaveral Air Force Station to meet with the RSS personnel—they were not NASA employees. By congressional law the protection of the civilian population from rocket mishaps was the responsibility of the Department of Defense, and DOD had given the job to the USAF. And the only way the air force could guarantee that protection was to place explosives on everybody's rockets, NASA's as well as all military and commercial missiles. (On the shuttle, the explosives were placed on each SRB and the gas tank. While there was none on the orbiter, detonation of the other explosives would also destroy the orbiter and kill the crew.) During every missile launch, USAF officers, who served as RSOs, monitored the machine's trajectory. If a rocket strayed off course, it would be remotely blown up to prevent it from falling on a city.

In multiple meetings I examined every aspect of the design of the RSS and the selection and training of the RSOs. (I would learn that RSOs routinely declined invitations to attend KSC social functions with astronauts. They did not want their launch-day judgment impaired by a friendship with crewmembers they might have to kill.) The system was as fail-safe as humanly possible. In these same meetings I also learned that the Range Safety Office was proposing some changes to shuttle launch abort procedures. They worried that in some aborts, pieces of the jettisoned gas tank could land in Africa. Their suggested solution was to have astronauts burn the OMS engines during these aborts. The additional thrust produced in the burn would result in an ET trajectory that would drop the fuel tank into the Indian Ocean.

When I brought this request to John Young, he became as hot as a reentering ET, arguing it was a dumb idea. The OMS propellant was the gas used for the final push into orbit, for maneuvers while in orbit, and for the braking maneuver to get out of orbit. The RSOs were asking us to burn gas during ascent that we might later need—just to put another zero behind their already conservative risk-to-Africans probability numbers. I agreed with Young. But then the trajectory planners at MCC did their own studies and found that igniting the OMS engines pre-MECO (burning them at the same time as the SSMEs) would actually improve nominal and launch abort performance. In other words, it would improve the crew's chances of reaching orbit or a runway. When I brought this data to Young, I expected him to enthusiastically endorse it, but I was stunned when he didn't. His position was that we would never do an OMS burn on the uphill ride. I assumed I hadn't made myself clear and tried again. "John, I'm not suggesting this be done to satisfy the RSO. This is our own FDO recommending it. The data shows it will improve performance during the abort." John would hear none of it.

Over the next several weeks, in multiple meetings in Young's office, I continued to bring him the results of various meetings on the pre-MECO OMS burn issue. The ball was rolling. It was going to happen.* Young was beyond angry at this news and focused his anger at me. Again and again I tried to make him understand the pre-MECO OMS burn was something FDO wanted to do to protect the crew. But he was deaf to my logic. Instead he remained focused on the fact the Range Safety Office wanted the OMS burn to keep the ET off Africa.

I appealed for help from the JSC office pursuing the OMS burn change, the office of Flight Director Jay Greene. Jay had cut his teeth as a young MCC flight controller during the Apollo program. I held him in great esteem. He was heart-and-soul dedicated to crew safety. If he and FDO were saying that an uphill OMS burn was going to make things safer for the crews during some aborts, then it would. I asked him to come to Young's office with the supporting engineers to make their case to Young. He would be happy to was his reply. I felt good about what I had arranged. Jay was a well-regarded flight director. John would *have* to listen to him.

At the appointed hour I rendezvoused with Jay and his entourage of engineers and we walked to Young's office. It was empty. When I asked where he was, his secretary sheepishly replied, "He went to get a haircut." I wanted to scream. He had stiff-armed me. His mind was made up. He didn't want to hear any contrary arguments from anybody.

In an attempt to gain the support of other astronauts, I presented some data on the RSS situation at the September 15, 1986, Monday morning meeting. I was hardly able to finish a sentence. Young heckled me at every turn. I was humiliated. Over a beer I mentioned my travails to Hoot Gibson. Hoot exploded, "I've had the same problem with him on the issues I'm working and I've just quit listening and talking to him."

As the weeks passed I fell further into the depression that had started with *Challenger*'s loss. I had lost friends. I had lost a mission into polar orbit. Now the core of my professional life, my work ethic, was slipping away. All my life I had been intent on getting the job done. When the first psychiatrist of my TFNG interview had asked me what

*It did happen. Pre-MECO OMS burns are now regularly done during nominal ascents and are part of shuttle launch abort procedures.

my personal strength was, I had truthfully replied, "I always do my best." It was my hallmark. I knew I wasn't the smartest astronaut. But I was solid, reliable. I always got the job done . . . until now. I hated my job. I hated my boss. When I slept, which wasn't much, I had dreams of Judy's necklace and exploding shuttles and writhing SRBs and walking through the gore of a crash site.

My distress had long been known to Donna. Every evening I would recount my stories of abuse to her. As always, she listened and lent her support . . . and lit more bonfires of votive candles to send her prayers heavenward for my delivery from Young. We talked about leaving NASA. I could return to the air force, but I knew I wouldn't be happy there. The only thing awaiting me in the USAF was a desk. I would never see the inside of a cockpit again. I was too old and too senior in rank. I would end up buried in the bowels of the Pentagon. I didn't want to leave NASA. I wanted to fly again. My *Discovery* flight couldn't compare with what some of my peers had done on their missions. Pinky Nelson, "Ox" van Hoften, Dale Gardner, and Bob Stewart had all done tetherless spacewalks. They had donned MMUs and, like real-life Buck Rogerses, had jetted away from their shuttles into the abyss of space. Kathy Sullivan, Dale Gardner, Dave Griggs, and Jeff Hoffman had done traditional tethered spacewalks. Sally Ride had used the robot arm to deploy and retrieve a satellite. Rhea Seddon had used the robot arm in an attempt to activate a malfunctioning satellite. I wanted to do similar things that challenged my skills as a mission specialist. I wanted a spacewalk flight. I wanted to fly a mission with an RMS task. I wanted a high-inclination orbit so I could see my Albuquerque home from space. And there was only one place on Earth I could do these things . . . at NASA. As much as I wanted to walk into Young's office and tell him, "Take this job and shove it!" I couldn't. There was no place else to go and ride a rocket into space. I would have to endure.

On September 19, astronauts celebrated for the first time since *Challenger* with a party at a local club. I had been looking forward to it. After nine brutal months, it would be good to erase my brain with a few drinks and have some fun with my fellow TFNGs. Donna and I sat at a table with the Brandensteins, Coveys, and Boldens (class of 1980). After dinner, Bob Cabana, the class leader of the latest group of astronauts to arrive at JSC (class of 1985), walked to the stage and invited George Abbey to step forward and receive an autographed photo of their class. The image immediately brought to mind our TFNG efforts to brownnose Abbey back in 1978. Kathy Covey let out a "woo, woo, woo" catcall. We all understood her sarcasm. Every astronaut class prostituted itself to Abbey thinking it was going to help them. The class of 1985 would soon learn what we had all learned—shoving your nose up Abbey's behind didn't get you anywhere. In a year they would all be cursing him and Young like everybody else.

Others at the table were soon speaking of their disgust with our leadership. At that, Kathy, a very successful and hard-nosed businesswoman, began to mock our impotence. "I've listened to this shit for years. You guys are so gutless you deserve what you get" was her message. Of course she was right. But it all went back to that incontrovertible fact . . . there was no other place on the planet where we could fly a rocket. If Satan himself had been our boss and demanded we take a fiery pitchfork up the wazoo before we could climb into a shuttle cockpit, all of us would have long ago become acrobatic in our ability to bend over and spread our cheeks.

I grabbed another beer and then another. I didn't want to hear any more of this. I was burned out. But it was impossible to escape. As the party was breaking up, Ron Grabe (class of 1980) took me aside. "Mike, you better watch your six o'clock." It was fighter pilot lingo; I had an enemy on my tail. "This week I was waiting to see Young and I heard him on the phone. I don't know who he was speaking with, but I assumed it was Abbey. He was saying, 'Mike Mullane is one of the enemy. He's a nice kid and all that, but he's on the side of the Range Safety people.'"

A "nice kid"? I was forty years old. And what crime had I committed to earn the label "enemy"? I was guilty of doing my assigned job.

I thanked Grabe for the warning and added, "I guess I'll talk to P.J." P. J. Weitz was a Skylab-era astronaut working as Abbey's deputy. He was well regarded, and I considered him the only manager I could trust within all of NASA.

Grabe added, "Don't bother with P.J. I've already spoken to him about Young. I told him John has become unbearable. Nobody can make an objective presentation on any subject. He has made up his mind on everything. I used your Monday morning presentation on the RSS as an example. P.J. was sympathetic but said he couldn't do anything."

I reached a new nadir of depression. It was never clear how Young influenced flight assignments. Most of us believed he had nothing to do with them, which, if true, was absolutely amazing given the title on his office door: Chief of Astronauts. But none of us knew for sure. Maybe Abbey did listen to his input. I couldn't just dismiss Grabe's warning. I did need to watch my six.

The following week I made an appointment to see Abbey. I had been at NASA for eight years and had only met with George on a handful of occasions and always in the company of others. I had never had any real one-on-one time with him. I approached his desk with the same trepidation I imagine a departed soul experiences while being escorted by the seraphim to the judgment seat of God.

He motioned for me to take a seat and I began to explain my problems with Young. Abbey wouldn't look me in the eye. As I spoke he continued to shuffle through papers on his desk as if my problem were the merest of trivia. I was only a couple sentences into my rehearsed speech when he saw where it was going and mumbled, "Don't worry about that," to his ink blotter. "John is just frustrated he can't do more." I kept talking. I needed resolution. I was still working the RSS and OMS burn issues and being savaged by Young in the process. I couldn't go on like this. Abbey interrupted me with a dismissive wave. "Don't worry about it. You'll be getting too busy with DOD affairs in the next six months." I was silenced by that comment. What was he suggesting? Was he hinting I was in line for a Department of Defense shuttle mission? There were several DOD payloads ready to go on the shuttle—satellites so optimized for the shuttle cargo bay they could not be easily switched back to the air force's unmanned boosters. Or was Abbey implying I would soon switch jobs from Range Safety to review the safety of DOD payloads? Or was this a polite warning that my career at NASA was being terminated and I would be going back to the USAF? There was no divining what George Abbey meant.

I came away from George's office only slightly unburdened. His manner suggested my career was intact. But in the same breath he also told me to basically ignore John Young—his chief deputy. That command was more proof NASA's leadership structure was a joke. How could I do that? Young was my immediate boss. He signed my air force performance reports. If any generals ever called to discuss my promotion potential, they would be talking to Young. Besides, there were some serious range safety issues that needed to be addressed. Was I supposed to "not worry about those"? There was also the possibility that perhaps John hadn't been talking to Abbey when Grabe overheard him. Maybe he had been bad-mouthing me to one of his champions, a champion far above Abbey's level who held veto power over Abbey's crew selections and was ready to redline me from any list based on Young's input. I hated the position I was in. I couldn't just ignore Young. My prayer was that Abbey would discuss the situation with Young and he would become rational on the OMS burn issue. But all hope in that regard was dashed a couple months later when I was warned a second time that there was somebody at my six o'clock. This time the messenger was Hank Hartsfield. "Mike, John has a real hard-on against you about the way the pre-MECO OMS burn issue has played out. I heard him mumbling that maybe you should be replaced. I hope your career hasn't been damaged."

I was blind with fury. At every meeting on the topic I had dutifully represented Young's position that he disapproved of burning OMS fuel during powered flight. But Young never personally attended these meetings to defend his position. He never used his bully pulpit as a six-time astronaut, moon-walking hero, and chief of astronauts to formally make his case. When I suggested to him in writing that he attend a Flight Techniques Panel meeting at which Jay Greene was going to "press forward with implementation [of pre-MECO OMS burns]," John shot back his written answer: "NO! We are <u>NOT</u> going to a forum or voting on this issue!"

I thanked Hank for the warning, suppressing the urge to ask, "Who do I appeal to for justice? Who runs this asylum called Johnson Space Center?"

But Hank's warning was the final straw. I broke. Mike Mullane, the man who prided himself on being able to hold it all inside, be it an enema in the colon or an agony of emotions in the soul, the man who had lived a life in abject fear of doctors, the man who thought psychiatry was for the feminine and weak . . . that iron man, Mike Mullane, called Dr. McGuire's office and made an appointment. I was losing my mind.

On the day of the meeting I picked up the phone several times to cancel. I was certain that if I walked into McGuire's office I would be recording a new astronaut first. I would become the first astronaut in the history of NASA to voluntarily see a shrink. I would be admitting failure. I would be violating the "Better dead than look bad" commandment. I could imagine how the office grapevine would carry the news if my dark secret was ever discovered. "John Young put Mullane in tears. He ran to the shrink like one of those weepy women on *Oprah*." My finger hovered over the phone keypad as this image of personal failure filled my mind. I would cancel the appointment. But I would always come back to the question, "What choice did I have?" I was going freakin' nuts. I would hang up the phone only to immediately snatch it back and begin to question myself all over again.

Somehow my resolve triumphed. I made it to zero hour. I told my secretary I was going to the gym and then took a circuitous route to McGuire's temporary office. He merely consulted for NASA. His primary job was in San Antonio with the University of Texas. I found the unmarked room and walked by it several times, checking the hallways for any prying eyes. A Baptist preacher on a clandestine rendezvous with a prostitute could not have acted more suspiciously. The hallway was deserted. Finally I grabbed the door handle, took one more hurried glance in all directions, rushed into the room, and immediately closed the door. That entry alone was probably enough for McGuire to make a diagnosis: paranoid.

As he had ten years earlier, Dr. Terry McGuire met me with a broad smile and enthusiastic handshake. "Come on in, Mike. Have a seat. What can I do for you?" He was largely unchanged from how I remembered him—tall, trim, yielding to baldness, clean shaven. He had the perfect voice for his job—deep, melodious, and soothing.

While I didn't think there was a damn thing he could do for me, I cut to the chase. "Young and Abbey are driving me fucking nuts."

McGuire laughed at that. "I've heard that from a number of your fellow astronauts."

I'm sure he noticed the shock on my face. "Are you saying I'm not the first astronaut to meet with you?"

"Not at all."

The revelation was like a giant weight being lifted from my shoulders. Misery loves company and now I was being told I was part of a miserable crowd. Suddenly, I wanted to interview him, to find what others were saying, but he quickly steered me back to topic. "So, tell me what's happening with Young and Abbey."

For the next hour the demons of anger and disgust flew from my soul like bats from a cave. Emboldened by the thought that others had sat in this same seat, I didn't hold anything back. I told of my ongoing head-butting with Young on range safety and OMS burn issues and the warnings other astronauts had given me that my career was in peril for doing my job. "We're all afraid to speak up for fear it will jeopardize our place in line. There's no communication. Nobody understands how crews are chosen or even who chooses them or who has veto power over them—and that's all that matters to us, flight assignments. Fear dominates the office."

I recounted how astronauts had recently attempted to forecast flight assignments based upon where people would be parking. "Hank Hartsfield put out an updated parking lot map. Some astronauts jumped on it as if it were the Rosetta Stone, which could decipher their place in the flight line. They assumed those with the closest parking spaces to the simulators were to be on the next missions. It's sick. It just shows our desperation for some insight into the flight assignment process."

I told him of something Hartsfield had related. "Hank is working as Abbey's deputy and told me that George is resisting bringing computer links into the astronaut office. It's Hank's opinion that Abbey doesn't want us to have a communication path he can't control."

I told him of a revealing incident in which one astronaut suggested, "There will be one hundred suspects, all astronauts, if Abbey was ever to die from foul play." Another astronaut offered, "No, there won't be one hundred suspects. There will be one hundred astronauts clamoring to take responsibility . . . 'I'm the hero . . . I did it.'" A hundred astronauts were on the verge of going postal.

I explained the profound us-versus-them attitude that had come to dominate our relationship with Young and Abbey. I told of astronauts who were perceived as spies for the duo. "Whenever they enter a conversation, everybody watches what they say for fear it will come back to haunt them. It's like being in a prison yard and worrying about the warden's stoolies overhearing an escape plan."

I told him how we had all hoped the *Challenger* disaster would be the catalyst for change in management, but nine months had passed and nothing had changed and, with each passing day, it became more evident nothing ever would change. I offered my opinion that "Nobody runs NASA. Young and Abbey don't answer to anybody. They're bulletproof."

Throughout my diatribe, I couldn't get it out of my head that I was engaged in an exercise of futility. What was McGuire going to do? He wasn't even a NASA employee: He was a consultant. I was wasting my breath.

I finally stopped and thanked him for listening. "I know there's nothing you can do on this, but it's been helpful to get it off my chest. Knowing others have been driven to you in their frustration is definitely helpful."

McGuire said nothing to dissuade my belief that he was powerless to effect any management changes. He would have been lying if he had, was my certain opinion. He encouraged me to stick it out. Changes might be in the works, brighter days might be ahead, blah, blah, blah. It was what I had expected. He was as impotent as the rest of us. He was a good listener but he had no cure for what ailed me. I wanted to fly in space again and my immediate boss had twice indicated, if he had anything to do with it, that was never going to happen. I had long exhausted myself wondering if Young's opinion mattered at all.

As I rose to leave, McGuire handed me a ten-page, single-spaced document. "You might want to read this sometime. It'll help explain the situation you're in."

I wanted to say, "I don't need to read anything to know the situation I'm in . . . It's called deep shit," but held my tongue. I glanced at the cover page, *Leadership as Related to Astronaut Corps, by Terence F. McGuire, M.D., Consultant in Psychiatry.* It was undated. My curiosity was piqued by the title. Why was McGuire writing about astronaut leadership? I could only assume it was a self-initiated private work. "Publish or perish" was the order of the day for university professors. I rolled the document into my hand, thanked McGuire for listening, and departed.

I wasn't about to be found at my desk reading anything with McGuire's name on it, so I put the document in my briefcase and took it home. That evening I popped a beer and began reading. "One of the more operationally practical ways of viewing personality subdivides the population into six basic clusters of characteristics that define distinct personality types. . . ." Yawn. I felt like I was back in high school reading *Moby-Dick* (a book I'm convinced nobody has ever completed). But as I read further I realized McGuire did have an extensive knowledge of what was happening in the astronaut corps.

"In the last eight years or so, the dissatisfaction level relative to management style has risen significantly, if I am to judge from all the unsolicited comments offered by astronauts and their spouses. The level of dissonance is much higher than I experienced in my military career as a flight surgeon-psychiatrist working almost exclusively with pilots and their families. Nor have I seen its equal with the elite flying units or special projects air crew for whom I was a long-term consultant. . . . Though they are exceedingly careful about the setting in which they [astronauts] give voice to their dissatisfaction, there is no doubt that the current managerial style constitutes an important morale issue with the astronaut community and, for many, has a stultifying effect on creativity and open discussion."

The writing was couched in scientific mumbo jumbo and not a single manager's name was mentioned. The opening paragraph implied it was nothing more than a technical paper. "This is a background document on leadership as it relates to the astronaut group, more specifically, on the impact of various leadership styles upon the morale, creativity and productivity of the astronauts." But much of the remainder of the work focused on a particular leadership style as it related to astronauts—the autocratic power merchant. It didn't take a rocket scientist to see the similarities between that style, as described by McGuire, and the way Abbey operated:

"Like a good Pavlovian psychologist, he has learned not only that rewards and punishments reinforce behaviors, but also that there are times when an inconstant system of incomplete rewards can evoke even stronger adherence to desired behaviors than a predictable full-scale reward system. This second approach also lowers his predictability and keeps people off balance."

". . . No one in the cadre is allowed to know all parts of the grand plans the power merchant may have."

"Inconsistency, ambiguity, silence, evasion . . . all have their place in his studied unpredictability."

Later in the document, McGuire makes the point, "The autocratic managerial style is the most antithetical to the needs of the astronaut corps, a group who in most settings would be chiefs rather than indians [*sic*]. Though calculated to be the least effective, if the autocrat is open, non-devious and fair, he can still be acceptable to the prototypical astronaut. But if he is the type of leader who is oriented toward the accumulation of personal power for power's sake, rather than for the good of the company, his impact on the corps will be destructive. Men of such inclination are drawn by nature toward the autocratic style. So I have gone to special pains to identify the hard-core autocratic power merchant because, of all leadership approaches, it has the greatest potential for negative impact not only on personnel clusters such as the astronauts, but also within the total institution. . . ."

He continued, "For many years astronaut morale has been, I believe, considerably below its potential. Many of the fine men who have moved on from its ranks to other endeavors have told me of the negative role astronaut management has played in their decision to leave. As is so often true, the most capable men, those with more options and more confidence, are the most at risk to depart NASA for new challenges. Usually they elect to leave with as little surface disturbance as possible, out of deference to NASA as an organization. In my several decades of association with NASA, I have never seen a more propitious time to institute change, nor a time in which the morale-boosting effects of realistic positive change would be more welcome."

I set aside the document completely befuddled. What did it all mean? Had someone in management commissioned McGuire to document astronaut frustrations and scientifically show how Abbey's leadership style was the direct cause? If so, to what end—as justification to get rid of George? His comment ". . . never seen a more propitious time to institute change . . ." certainly sounded like a recommendation to somebody. It certainly wasn't the type of statement you would expect to find in a technical paper written for publication in a medical journal.

But I wasn't about to go back to McGuire and question him. More than ever I felt like I was living in medieval times with plots swirling about. I had one objective . . . not to get burned by any castle intrigue. If somebody was attempting to assassinate John and/or George I wished them luck, but I didn't want to be a participant. Like a serf in the field, I wanted to be invisible when the opposing armies swept past. I just wanted to hold on long enough to fly another space mission and then I would be gone from this madness.

CHAPTER 28

Falling

In mid-October 1986, astronaut medical doctors Jim Bagian and Sonny Carter presented the *Challenger* autopsy results. We expected to hear the answer to the question that had tormented all of us since the moment of *Challenger*'s destruction. Had the crew been alive and conscious in their fall to the water? I had been certain they had not. I had said it a hundred times to Donna, "At least they died or were knocked out instantly." That belief was my security blanket. It was too great a horror to think they may have been conscious in the two-and-a-half-minute fall to water impact. New revelations throughout the investigation had given me momentary doubts but I had always managed to build a new scenario to hide behind. My "cockpit-shredding explosion" theory had long been proven wrong. The fire leaking from the right-side SRB had weakened its bottom attachment to the ET. As the SRB pulled free it ruptured the external tank, and the aerodynamic forces and the G-loads of the moment caused the catastrophic breakup of the stack. There had been no high-power detonation. The enormous "explosion" seen in the sky was merely tons of liquid oxygen and hydrogen vaporizing and burning. NASA cameras had picked out the cockpit module as a piece of the fragmentation. It trailed some wires and tubing but otherwise appeared intact, suggesting it was bearing a live, conscious crew. But I created a scenario in which the cockpit G-forces at the moment of breakup had pulled the crew seats from their floor attachments and hurled them against the interior of the cabin, killing the occupants instantly, or at least knocking them unconscious. When engineers later determined that the cockpit G-loads were not incapacitating, much less fatal, I created a scenario in which a window had broken or, in some other manner, the pressure integrity of the cockpit had been explosively compromised, causing crew unconsciousness within seconds.

The first fact that Bagian and Carter put on the table was that the time of death could not be determined from an examination of the crew remains. I was not surprised. High-performance-vehicle crashes typically leave little of the human body for pathologists to work with. In the case of *Challenger*, the weeks of immersion in salt water had resulted in additional deterioration of the remains. The story of crew survivability and consciousness would have to be told by the remains of the machine, not the crew. Bagian and Carter began that story.

There was proof, in the form of Mike Smith's Personal Emergency Air Pack (PEAP), that the crew had survived vehicle breakup. PEAPs were portable canisters intended to provide emergency breathing air to a crewmember escaping through toxic fumes in a ground emergency. They were not intended for use in flight. But Mike Smith's PEAP, stowed on the back of his seat and only accessible in flight by mission specialist 1 or 2, had been found in the on position. Either El Onizuka (MS1) or Judy (MS2) had to have thrown the switch and there would have been only one reason to do so—they were suffocating. Breakup had ripped away their only source of oxygen, the tanks under the cargo bay. Within a couple breaths, the residual oxygen remaining in their helmets and in the feed lines would have been consumed. The fierce urge to breathe would have immediately driven all of the crew either to turn on their PEAPs or raise the faceplates of their helmets. Judy or El had turned on Mike Smith's PEAP knowing he could not reach the switch himself. (Onizuka, sitting directly behind Mike, had the easiest access to the switch, though Judy, sitting to El's left, could have reached it with some difficulty.) Enough pieces of one other PEAP were recovered to determine it had also been in the on position, but crash damage made it impossible to establish the seat location of that canister. The fact that two PEAPS had been turned on was proof the crew had survived *Challenger*'s breakup. I would later learn that some of the electrical system switches on Mike Smith's right-hand panel had been moved out of their nominal positions. These switches were protected with lever locks that required them to be pulled outward against a spring force before they could be toggled to a new position. Tests proved the G-forces of the crash could not have moved them, meaning that Mike Smith made the switch changes, no doubt in an attempt to restore electrical power to the cockpit. This is additional proof the crew was conscious and functional immediately after the breakup. Mike Smith's PEAP also provided proof the crew had been alive all the way to water impact. It had been depleted by approximately two and a half minutes of breathing, the time of the cockpit fall.

The question remaining was whether the crew had stayed conscious beyond the few seconds needed to activate their PEAPs and for Mike Smith to throw some switches on his panel. They could not have remained conscious if the cockpit had rapidly depressurized to ambient (outside) air pressure. Breakup occurred at 46,000 feet, an altitude 17,000 feet higher than Mount Everest, and the nearly Mach 2 upward velocity at breakup continued to carry the cockpit to an apogee of approximately 60,000 feet. To stay conscious in the low atmospheric pressure of these extreme heights, the crew would have needed pressurized pure oxygen in their lungs and the PEAPs only supplied sea level *air,* a mixture of about 80 percent nitrogen and 20 percent oxygen. But had there been a cockpit depressurization?

An explosive depressurization—due to a window breaking, for example—would have been a blessing and I prayed fiercely that had been the case. But *Challenger*'s wreckage said it didn't happen. If it had, the cockpit floor would have buckled upward as the air in the lower cockpit rapidly expanded. The wreckage revealed no such buckling. That news was a dagger in my heart.

Bagian and Carter explained there was still the possibility of a nonexplosive but rapid enough depressurization to cause quick unconsciousness. Such air leakage could have occurred due to numerous penetrations at the rear cockpit bulkhead. These provided pathways for wire bundles and fluid lines to pass between the cockpit and the rest of the orbiter. At breakup those wires and tubes were violently ripped apart, and it was possible the pressurization sealing for those manufactured penetrations could have failed. There was also evidence of breakup debris striking the cockpit from the outside. A piece of steel had been found jammed into a window frame. While that particular piece of debris did not penetrate the cockpit, other debris might have, resulting in a depressurization rapid enough to cause unconsciousness.

But it was all conjecture. There was no way to know the pressure integrity of the cockpit and, therefore, the state of crew consciousness. Bagian and Carter did have some ancillary evidence suggesting crew inactivity, which some thought could be a signature of crew blackout. Every piece of paper recovered from the wreckage was examined to see if any crewmembers had written a note. Nothing had been discovered.

Neither had the cockpit overhead emergency escape hatch been blown. Some astronauts had suggested they would have jettisoned it as they neared the water to facilitate escape if impact was survived. The status of Mike Smith's PEAP also hinted at crew inactivity—the canister was only depleted by two and a half minutes, which meant his visor had remained closed during the fall. If it had been open, all five minutes of PEAP air would have leaked out. But, if the crew had been conscious, wouldn't they have *raised* their visors to talk to one another in their fight for survival? That was Bagian's and Carter's hypothesis. After all, we were trained to react to emergencies as a team and that required communication. At breakup the intercom failed, leaving visors-open, direct speaking as the only means of communicating. (Crash damage had obliterated all the helmets and all but Mike's PEAP, making it impossible to know if the visors of the others were up or down.)

After the presentation was concluded, someone put the escape question to Bagian and Carter. "If this had occurred during OFT [the first four shuttle flights in which the two-man crews had ejection seats], do you think the crew would have been able to bail out?" Their answer was a definite yes. The OFT crews had worn pressure suits. Even if cabin pressure had been lost, those suits would have kept the crew conscious and they would have been able to pull an ejection handle.

Carter next informed everybody that the flight surgeon's office was going to archive a clip of our hair and a footprint to facilitate our identification in the event of a future shuttle loss. That comment suggested how difficult identification of the *Challenger* crew remains had been. Even dental records hadn't been enough. John Young sagely observed, "When extraordinary methods are being taken to make sure you can be identified after you're dead, everybody ought to think twice about the job they're in." He was right.

He was also right when he added, "We shouldn't fly again until we have an escape system." Already some NASA managers were suggesting we should return to flight as quickly as possible and the escape system modifications could catch up. As much as I disliked Young for his attempts to torpedo my astronaut career, the position he took on some issues were the right ones.

Carter reminded everybody, "Keep the information you just heard to yourselves." At that Dick Richards (class of 1980) lashed out, "Who does NASA think it's protecting? The families? They don't care if the information is released."

John Young answered him. "NASA is protecting NASA." He had it right again.

After the meeting broke up, I went to the gym for a run. It quickly became a sprint. I wanted to punish myself. I wanted the agony of burning lungs and a pounding heart and aching legs to overwhelm me so I wouldn't have to deal with the reality of what I had just heard. Sweat stung my eyes but I made no effort to wipe it away. I had become a self-flagellating penitent. Pain was good. I relaxed my jaw to its limits and tilted my head back, trying to form a straight pipe to my lungs. Strings of saliva grew from the corners of my mouth and were jerked away by the pounding of my legs. My respiration took on the sound of an emphysemic wheezing and gasping for breath. Several NASA employees passed opposite me and I caught the question in their eyes: "What's he running from?"

I was running from my thoughts . . . and predictably losing. Judy or El had flipped Mike's PEAP to on. There had been no cockpit floor buckling, therefore no explosive decompression. Those twin facts opened the door to the possibility that the cockpit had held its pressure as tightly as a bathysphere and the crew had been conscious throughout the fall. Try as I might I could not find shelter in the evidence of crew inactivity. Would I have written a note? I seriously doubted it. Would I have jettisoned the overhead hatch? No. Of that, I was certain. Had I been Scobee or Smith, I would have been fighting to regain vehicle control all the way to the water, knowing that if I didn't, death was certain. The hatch on-off status was irrelevant to survivability. I also thought it was a big leap to assume the crew would have felt it necessary to raise their visors and communicate by direct speaking. I had been in the backseat of F-4 jets when the intercom had failed and hand signals had worked fine. And Scobee and Smith, sitting side by side, had the advantage of being able to see exactly what the other was doing. If I had been in their position and was conscious with the helmet visor down, would I have taken the chance of raising it to communicate? To do so would have meant overcoming years of jet crewmember training, which emphasized keeping your mask *on* during flight, particularly if there was any hint cockpit pressure integrity might be compromised. The mask in this case would have been the helmet visor. I would have kept it down.

Like everyone else I wanted to believe the crew had been unconscious but there was no hard proof. I kept seeing the horrific *other* possibility, that they had gone down in *Challenger* like galley slaves chained to the benches of their sinking ship and been aware of every torturous second. They had been trapped. They had no escape system. They were flying an *operational* space shuttle.

I couldn't go on. I hit my wall. I slowed to a walk, steered off the track, found a tree, and collapsed against it. There would be no escaping the projector of my mind as it played what might have been the last moments of *Challenger.*

"*Challenger,* you're go at throttle up."

Scobee answered, "Roger, Houston. Go at throttle up."

Mike Smith watched the power tapes climb toward 104 percent. Even as he was doing so, the stack was disintegrating

The leaking fire had weakened the bottom SRB attachment strut. The right-side booster snapped free, rupturing the ET. Tons of propellant poured from the gas tank. The left-side booster ripped from its struts and joined the right-side SRB in chaotic, unguided flight. *Challenger*'s fuselage and wings were broken into multiple parts. The cockpit was torn from its mounts.

At the instant of vehicle breakup the crew was whipsawed under their seat harnesses. Checklists were snatched from their Velcro tabs and jerked on their tethers. Pencils and drink containers separated from their tabs and were hurled through the volume. The noise of debris crashing into the outside of the cockpit added to the chaos. Exclamations of surprise came from some of the crew's throats, but fell dead in their microphones. All electrical power had been lost at the separation of the cockpit from the rest of the fuselage.

The mayhem of breakup lasted only a moment before the equally startling calm of free fall began. While the cockpit and the other debris were still moving upward at 1,000 miles per hour, they were freely under the slowing influence of gravity. Like human cannonballs, the crew had experienced a momentary violence, followed immediately by the silence of gravity's grip. They floated under their harnesses. Pencils and pens spun in the air around them. Checklists floated on their tethers.

The crew was alive but suffocating. They turned on their emergency air packs. Judy or El switched on Mike Smith's PEAP.

Scobee and Smith were test pilots and reacted as they had been trained. Even the brief, wild ride through breakup would not have mentally incapacitated them. They had faced countless serious emergencies in their flying careers. They knew the situation was perilous, but they were in a cockpit with a control stick and there was a runway only twenty miles away. They believed they had a chance.

They snapped their attention to the instruments hoping to identify the problem, but the cockpit was electrically dead. Every computer screen was a black hole. Every caution and warning light was off. There were no warbling emergency tones. Every "talk back" indicator showed "barber pole"—its unpowered indication. The attitude indicator, the velocity, acceleration, and altitude tapes were frozen with OFF flags in view. They had nothing to work with. As they attempted to make sense of the situation, Scobee's hand was never off the stick. He fought for vehicle control, oblivious to the fact there was no longer a vehicle *to* control.

"Houston, *Challenger*?" He and Mike Smith made repeated calls to MCC, but those were into a lifeless radio.

With no instrument response and an apparently dead stick, Scobee and Smith mashed down on their stick "pickle buttons" to engage the backup flight system. It was the emergency procedure for an out-of-control situation. If the problem was due to a primary flight system computer failure or software error, the BFS computer would jump online and bring life back to the cockpit. Again and again their right thumbs jammed downward on the spring-loaded red buttons. Again and again they searched the instruments hoping to see life in them, hoping to have something, anything, to work with. But *Challenger* was now a blossoming cloud of debris. No switch was going to put her back together.

Very quickly the crew realized the futility of their actions. The upstairs crewmembers—Dick, Mike, El, and Judy—had window views of the disaster in which they were immersed. As the tumbling cockpit moved ever higher those views became more synoptic. They looked downward to see the white-orange cloud marking the place of *Challenger*'s death. They saw the billowing trails of the disconnected SRBs.

The downstairs crewmembers—Ron McNair, Christa McAuliffe, and Greg Jarvis—were locked in the most horrifying of circumstances. They had no windows, no instruments. They were totally dependent upon the upstairs crewmembers to keep them informed on the progress of the flight. But no words came. At the instant of breakup the intercom went dead and the mid-deck lights went out. They were trapped in a tumbling, darkened, silent room.

As the cockpit arced across its apogee, the upstairs crew saw the sky turn space-black, *Challenger*'s lost goal. The silence was nearly total, just the merest whisper of wind. Then, the two-minute fall to the sea began. The rippled blue of the Atlantic filled the windows. The noise of the wind rose to a loud rush as the cockpit quickly reached a terminal velocity of nearly 250 miles per hour. The upstairs crew watched the finer details of the sea become visible: rumpled wind-blown areas, the froth of whitecaps, and brighter splashes marking the impact of other pieces of their machine. The horizon rose higher and higher in their windows, the blue reaching toward them until . . . blessed oblivion.

CHAPTER 29

Change

On January 9, 1987, Abbey made a rare and impromptu appearance before the astronaut office. Since his prior visits had almost always included flight assignment announcements, there was a buzz on the walk to the conference room. I couldn't believe my name would be on any press release. I had significant doubts I would ever see my name on a crew list again. But hope springs eternal in the souls of astronauts. The fact that the meeting was unscheduled, on a Friday afternoon, no less, suggested something unusual was in the offing.

As always, Abbey spoke at low volume and everybody craned forward to listen. For ten minutes he discussed some changes in the management structure of HQ, a topic none of us believed was the reason for the meeting. We were right. He concluded his HQ remarks and then, almost offhandedly, mumbled, "The crew for STS-26 will be Rick Hauck as commander, Dick Covey as pilot, and Dave Hilmers, Pinky Nelson, and Mike Lounge as MSes."

For a long moment the room was gripped in a stillness that rivaled deep space. We were hoping Abbey would continue with more crew assignments, or at least tell us when those might happen. But there was nothing. Except for the lucky five, who wore embarrassed smiles, the rest of us slumped in crushing disappointment. Why didn't Abbey get it? If he was the power monger that many believed him to be, couldn't he see the power to be gained with one hundred faithful-unto-death astronauts? With a little communication, that's exactly what he would have had. But he didn't offer a single hint regarding the timetable for other assignments. In the enduring silence, I noticed some of the faces around me hardening into glares of something beyond anger. If I were Abbey I would hire a food taster.

The meeting broke up and I drifted back to my office. Several other TFNGs came by for a losers' commiseration session. The USAF contingent was angry that a navy astronaut, Rick Hauck, would be commanding the return-to-flight mission. Rick would be making his second flight as a commander while his PLT, fellow TFNG and air force colonel Dick Covey, had yet to command his first mission. Others were livid that Pinky Nelson had been assigned to the flight. While Pinky was well liked, he had taken a sabbatical to the University of Washington after *Challenger*. The rest of us had stuck around to do the dog work and be brutalized by Young in the process. In our minds Pinky hadn't paid the dues to have received such a prize as the first post-*Challenger* mission. It was also a sore point that his last mission had been the flight prior to *Challenger*, so he had the additional plum of having back-to-back missions. Norm Thagard was certain Abbey had picked Nelson just to show the rest of us how unfair and capricious he could be. I recalled a line from McGuire's astronaut leadership document, "Inconsistency, ambiguity, silence, evasion . . . all have their place in his studied unpredictability."

There were other aspects of this crew selection that would have angered us even further had we known about them. Years later, at our TFNG twentieth-anniversary reunion, Rick Hauck would tell me that Abbey had allowed him to select Dick Covey as his pilot. No other TFNG commander I ever spoke with had been given that responsibility. Abbey had always named the mission crews, the CDRs, the PLTs, the MSes, everybody. Hauck also revealed he had been told six months prior to the press release that he would command the return-to-flight mission but had been sworn to secrecy by Abbey. I wondered how many times during those six months other hopeful commanders had been in Rick's company wondering aloud who would command STS-26, and Rick had pretended to wonder with them. Deep secrecy. It was Abbey's style and it was killing astronaut morale.

My winter of discontent continued. As we had anticipated, the lightweight SRB program was canceled and, along with it, all Vandenberg AFB shuttle operations were terminated. I would never see polar orbit.

Challenger's wreckage—*all of it*—was sealed in a pair of abandoned Cape Canaveral missile silos. It was another head-shaking moment for me. Pieces of the wreckage should have been retained for permanent display in key NASA locations as reminders of the cost of leadership and team failure. At a minimum, displays of the wreckage should have been placed at NASA HQ and in every NASA field center's headquarters' building. The LCC and the MCC buildings needed a similar display. Even the astronaut office should have been the site of an exhibit. Other astronauts agreed. I heard Bob Crippen remark that every astronaut should be *required* to view the wreckage before it was sealed away, adding, "I don't think some of the civilian astronauts yet appreciate the risks they are taking when they climb into a space shuttle." But no such displays were established. *Challenger*'s broken body was sealed away as if the very sight of it was somehow obscene.

I continued to be beaten up by John Young any time I had anything to say about range safety or pre-MECO OMS burns. Every Monday rumors of his and Abbey's imminent removal swept through the office like blue northers out of the panhandle. But come Friday, nothing had changed. A good night's sleep had long become a memory. I would get up at weird hours and take walks or go for a run. Donna and I talked ad nauseam about leaving NASA. I had my twenty years with the air force. I could retire from it and NASA, go back to Albuquerque, and get a job. But every time I thought of giving up the T-38, of never hearing, "Go for main engine start," of never again seeing the Earth from space, I would get angry. I was doing my job. I was doing a good job. Why should I be driven away for that?

In the spring of 1987, I got a temporary reprieve from astronaut frustrations. With the shuttle grounded for at least another year, the USAF decided it would be a good time to reacquaint their astronauts with air force space operations. The navy planned to do the same for their astronauts. Both services referred to the program as a "re-bluing," a reference to the fact we would be back in our blue military uniforms. We would travel to various United States and overseas bases to be briefed on how military space assets were being used to counter the Soviet threat.

When word of this program reached the civilians in the astronaut office, one particularly bookish scientist challenged the fairness of it. "If the air force and navy are sending its astronauts on a re-bluing, what is NASA going to do for us civilians?" Mark Lee, an air force fighter pilot, looked at the whiner and replied, "You guys are going to get re-nerded."

West Berlin was the best place to get eyeball to eyeball with the enemy, so the air force flew us there. This was 1987 and the infamous Berlin Wall still had two years of life left in it. We attended various classified briefings and got a helicopter tour of the Iron Curtain, flying over death strips guarded from watchtowers and barricaded with razor wire.

One evening we donned our uniforms, passed through a border checkpoint, and walked into East Berlin for supper. The city was still considered occupied and the military personnel of the occupying countries could pass into one another's zones, although it was a one-sided passage. The East didn't allow their troops into the West, knowing they would never come back.

In our walk from West to East we traveled back to 1945. Color had yet to come to this part of the world. Everything was gray and drab, even the clothing of the women. Remote-control TV cameras mounted on buildings watched us and other pedestrians. The streets were heavily patrolled by Kalashnikov-toting East German and Soviet guards. They glared at us like we were the enemy, which, of course, we were. As we passed one pair of guards, I pointed to a medal on my chest and said to John Blaha (class of 1980) in an intentionally loud voice, "And I got this one for killing ten commies." The hostile expressions of the guards didn't change. Apparently they didn't speak English, which was probably a good thing for me.

Our air force host led us to his favorite East Berlin restaurant. I was prepared to be disappointed, but the place was clean, brightly lit, and staffed with young and beautiful East German fräuleins. As we entered, the rest of the patrons, all East German and Soviet military officers, gave us their best game face. We ignored them. Several tables were shoved together to accommodate our entourage and we got down to the business of drinking. We were soon a rowdy spectacle for the rest of the crowd. They stared at us with disapproving expressions, as if laughing and smiling were forbidden in the workers' paradise.

Later in the evening an intoxicated John Blaha grabbed a vase of daffodils and began to peer into each bloom with the focus of a horticulturist. I wondered if he had slipped into alcohol poisoning, but he whispered to me, "I'll bet the KGB has bugged this vase. They're probably in a back room listening to everything we're saying. Well, I'll give them something to think about." He lifted the flowers to his mouth like a microphone and began to speak loudly into their blooms: "Mike, wasn't that briefing about our new F-99 Mach 7 fighter really interesting?" Then he handed the vase to me.

I joined in the fun. "Yeah, and to think Mach 7 is its *single*-engine speed."

The others at the table picked up on our disinformation campaign and the vase of flowers went from hand to hand while the rest of our group made even more outrageous claims about secret weapon systems we had recently seen or flown. Meanwhile, the humorless commie diners stared at us as if we were mad. Since we were talking into daffodil blooms, I could understand their bewilderment.

When the vase finally made it back to Blaha, he closed the floor show by speaking into it in an exceptionally loud voice. "Why is it that visiting Soviet basketball teams never play the Celtics or Lakers? Whenever they come to the USA they always play some piss-poor university team. What are they . . . pussies?" We all wondered how that would translate back in the Kremlin.

Imagine my shock when, several months later, Blaha ran into my office with a newspaper article describing how the Soviets, for the first time in history, were going to allow their basketball team to play an exhibition game with an NBA team. "I told you that vase was bugged," Blaha shouted. We laughed at the image of an army of KGB spies hunting for that F-99 fighter.

Our journey into the heart of the enemy camp wasn't the highlight of that evening. Back at our hotel, four of us donned our bathing suits and headed for the sauna. There we encountered a middle-aged fräulein with a Mr. T physique who handed us towels and shower clogs and then pointed to our suits and said, "*Nein.*" The suits were not allowed. It was a nude spa. We exchanged a few self-conscious glances. But there were no other females present and only a saliva test would have confirmed our receptionist's gender. We stripped. What a photo that would have made . . . four of America's heroes marching to the sauna like we marched to our space shuttles, except we were marching completely bare-assed. We opened the door and entered a steamy room. When our eyes adjusted to the dim light we realized we were sitting with a half dozen naked women. The spa was coed. Oh well, when in Rome . . .

Later I was climbing out of a small pool when a very attractive and very naked German woman came to me. Someone in our group must have dropped the astronaut bomb because she wanted to ask a few questions about flying in space. I could barely understand what she was saying . . . not because her English was poor. On the contrary, it was excellent. Rather, it was because 99 percent of my meager mental powers were being used to force my eyes to look straight ahead. As she spoke, my brain was screaming, "*Don't look down! Don't look down!*" I felt it would be a serious breach of naked etiquette to talk to her breasts, something we denizens of Planet AD regularly did with clothed women. Given my struggles it was a wonder I could form a coherent sentence.

Meanwhile, as I did my best to be a naked gentleman, I noticed she had no qualms about looking at *my* body. As she spoke her eyes wandered up and down as if she were appraising a cut of beef. I felt *so* violated.

Even the naked ladies weren't the most memorable part of our rebluing trip. Events five thousand miles away trivialized everything we had encountered. We received word from Houston that John Young's tenure as chief of astronauts had ended. He had been reassigned to the position of JSC deputy for engineering and safety, a technical rather than team-leadership position. The celebration was immediate. Most of us had been looking forward to this day for a long, long time. My celebration was probably the most unrestrained. For the past year, John had made my life miserable. While I had heard of only two incidents in which he had suggested I was lacking as an astronaut and should be replaced, God only knew how many other times he had said it and to whom he had said it. Despite Abbey's "forget it" comment, I couldn't believe my reputation hadn't been damaged. Young had been my tormentor, and my joy at his departure was unalloyed. That's not to say I couldn't admire the man for his achievements in the cockpit. He had flown in space six times, including a moonwalk mission and the first space shuttle mission. The latter had probably been the most dangerous mission ever flown by any astronaut. While many of us questioned John's leadership abilities, no one doubted his flying skills and guts.

On April 27, 1987, TFNG Dan Brandenstein was picked to replace Young. I knew he would do a superb job as chief of astronauts. But at the same time I was angry that Abbey had screwed the air force again. The grapevine had it that the selection criteria for the position had mandated a TFNG who had flown as a shuttle commander. There were three navy TFNGs who qualified: Brandenstein, Hauck, and Hoot Gibson. There was only a single USAF TFNG veteran commander: Brewster Shaw. And why did such a disparity exist? Because of Abbey's longtime preferential treatment of the U.S. Navy astronauts. If a bomb went off under Abbey's car, the air force TFNGs would be at the top of the suspect list.

CHAPTER 30

Mission Assignment

With Brandenstein at the helm of the astronaut office, the summer of 1987 passed much more pleasantly. At the Monday meetings there were actual exchanges of ideas. Astronauts, me included, were able to get up and make a presentation without being blasted with criticism. Dan even addressed one of the criteria for crew assignments, a first in my nine years with NASA. "Crews will be picked not only on how they have performed in simulations and on past missions, but also on how well they perform their office duties." To imagine . . . someone in a management position at NASA was actually revealing something about the crew selection process. It was enough to make me want to step outside and see if a squadron of pigs was flying over. Actually, what Dan gave us wasn't much . . . and *couldn't* be much because Abbey was still God. But he was doing his best to be a real chief.

The days weren't all sunshine and roses. Along with the rest of the office, I remained in flight assignment limbo. Also, STS-26 was slipping into the summer of 1988, a year away. If and when I ever got another mission, it was moving in lockstep to the right, too.

During this period of recovery from *Challenger,* Abbey pressed ahead with a previously scheduled new astronaut class selection. Every astronaut, and probably every other thinking person in NASA, thought it was insane to be selecting another group of astronauts when it was obvious the future shuttle flight rate was going to be a fraction of what it had been. Why bring more superachievers into certain frustration? Astronauts speculated that Abbey wanted more people to expand his empire. Whatever Abbey's motivations, the selection was made and another group of fifteen astronauts, the class of 1987, walked into NASA that summer.

At an Outpost Tavern welcoming party for this class, I ended up alone with George. I turned from getting a beer and he was approaching me with purpose. *Uh-oh,* I thought. *I sure hope he doesn't ask me about a document bearing Dr. Terry McGuire's name.* I was still terrified that Abbey had hidden cameras around JSC, or had somehow put a homing device on all of us so he could keep track of where we went and who we talked to. Maybe he had listening devices in every office, including the ones that McGuire used. I regretted ever having seen that astronaut leadership document. Whether I liked it or not, it made me a co-conspirator in any possible plots against him.

From his mumbles I thought I heard "How are you doing, Mike?"

"Fine, George." My heartrate was at *Go for main engine start*–speed. That's what happens when God is speaking to you and you're hiding a mortal sin.

"Are you going to be around this week?"

Here it comes, I thought. He wanted to see me in his office . . . with McGuire's treatise in hand. I was ready to blurt out, "I'm innocent! I didn't have anything to do with it! McGuire wrote it before I ever spoke to him. The others are evil, not me. Kill them. Mercy, my liege, mercy!" But all I croaked was, "This week? Well, yeah . . . I'll be here." At the moment I was very glad George never made eye contact with his audience. If the conversation continued in the direction I thought it was going, I wouldn't have to worry about him discovering any lies in my eyes. We were both talking to our shoes.

"That's good. There are some things we need to discuss."

Oh, God. I'm screwed.

Abbey continued. "The SRB testing is going well. More flight assignments will have to be made. We'll need to talk about that." I almost dropped my beer. The topic of conversation wasn't McGuire! While I couldn't be certain (nobody could be certain with Abbey about anything), I sensed he was teasing me about an imminent flight assignment. I looked at him and sure enough there was a coy smile on his face. He was actually relishing his godly role as the bearer of good news.

I immediately went to Donna to tell her about the exchange. I could see she was conflicted. She was happy that I might be on the verge of drawing a second space mission, but terrified I would die flying it. Several of the *Challenger* widows were at the party and every spouse, Donna included, was watching them and thinking, *That could be me.*

The next week I sat in my office, snatching up the phone on the first ring hoping to hear Abbey's voice, but the call never came. My paranoia began to ratchet upward. Maybe I had read too much into Abbey's words. Maybe the coy smile I thought I had detected had been nothing more than a gas pain grimace. Maybe George knew of my treasonous McGuire visit and was playing with me.

The week after also came and went with no call, and I was certain I had been toyed with. If a bomb went off under his car now, I would be alone at the top of the suspect list.

Finally, on September 10—my forty-second birthday—I landed from a T-38 mission and found a note on the crew lounge door asking me to call Abbey . . . at home. I was sure this was the call in which I would learn of my assignment to a second mission. Why else would Abbey want me to disturb him at 10:15 P.M.? What a birthday present this was going to be! I dialed the number.

But it was another disappointment. George acted as if there had been no reason to call him at home. All he wanted to know was if I had seen a letter written by a New Mexico congressman on the shuttle program. I was certain, now, that Abbey was the cat and I was the crippled mouse. He was playing with me. There was no pending flight assignment.

On Saturday night I was able to momentarily forget about mission assignments. The class of 1987 hosted its first party and provided some great escapism entertainment in the form of a skit modeled after the TV show *The Dating Game*. Dan Brandenstein played the eligible bachelor. He was onstage and screened from several women . . . or rather class of 1987 men in drag, who were vying for his affection. The only real female participant in the skit was Mae Jemison, the first black woman astronaut. She was introduced as "celebrity host Vanna White." I'm sure Johnny Cochran could have found a lawsuit in that. One of the men in drag was new astronaut Mario Runco. Imagine a tall, muscular Klinger from *M*A*S*H* and you have an image of Mario. He had a classic Roman nose, a perpetual five o'clock shadow, and a regional New York accent—Mario spoke Bronx. For the skit he squeezed into black fishnet stockings, a low-cut dress, and high heels. It was an ensemble that revealed enough hair to have generated a Sasquatch sighting. He was, without question, the ugliest drag queen to have ever put on lipstick.

The class of 1987 gave Dan the list of questions he was required to ask of the prospective dates. Since the military personnel from the new class were also from Planet AD, many of the questions were sexually suggestive. One was an obvious play on the psych questions being asked in the astronaut interview. Apparently those hadn't changed in the past decade. "If you died and could come back as any animal, what would it be?"

Mario appeared to fall into deep thought on such a complex question. Finally he answered, "I would like to come back as . . . a beaver." As if the double entendre needed emphasis, he casually spread his legs. It was a move Sharon Stone would make famous years later in the movie *Basic Instinct,* but Mario did it first. It was also a move that is irrevocably burned into the synapses of my brain, where memories of my Most Terrifying Sights are stored. Even today, when I look at a blank white wall, I see that hair-way up his skirt and shiver in terror.

The remaining questions and answers were scripted to ensure Dan selected Mario's character as his date. When Mario came from behind the screen, he went to Dan, grabbed him, twirled so that his back was to the audience, and planted a kiss on Dan's lips . . . or so it appeared. Actually he clamped his hand over Dan's mouth and kissed the back of it. Mario was a hell of a thespian.

The skit continued with a "word from our sponsor." Two members of the 1987 class came onstage dressed as the hayseed spokesmen for Bartle & James wine coolers. The real B&J television advertisements were laugh-out-loud funny. They featured one character with a boring, monotone voice explaining some bizarre use of the product beyond its intended purpose as a beverage. As he did so, his doofus-looking silent partner, Ed, would give a demonstration in the background.

The B&J advertisement the class of 1987 presented was definitely not ready for prime time. One astronaut adopted the deadpan voice and mannerisms of the B&J protagonist and explained how the wine coolers could be used to prevent the spread of STDs. Silent Ed rolled a condom onto a B&J bottle and vigorously shook it. The carbonation in the drink inflated the latex into its hotdog shape. Ed peered closely at the phallus, searching for leaks. As if that weren't suggestive enough, the advertisement spokesman continued, "The alcohol in Bartle & James wine coolers can also be used to disinfect body parts that might be exposed during intimate relations." Ed used that as his cue to pour some of the B&J into his palm and splash it on his face like aftershave. Political correctness might have subdued the office parties of the rest of the country, but it had yet to wet-blanket astronaut parties.

The following Monday I walked into my office still thinking about the skit. It had been a great party and I intended to tell the new arrivals how much I enjoyed their antics. But those thoughts evaporated when I arrived at my desk. A note from my secretary read, *Please meet Dan Brandenstein at 8:15 A.M.* My office mate, Guy Gardner, had the same note on his desk and I quickly discovered three other astronauts were also notified of the meeting: Hoot Gibson, Jerry Ross, and Bill Shepherd. With two pilots and three MSes, the notification certainly suggested a flight assignment announcement. But I wasn't about to cheer yet. John Young had never announced flight assignments. That had always been exclusively Abbey's job. The fact that Dan Brandenstein's office, and not Abbey's, had called put a lid on my simmering anticipation. There were certainly other things Dan might want to see us about. Again, I prayed it wasn't anything associated with Dr. McGuire, as in, "Which one of you idiots has been talking to the shrink?"

We walked into Dan's office. It was still strange to see a TFNG in the big-time. As a navy pilot, Dan had been firmly in the grip of Planet AD's gravity. No more. His new management position had blasted him to escape velocity. We would all miss him.

Dan welcomed us with a smile, which I immediately interpreted as a good sign. "Abbey wants to see you guys. I'll walk over with you." There it was, the Abbey connection. More and more it was looking as if September 14, 1987, would be a special day for me. As we walked to the JSC HQ building my heart was a-flutter. It had been three years since I had stepped from *Discovery*. By far, the last twenty months had been the worst in my life. I had buried four TFNG friends killed in a preventable tragedy and had endured John Young's abuse. I couldn't wait to get back in space. *Please, God,* I prayed, *let this be what I think it is.*

Abbey, too, was ready for us with a smile. After a moment of small talk he relieved our suspense. "I was wondering if you guys would like to fly STS-27?"

Is a crab's ass watertight? was one rejoinder that came to my mind. *Hell, yes* answered both questions.

Our group immediately broke into jokes and giddy laughter. No one really answered Abbey's question, but, of course, we didn't have to. He was offering us gold and nobody ever turned that down. I was now officially a crewmember for the second post-*Challenger* mission. It was a classified Department of Defense mission so nobody yet knew exactly what we would be doing, but it didn't matter. We were an assigned crew. That was *all* that mattered.

As I floated in weightless joy back to my office, I considered for the billionth time that strange man known as George Washington Sherman Abbey. He defied analysis. To borrow a quote from Winston Churchill, George was "a riddle wrapped in a mystery inside an enigma." It seemed he went out of his way to drive astronauts to loathe him. Even in this STS-27 crew assignment some would be rightly embittered. Bill Shepherd was class of 1984 and would be flying his first mission before two mission specialists from the class of 1980, Bob Springer and Jim Bagian, would fly their rookie flights. And STS-27 would mean Hoot Gibson would be flying his second mission as a commander before eight other TFNG pilots had yet to command their first mission. The STS-27 crew assignment press release was going to be a bitter pill for many in the office to swallow.

Hoot would later tell me Abbey had informed him several weeks before the official announcement that he would be the CDR of STS-27. Hoot had replied, "George, it's not my turn." Abbey had said, "Turns have nothing to do with it." He might as well have said, "I don't give a shit about astronaut morale." The statements were identical.

While sitting in Abbey's office, though, I had never seen him as jolly as he had been while telling us of our new mission assignment. It was as if he was high on our happiness. Why couldn't he understand it could be like that 24/7/365? All he had to do was understand that turns did matter, that visibility into flight assignments mattered a hell of a lot, that open communication mattered, that being positively stroked once in a while mattered . . . hell, being *negatively* stroked once in a while, getting *ANY* performance feedback once in a while, mattered. During those ten minutes in his office I loved George Abbey, but the moment passed. Now, if there were conspirators somewhere in NASA's hierarchy preparing to strike, I wished them all the luck in the world.

That evening, as I told the kids about the flight, my sixteen-year-old daughter, Laura, said, "You're not going to die on me, are you?" She said it with a smile, trying to make a joke out of it—a chip off the old block—but I knew she was worried. So were Donna, Pat, and Amy. And I knew, as soon as STS-26 was on the ground, I would be worried. Just as it had been with STS-41D, I knew Prime Crew night terrors awaited me for STS-27. But I had to do this. I couldn't stop or turn away from a flight into space any more than a migratory bird could ignore the change of seasons. It was in my DNA, beyond rational understanding.

CHAPTER 31

God Falls

We were the last crew ever to receive news of our mission assignment from George Abbey. On October 30, 1987, George was reassigned to NASA HQ in Washington, D.C., to assume the job of deputy associate administrator for spaceflight. Hoot would later tell me Abbey had hinted he didn't want the "promotion." That was easy to believe. For ten years George had wielded enormous power at JSC and now it was being taken from him. His loftier-sounding HQ title came with about as much power as one of those twenty or so vice president positions on the staff of a local bank. Every astronaut was of the opinion that this had been an assassination. But by whom? Many astronauts suspected Admiral Dick Truly. Dick was now the number-two man at NASA HQ and being groomed for the NASA administrator position (which he would assume in 1989). As a former astronaut, he certainly knew of George's leadership style, so he had motive. I wasn't about to ask Dr. McGuire if he had a hand in it. I doubted he would have told me and, besides, I didn't want any further association with the matter. It was the kind of office intrigue that could only hurt a career.

I didn't struggle with my emotions when Young was removed, but, with Abbey, it was different. Abbey had raised me from the huddled masses to be an astronaut. He had picked me to fly on two space shuttle missions. Besides my own father, no other man had influenced my life as much as George Abbey. And any unbiased outsider would say he had treated me fairly. He had selected me to fly my first mission before seven other TFNG MSes got their rookie rides. He had selected me to fly my second mission before eight of my peers got their chance at a second mission. With STS-27 I was assigned to an important flight with a robot arm task. Yes, I was definitely conflicted about Abbey's bureaucratic demise. I was disturbed enough to wonder if it had really been *my* problem all along. Maybe a tougher man could have accommodated Abbey's Machiavellian managerial style. But deep down I knew that wasn't the case. McGuire had said I wasn't the only astronaut to have come to him, and given the time it would have taken him to acquire the data and write his astronaut leadership document, he must have been talking to astronauts who had "lost it" long before I did. Even as I questioned my emotional mettle as the source of my Abbey problem, other astronauts were celebrating. One commented, "If we had any lamp shades in our offices we'd be running up and down the halls with them on our heads." Another celebrant was heard in the hallway singing, "The wicked witch is dead. The wicked witch is dead." Still another bitterly offered, "I hope one day to see Abbey working as a grease monkey on the flight line at the Amarillo airport. It's all he's good for." I doubted there were any astronauts, even those who had benefited most from his largess, who were sorry to see Abbey go.

No, the problem wasn't mine. But, still, I didn't like the way I felt. I had wanted to love this man unconditionally. Like Donna, he had been integral to my dream fulfillment. If somebody else had been leading astronaut interviews in October 1977, would I have made the cut? I doubted it. Just as a different wife would have meant a different life, I suspect a different chief of FCOD would have had a different criterion for astronaut selection, most likely one I wouldn't have met. I had wanted to render Abbey a lifetime of fealty. At our welcome to JSC in 1978 there had been no more loyal TFNG on that stage. But over the years, his Stalinist-like secrecy, his indifference to the fear that dominated the astronaut office, his unfairness to the air force pilots, his gross inequities in flight assignments, and his abysmal lack of communication had drained my allegiance completely.

As the office celebration continued, one astronaut commented, "Until someone drives a wooden stake through his heart, I won't believe he's really gone." The comment proved prophetic. Abbey's JSC days weren't over by a long shot. He used his time at HQ to ally himself with Dan Goldin, who, in 1992, would become NASA administrator. In 1996 Goldin appointed George director of Johnson Space Center, arguably the most powerful position within NASA. Ultimately, George got it all, proving what every TFNG had believed for so many years, that

Abbey was unsurpassed in his ability to manipulate a byzantine organization like NASA's. It was a talent the CIA could have employed. If, during the darkest days of the Cold War, they had parachuted Abbey into Moscow's Red Square, naked, not knowing a word of Russian, and without a single kopeck in his hand, George still would have become a member of the Politburo within a year and Soviet premier within two. We could have ended the Cold War decades earlier.

On December 5, 1987, the astronaut office held a going-away party for Abbey at Pete's Cajun BBQ. It was billed "George Abbey Appreciation Night." About a third of the office were no-shows. Two astronauts told me they were boycotting the event. I suspected some of the other absentees were of a mind that they would rather fly a night TAL into Timbuktu than show any appreciation for George. I attended merely to celebrate the fact he was gone from our lives. There would be no weepy singing of "*Auld Lang Syne*" at this party. Mark Lee, class of 1984, was the MC and did a terrific job, donning a buzz-cut wig to mimic Abbey. It was obvious Mark was of the opinion that George was forever powerless over astronauts, because his humor was "I'll-never-see-this-man-again" acidic. He made reference to Abbey's preferential treatment of the navy astronauts, of Dick Truly's suspected role in removing him from the FCOD position (Dick Truly and the new JSC director, Aaron Cohen, were both noticeably absent from the event), and of George's tendency to suffer episodes of narcolepsy when he didn't want to hear an opinion contrary to his. Then, to everybody's astonishment, Mark took off his wig and grew sentimental. "You can't have a boss for so many years and not be choked up about his departure." I couldn't believe it. Others around me held similarly incredulous looks. *Choked up about George's departure?* Only if one of Pete's Cajun rib bones got caught in my throat. But Mark continued his endearing comments, glibly flowing to a conclusion, "And for those of you who might be feeling a lump in the throat and getting all misty-eyed thinking about George leaving . . . just remember what an asshole he can be!" Mark had flown well beyond the edge of the envelope. Everybody cheered and applauded. Abbey smirked. What was behind that smirk? I wondered. It could have been anything from *I'm proud of these guys* to *I'll get even with these traitorous dickheads if it's the last thing I ever do.*

Abbey was dead to me. The other significant man in my life, my father, would die unexpectedly in his sleep a few months later—February 10, 1988. He was sixty-six. He had lived exactly half his life with legs and half without. Only in the dim recesses of my mind, far back to my eighth and ninth years of life, could I see him standing next to me on rock-solid legs showing me how to turn into a fastball and lay down a bunt. Those memories were all but obliterated by the thirty-three years of images from his post-polio life. In those, I always see him in his wheelchair, as if the device were an organ of his living body. And in those scenes I see a man who stood taller than most of us will ever do on two good legs. My dad was my hero.

I chose to drive the eight hundred miles to Albuquerque for his funeral so I would have some quiet time for reflection before being plunged into the administrative details of his death. Two of my brothers lived in Albuquerque so my mom had plenty of help to deal with the situation, though I knew she wouldn't need it. Grief didn't stand a chance with my mom. Two days later, when I entered the house, she pulled me aside and said, "Mike, I loved your dad very much, but you won't see me cry. It's not in a Pettigrew to cry." And she didn't.

Like many children who experience the sudden loss of a father, I regretted all the things I hadn't said while he had been alive to hear them. I wished I had told him how much it had meant to me that he had stayed in our kid-games even after polio. My brothers and I would carry him to the mound and he would be our permanent pitcher in games of baseball. He would referee our driveway basketball games. He would play water polo with us, his shrunken legs trailing in his wake like two ribbons of flesh.

He had endowed me with an easy and spontaneous sense of humor and I wanted him back to hear me say, "Thanks, Dad." I could recall him sitting in his wheelchair, fully clothed, smoking his pipe and deciding it would be fun to go swimming. He bellowed, "Banzai," and raced his wheelchair straight into the pool. It sank to the bottom, where he sat for several seconds with his pipe still clamped in his teeth. My brothers and I laughed at his underwater pose. We begged him to do it again and, after hauling him and the wheelchair from the water, he did.

I wished for a chance to tell of the thrill I felt watching my rubber band–powered gliders soaring across the desert. Dad had shown me how to build them and I could still see his hands, made huge and calloused from the use of his crutches and wheelchair, guiding mine in the sanding of the balsa-wood propeller spinner.

I wanted him back to tell how much it had mattered that he had kept me home from school to watch Alan Shepard ride his candle into space. I wanted to tell him thanks for driving me to Kirtland AFB, where I watched the weather officer launch his radiosondes. Dad would then hurry me home so I could follow the helium balloon with my Sears telescope. I would hope against hope the instrument package hanging beneath it would parachute into my yard. Just the thought of touching something that had been in the stratosphere would turbo-boost my heart.

I remembered him buying me my first rocket—a plastic device powered by vinegar and baking soda. And then there were his gifts of Tinkertoys, chemistry sets, Erector Sets, a Heathkit crystal radio, and my *Conquest of Space* book. I ached to tell him how all of those had empowered my dream.

Dad's death unleashed long-dormant memories of the minutiae of my childhood. By itself each memory seemed inconsequential but connected together they revealed my pathway into space. George Abbey may have selected me as an astronaut but my dad made me one.

Dressed in military uniform, with his medals and wings on his chest, Dad was laid to rest in the Veterans Cemetery in Santa Fe, New Mexico. A rifle salute was fired and the honor guard lifted the American flag from the casket, folded it, and presented it to my mother. "Taps" sounded as I came to attention and rendered a salute to the greatest man I have ever known. Tears finally came to my eyes. There are some things that will make even a Pettigrew cry.

Later, I would return to his grave and place the mission decals of STS-41D, STS-27, and STS-36 on the marker. Each patch had the name "Mullane" on it. They were Dad's missions, too.

CHAPTER 32

Swine Flight

STS-27 was a classified DOD mission. I wouldn't be able to share much with Donna. I had entered the "black" world of the Cold War, where I would be taking trips to locations I couldn't discuss. I would study checklists in an underground vault. At parties Donna wouldn't be able to ask our contractors and support team about their work. She wouldn't even be able to ask in what city they worked. To complicate the spying efforts of Russian ships, the launch date wouldn't be announced until twenty-four hours prior to the planned liftoff. That little detail would seriously complicate family travel arrangements. As Donna would later say, "It's like making plans for a wedding where the date is kept secret." The mission photography would be classified. There would be no photos of me with my payload. When asked what the mission was about, I would have to borrow a line from *Top Gun*: "If I told you, I'd have to kill you." (Four years after the mission some aspects of it were declassified. I can now say I used the robot arm to deploy a classified satellite into space. I am forbidden to describe the satellite or its intended function.)

I was thrilled with my crew. Hoot Gibson was a natural-born leader. He didn't micromanage as some commanders did. (One was known to reach completely across the cockpit to make a switch change rather than allow the crewmember at that position to do it.) Hoot gave each of us our duties and set us free to be creative to get the job done. He was also a blood brother from Planet AD. The office secretaries quickly named STS-27 "Swine Flight," and gave each of us strap-on novelty pig snouts because of our animal "snorting" sounds whenever an attractive woman came within eyeshot (as in, "I'd like to snort her flanks").

Guy Gardner, Jerry Ross, and I had trained together on the canceled STS-62A mission so we were already teamed. Rookie Bill "Shep" Shepherd was a soft-spoken, powerfully built Navy SEAL who specialized in underwater demolition. Like Hoot and I, Bill was from Planet AD. He was also a bachelor astronaut, which meant he had achieved a higher state of earthly rapture than the Dalai Lama.

Everyone was excited to be doing something warriorlike on the mission, instead of commercial or scientific. It felt good to don a military uniform again and pose for our crew photo. We were going to stick it to the godless commies in space. There were no press releases on our mission preparations. The air force wanted us to remain as invisible as possible, which proved easy to do. Rick Hauck's STS-26 mission, aka "The Return to Flight Mission," was so hyped by NASA that it provided a very dark shadow in which we could hide. But the overarching importance attached to STS-26 grated on us and the rest of the astronaut office. We felt Rick and his crew were wearing their fame too conspicuously, which was a grievous violation of astronaut commandment number two, "Thou shalt not glory in public adoration." While the press marveled that anyone could be so brave as to fly on the first mission after *Challenger*, every astronaut knew it would be the safest space mission ever flown. Not only had the SRBs been completely redesigned and retested, but every shuttle system had been put under a microscope, and appropriate changes had been made. Also, the STS-26 mission objective was relatively trivial, the release of a TDRS communications satellite, something that had been done several times in the past. Hoot called it "The Quiche Mission." Another AD pilot observed, "We even let the girls release TDRSes." But it was obvious the STS-26 crew thought their mission was the most important spaceflight since Angel Gabriel flew to the Virgin Mary. We all wearied of listening to their Monday morning pontifications on the criticality of STS-26 issues. When the limits of forbearance were finally exceeded, the glorious "Return to Flight" crew became a target of satire for the invisible "Swine Flight" crew.

The first public act of rebellion occurred at an astronaut reunion party. Shep and I were sitting at a bar when the helium balloons anchored as table decorations caught our eye. We grabbed a brace of these, ripped the nozzles open, and inhaled the gas. With squeaky falsetto voices we wandered through the audience introducing ourselves to legendary astronauts from the Apollo program. "Hi, I'm Rick Hauck, commander of the STS-26 crew. Would you like my autograph?" Meanwhile Buzz Aldrin, Pete Conrad, and other celebrity astronauts looked at us with expressions reading, "The astronaut corps has sure gone to hell." Again and again we would run back to the bar for a swallow of beer and a hit of helium, and then it was off to another moonwalker. During one of these refills Shep must have gotten some bad gas and experienced a flashback to some combat event. His eyes glazed over and he fell into a thousand-yard stare and then, without provocation, he grabbed the collars of my golf shirt and ripped it open. I glanced around to ensure there were no knives on the bar, then retaliated. I grabbed his shirt and ripped it apart. A handful of TFNGs gathered around to watch me die. The 145-pound weakling had just kicked sand in the face of a knife-skilled SEAL. Fortunately for me, Shep's post-traumatic stress passed quickly and he merely laughed at the tatters of his shirt. We drained our beers, took another helium hit, and headed back into the audience. I found Jim Lovell and in a Donald Duck voice repeated my lie, "Hi, I'm Rick Hauck, commander of the STS-26 crew. Would you like my autograph?" Lovell looked at me as if I were a derelict.

Swine Flight's most outrageous assault on the sanctity of the STS-26 mission came after a fund-raiser for a *Challenger*-related charity. This was a black-tie affair and most astronauts and spouses were present. The venue was the downtown Houston performing arts center, the Wortham Center, and hundreds of local dignitaries and their spouses were in attendance. As the program drew to a close, the master of ceremonies brought a young girl on the stage to sing Lee Greenwood's popular song, "I'm Proud to Be an American." As she was belting out this arrangement at 150 decibels, the MC screamed into his microphone, "Ladies and gentlemen, I give you the crew of STS-26, the Return to Flight Mission!" At this cue the orchestra pit platform began a slow rise. Artificial smoke swirled about it and spotlights flashed through the vapor. And there, to the astonishment of every astronaut, were Rick Hauck and Dick Covey. They stood like carvings on Mount Rushmore: chins jutted out, chests puffed up, arms rigidly at their sides, steely eyes straight ahead. The public crowd around us went wild with their cheers and applause. You would have thought the platform bore Jesus Christ, Himself. Meanwhile every astronaut and spouse wanted to vomit. Shep looked at me and made a finger-in-the-mouth gagging pantomime. *What next,* we wondered—*the STS-26 crew driving to the launchpad in a convoy of pope mobiles, each man waving clinched hands over his head in self-congratulations?* It was too much.

The very next evening our crew had a party during which, no surprise, the favorite topic of conversation was Rick Hauck's ascension into heaven. Shep plotted a "let's get 'em" mission with the same intense focus a SEAL might plot to blow up an enemy fortification. Early Monday morning he and I smuggled two fire extinguishers into the astronaut conference room. We wrapped them in our jackets and placed them directly behind Hoot Gibson's and Guy Gardner's seats. Jerry Ross brought a tape player loaded with Lee Greenwood's "I'm Proud to Be an American." Hoot and Guy secreted black bow-ties in their pockets.

After Rick covered his STS-26 issues, Brandenstein asked Hoot if he had any STS-27 items to discuss. That was Jerry Ross's cue to trigger Greenwood's tune. Shep and I fired off the fire extinguishers for the smoke effect and Hoot and Guy clipped on their ties and slowly rose from their chairs in a mimic of Hauck's and Covey's rise from the Wortham Center orchestra pit. The conference room exploded in laughter. There was thunderous applause. I looked at Rick. He had a smile on his face but his flexing jaw muscles said more about what he was really feeling. He had just been lampooned and was dying to issue a rebuttal, but he knew he couldn't. To do so would be a violation of astronaut commandant number three, "Thou shalt not show a weakness." A three-legged gazelle limping across the Serengeti would survive longer than an astronaut exhibiting a wounded ego among his peers.

As our STS-27 training progressed we were introduced to a new shuttle design feature, a bailout system. It wasn't what we had hoped. The best design would have had the entire cockpit being blasted away to parachute into the water. But this option would have required a complete redesign of the orbiter and there was insufficient money for that. Our second preference had been ejection seats. The shuttle was originally designed to include two of these for the two astronauts flying the first test flights. But two ejection seats were all that would fit in the upstairs cockpit and none could be added to the mid-deck. While it would have been relatively easy to reinstall the two upstairs seats, such a modification was also rejected. No mission specialist was going to climb aboard a shuttle in which the two pilots had the only escape capability.

With the elimination of a cockpit pod and ejection seats as potential escape systems, the engineers gave us the only thing they could give us, a backpack parachute. We would jump out the side hatch just like B-17 crewmembers did in WWII. Good freakin' luck! We'd be juiced against the wing like a grasshopper on an automobile windshield. But the engineers had a solution to get us clear of the wing—tractor rockets. A bundle of small rockets would be installed in the cockpit above the side hatch. After blowing the hatch, astronauts would lie on their backs on a table in the hatchway, attach their harness to a rocket, and then pull a lanyard, which would fire a mortar, hurling the rocket outward. A cable connecting the rocket to the astronaut would unreel for twenty or so feet before the rocket would ignite and then jerk the astronaut by the scruff of the collar out of the hatch and clear of the wing. When astronauts saw movies of this system being tested with anthropomorphic dummies there was grim laughter. It looked like something Wiley Coyote had ordered from Acme Rocket Company to catch that speedy Roadrunner. Fortunately, a more practical design was adopted from a suggestion by flight surgeon Joe Boyce—a slide pole. A banana-shaped, telescoping pole was installed on the ceiling of the mid-deck cockpit. After blowing the hatch, astronauts would throw a handle that would release springs to slam the pole outward and downward. Astronauts would then clip their harnesses to rings on the pole and slide out. When they came free of the pole, they would be underneath the wing. But even this design was a joke. The very reason ejection seats were invented was because aircraft crewmembers were being pinned inside cockpits by wind pressures and the G-forces of an out-of-control craft. Ejection seats overcame these forces by blasting the crewmembers out of the cockpit. The idea of getting out of an upstairs shuttle seat wearing nearly ninety pounds of equipment and encumbered by an iron-hard pressurized Launch-Entry Suit (LES), then climbing down the narrow interdeck ladder and making it to the side hatch while the shuttle was in powered flight and/or tumbling out of control (the two most common conditions of aircraft ejections) was a fantasy. The only scenario in which a backpack parachute would save a crewmember would be in controlled, gliding flight at subsonic velocities, and below 50,000-feet altitude. It was difficult for astronauts to imagine a failure that would put us in those conditions. Astronauts were still living with the consequences of an *operational* shuttle design.*

Many of us placed the slide-pole bailout procedures in the same category as the pre-*Challenger* contingency-abort procedures—busywork while dying. But we all completed the training. A mock-up of the side hatch and pole were installed on a platform over the WETF pool, and we practiced sliding down the pole to smack into the water.

While most of my time was spent in STS-27 classified training, I was occasionally required to perform other short-term duties. One proved enlightening. Pinky Nelson and I were given the task of polling the spouses for suggestions regarding the family escort policy. One wife responded that she wanted to sleep with her husband on the night before launch. Pinky and I short-stopped that recommendation. We could not imagine any astronaut wanting to be with their spouse in those hours. I certainly didn't want Donna in my bed. The beach house good-bye was agony enough—I couldn't imagine enduring an all-night good-bye. If the wife making the recommendation was alluding to having prelaunch sex, then her husband was a better man than me. Even a doughnut-size Viagra pill wouldn't help me at T-12 hours.

Many of the wives were extremely critical of what had happened to the *Challenger* spouses after the disaster. June Scobee and the other widows had been held at KSC so Vice President Bush could fly down and meet them. The wives were of one voice—in the event of a disaster they wanted to immediately return to Houston with their children. Screw the politicians. Pinky and I couldn't have agreed more and said so in our recommendations.

The spouses also complained that the part-timer politicians and Abbey's "buddies" were exempt from the rules; for example, the number of family guests allowed on the NASA buses. One wife wrote, "Senator Garn had no problem getting anybody he wanted to attend any of the dinners, ride on the family buses, etc." It was just another example of how NASA management had allowed the politicians to have their way with us. We recommended there be no exceptions to family

*Some astronauts believe even the backpack parachute arrangement might have enabled *Challenger*'s downstairs crewmembers to escape. However, there probably wouldn't have been enough time for the upstairs crewmembers to make it out.

escort policy for any crewmembers regardless of their pedigree. Theoretically this recommendation was irrelevant since *Challenger* had ended the passenger program. But neither Pinky nor I believed NASA HQ would ever say no to any politician who still wanted to fly (and we were proven right when, years later, Senator Glenn asked for a flight).

Another major source of irritation was the fact that wives could fly to the launch aboard NASA's Gulfstream jets at government expense, but their children could not. Since NASA required the spouses *and* children to be on the LCC roof for the launch, many of the wives felt Uncle Sam should pick up the transportation tab for the children, too. They also wanted their lodging needs handled by NASA. In prime tourist seasons and around holidays—where no-vacancy signs were the rule—a mission slip could make reservation extensions problematic and a terrible additional strain on the family. The spouses wanted NASA to take on that burden. Pinky and I agreed and added that to the recommendations.

These modifications to the family escort policy were adopted and, beginning with STS-26, NASA transported crew children to KSC via its Gulfstream jets. The agency also assumed control over all lodging issues. As for any wife who wanted to have KSC sex with her husband, she would just have to be satisfied with a quickie at the beach house or behind closed doors during a visit to the crew quarters. And she better not be a screamer—the rooms weren't that soundproof.

As the summer of 1988 was drawing to a close, John Denver came to the astronaut office to brief us on his plans to fly with the Soviets. Before *Challenger,* we had frequently heard Denver's name mentioned as a potential participant in NASA's passenger program. That program had been terminated by the disaster, so now the singer was pursuing a trip into space via a Russian rocket. On a visit to Houston, he made contact with JSC and was invited to the astronaut office to discuss his mission plans. He received a chilly reception. Most military astronauts harbored a severe dislike for all things associated with the commies. Russian bullets had been aimed at our planes in Vietnam. Our friends had been killed or imprisoned by their surrogates, the North Vietnamese. The idea that anybody would cozy up with those assholes for any purpose was an outrage to many of us. Denver was peppered with criticism. One Vietnam vet told him that his Russian plans "sucked." Denver argued that he wasn't being any more cooperative with the Russians than others in the past had been. "Like Jane Fonda" came a rejoinder from the back. Several astronauts applauded at that. Denver continued to defend himself, explaining that he had always been a big supporter of the space program and it had been a lifetime dream to fly in space. In fact, he said, "I was the one who first suggested to NASA they have a passenger program on the shuttle." That comment didn't win him any friends—many in the office were still silver-pinned astronauts because of the passenger program and the seats into space it had consumed. New tracers of criticism shot his way. One astronaut made the observation that when Denver returned from his mission, the press and public would elevate him to the status of "expert" on the space program just because of his celebrity. He would end up on every blue ribbon panel and space policy committee for the next decade, while the real experts, astronauts and others at NASA, would be forgotten. The meeting definitely didn't give Denver a Rocky Mountain High. He later dropped his Russian flight plans because of the cost, rumored to be $20 million . . . or maybe he was afraid of what an astronaut would do to him if he made the trip.

On September 29, 1988, STS-26 put America back in space. Four days later *Discovery* streaked out of the Pacific sky to touch down at Edwards AFB. The mission was virtually flawless. At Rick Hauck's call, "Wheel stop," I was once again part of a Prime Crew. With the title came a reserved parking place, euphoric joy, and intestine-knotting fear.

CHAPTER 33

Classified Work

December 2, 1988, found me and the rest of the STS-27 crew strapped into *Atlantis* waiting out a weather delay at T-31 seconds. We had already scrubbed the day before due to out-of-limits high-altitude winds. With the potential of a second scrub hanging over us, the mood in the cockpit was gloomy. I was beginning to think I was cursed. This was my sixth launchpad wait for only a second mission. The problem today was the weather at our transatlantic abort sites in Africa. They were below minimums. The launch director was on the phone with the astronaut observer in Morocco. With the APUs running, a decision had to be made quickly.

I heard the range safety officer speaking on the LCC voice net and used the opportunity to joke, "The RSO's mother goes down like a Muslim at noon." (The term *going down* was from Planet AD and referred to an act of oral sex.) I didn't worry about the RSO hearing me. He didn't have access to our intercom.

The crew cringed . . . and laughed. I was slandering the mother of the man who was just two switches away from killing us. Hoot shouted, "Mullane, don't joke about the RSO's mother! Pick on the pope's mother. Hell, pick on *Christ's* mother. Anybody but the RSO's mother!"

The laughter faded and the intercom fell quiet. I thought of Donna on the roof of the LCC. I knew the delay was killing her. Every wife-mother said the same thing: *Watching a husband being launched into space is like being in a never-ending difficult childbirth . . . without any pain medication.* My mom certainly thought so. She greeted me after my first mission with a sign reading, "September 10, 1945 was easier."

My fear was "off-scale high," an astronaut expression meaning the needle of a cockpit instrument had soared past the highest reading and had pegged itself against a physical stop. The fear wasn't any greater than it had been on my first launch. *Challenger* had changed nothing in that regard—I had known before STS-51L that flying the shuttle could kill me, and I knew this flight could kill me. On the drive to the pad I had passed the same rescue vehicles I had passed on my way to *Discovery* and had, again, thought of the body bags they certainly contained. I thought of the full-mouth dental photos and clip of hair and the footprint the flight surgeon had archived in Houston. In another ten minutes would somebody be pulling those from my medical file to send to a Florida pathologist? If it was to be so, I prayed I would be brave in whatever form of death *Atlantis* might serve me. *Challenger* had convinced me there would be no merciful, instantaneous deaths granted to any shuttle crews. The cockpit was a fortress; at least it was a fortress until it slammed into Earth. If it had kept the crew alive through *Challenger*'s destruction, it would keep me alive through any breakup of *Atlantis*. The shuttle engineers had done what engineers always do . . . built their machine to spec and then added their own margins. The seat I was strapped to would survive twice the number of Gs my body could withstand. The windows (each *triple*-paned) and the walls around me would remain intact until the Earth crushed them. I was strapped into a fortress that would keep me alive long enough to watch Death's approach. If fire was to kill me, I would have time to watch the flames. If a multimile fall was to kill me, I would watch the Earth rushing into my face. Even a cockpit decompression would no longer mercifully grant us unconsciousness, as it might have spared the *Challenger* crew. We now wore full-pressure suits that would keep us alive and conscious through any cockpit rupture.

As I stared at the countdown clock, still frozen at T-31 seconds, my prayers covered a spectrum of needs. *Please, God, let the TAL weather clear so we can launch . . . Please, God, let us have a safe flight . . . Please, God, don't let me screw up . . . Please, God, if I'm to die, let me die fighting, joking, helping the CDR and PLT with a checklist, reaching for a switch. Please, God, let me die as Judy and the others died . . . as working, functioning astronauts to the very end.*

Since I had first heard Jim Bagian and Sonny Carter reveal that Mike Smith's PEAP had been turned on by Judy or El, I wondered if I would have had the presence of mind to do the same thing had I been in *Challenger*'s cockpit. Or would I have been locked in a catatonic paralysis of fear? There had been nothing in our training concerning the activation of a PEAP in the event of an in-flight emergency. The fact that Judy or El had done so for Mike Smith made them heroic in my mind. They had been able to block out the terrifying sights and sounds and motions of *Challenger*'s destruction and had reached for that switch. It was the type of thing a true astronaut would do—maintain their cool in the direst of circumstances. "Better dead than look bad." My greatest fear was that I would fail if I was ever faced with a similar disaster, that I would die as a blubbering, whimpering, useless coward, an embarrassment to my fellow crewmembers and, worst of all, it would all be captured on the voice recorder to be played in a Monday morning meeting.

The launch director's voice put a stop to my depressing thoughts and pleading prayers. "*Atlantis,* the TAL weather is acceptable. We'll be picking up the count." There were audible sighs of relief on the intercom. Now . . . if only *Atlantis* would cooperate and keep humming along without a problem.

There was a short count by the LCC and the clock was released.

"Thirty seconds."

Hoot reminded everybody to stay on the instruments. An unnecessary order. If a naked Wonder Woman had suddenly appeared in our midst, nobody would have been able to pull their eyes from the displays. Well . . . maybe I would have taken a quick peek.

"Ten seconds. Go for main engine start." I wondered how many times I would have to do this before I could do it with a heart rate below 350 beats per minute.

The engine manifold pressures shot up. Fuel was on the way to the pumps.

Engine start. The now familiar vibrations of more than a million pounds of tethered thrust rattled me. I watched shadows move across the cockpit as *Atlantis* rebounded from the start impulse. When she was once again vertical, the SRB and hold-down bolt fire commands were issued. Seven million pounds of thrust rammed me into my seat. I was on my way into space for a second time.

"Passing 8,500 feet, Mach 1.5."

We came out of the other side of max-q and the vibrations noticeably lessened.

"*Atlantis,* you are go at throttle up."

"Roger, Houston, go at throttle up." At Hoot's call, I knew everybody was thinking the same thought. Those had been the last words heard from *Challenger.*

"Ninety thousand feet, Mach 3.2." Hoot gave the markers.

"P-C less than 50." The SRBs were done. A flash-bang signaled their separation and we all cheered. Someone added, "Good riddance." We wouldn't know it until we were in space but the right SRB had already placed us in mortal danger . . . not because of an O-ring failure but because the very tip of its nose cone had broken off and hit *Atlantis.* The number-three SSME was also running sick, a fact we wouldn't learn until after the mission. The inner-bearing race on its oxidizer turbo-pump had cracked. We were blissfully unaware of these two threats to our lives. The cockpit instruments were all in the green.

The rest of the ride continued smoothly. The sky faded to black while the flare of the sun painted the cockpit. We listened to the cadence of the abort boundary calls. With each call, we breathed a little easier.

"Here it comes, rookies . . . 40 . . . 45 . . . 50 miles. Congratulations, Guy and Shep. You're now astronauts." They cheered and the rest of us added our own congratulations. I thought again of the ridiculousness of the fifty-mile altitude requirement. Guy and Shep had earned their wings as we all had . . . at the instant the hold-down bolts had blown.

Hoot's calls continued. "Sixty-one miles, Mach 16 . . . a little over 2-Gs." We were paralleling the East Coast of America. No doubt *Atlantis* was generating some UFO reports. Even though the sun was up, the blue-white flare of our SSMEs would be visible all the way to Boston. We were steering for an orbit-tilted 57 degrees to the equator. Until launch that fact had been classified. But it was impossible to hide our orbit parameters after liftoff. Russian spy ships were most likely already sending our trajectory data to Moscow and their downrange radars would be picking us up as we came over their horizons.

"Twenty thousand feet per second and 3-Gs." Under the G-load Hoot's call was grunted.

"The engines are throttling." Guy watched his power tapes slowly drop toward 65 percent of maximum thrust to keep *Atlantis* at 3-Gs until MECO. If the engines failed to throttle, Guy was prepared to shut one of them off to prevent *Atlantis* from overstressing herself under higher G-loads. At this point she was nearly parallel to the Earth, running to the northeast with an almost empty gas tank, rapidly adding velocity. "Twenty-two thousand feet per second . . . 23 . . . 24 . . . 25 . . . here it comes . . . MECO." At slightly faster than 25,000 feet per second, about eight times faster than a rifle bullet, *Atlantis*'s computers commanded the SSMEs off. There was the *thunk* of ET separation, the *boom* of the forward RCS jets to get us clear of the tank, the noiseless squeeze of the OMS burn, and then we were in orbit. I started breathing again.

My stomach was flip-flopping like a hooked trout. It wasn't space sickness—I was still spared that malady. Rather, it was showtime jitters. It was time for me to deliver on the millions of dollars of training NASA and the air force had invested in me during the past year. I was to operate the robot arm to deploy our satellite payload.

Hoot and I faced aft toward the cargo bay, he at the starboard-side window with the orbiter controls at hand, I at the port side with the RMS controls. Our feet were jammed under canvas foot loops, anchoring our bodies so our hands would be free to grasp controls. Many science-fiction writers had assumed astronauts would wear magnetic or Velcro or suction-cup shoes to keep them anchored while working. The reality was much less sophisticated, just loops of canvas duct-taped to the steel floor in front of the control panels.

I opened the locks that held the RMS to the port sill of the cargo bay, prayed the astronaut's prayer one more time, "Please, God, don't let me screw up," then grabbed the Rotational (RHC) and Translational Hand Controllers (THC) used to "fly" the robot arm. For once, the incredible beauty of the Earth passed unseen beneath me. I had eyes only for the payload, *Atlantis,* and the robot arm. I focused on each one with the intensity of a doctor doing open-heart surgery. I steered the end of the arm over the payload grapple fixture and fired the snare, which rigidly latched the payload to the arm. Jerry Ross then released the cargo latches. My eyes now moved in a constant scan between the out-the-window view and the views on two cockpit TV screens. There were cameras in each corner of the cargo bay as well as cameras at the end of the robot arm and at its elbow joint. At any time I could select the view of two of these six cameras to better determine the proximity of the satellite to *Atlantis*'s structure. I also had Shep in the airlock watching from its outer-hatch porthole and Jerry watching the TV views over my shoulder, both men ready to scream, "STOP!" if contact looked imminent. The tolerances were exceedingly tight and I finessed the controls with the deliberation of a soldier probing the dirt for a booby trap. The payload, like all satellites, was as delicately constructed as fine crystal. Any mistake that caused a satellite-to-*Atlantis* impact could damage a critical component and turn the object into a billion-dollar piece of space junk and win me an open-ended assignment to Thule, Greenland, where I would get to hone new skills as a urinal scrubber. An impact could also foul the payload bay-door closing system, a mistake that could kill us. Needless to say, the other members of the crew were as focused as I was.

All went well. The Canadian-built arm handled like a dream. Within an hour I had lifted the payload clear of the cargo bay and had flown it to its release attitude. I called to Hoot, "We're there." He was all smiles and I knew that the rest of the payload team watching from the ground was wearing the same smiles. I had delivered for them. No Super Bowl–winning quarterback has ever felt more satisfied.

Hoot double-checked that his orbiter hand controllers were on and got a "Go for payload release" from MCC. On his cue I squeezed the grapple release trigger and pulled the arm off the payload. The satellite was now flying free in a 17,300-mile-per hour formation with *Atlantis*. Hoot quickly executed the fly-away maneuvers and we watched the satellite slowly recede in the distance until it was the brightest star in our windows. I parked the robot arm in its cradle thinking this would be the last time in the mission it would be needed. I would be wrong.

Time to celebrate. For all intents and purposes our mission was over. As orbiting astronauts were prone to say, "It's all downhill from here." We raided our pantry, ignoring the dehydrated broccoli the NASA dietician had included to grab some M&M candies and butter cookies. Soon a baseball game was in full swing. I would pitch an M&M to Guy and he would bat it across the mid-deck with a locker tool. Jerry and I would then field it with our mouths. (Astronauts never play with their food like this while other crewmembers are vomiting.) Hoot filmed the fun, something NASA was not going to be happy about. HQ had relayed to the astronaut office their growing displeasure with astronauts filming their weightless games. It was all the press would show and they felt it trivialized our missions. The press had ignored video of the STS-26 crew deploying their quarter-billion-dollar TDRS satellite and instead showed them dressed in Hawaiian shirts engaging in 0-G surfing.

Hoot next liberated a football from a locker. NASA was going to be honored during the January Super Bowl halftime show and HQ wanted a space-flown football to give to NFL Commissioner Pete Rozelle. The ball had been deflated to save space, but using a food rehydration needle, Hoot was able to blow enough air into it to give it a useable shape and we paired up for a hilarious weightless football game. As with the baseball game, we filmed our Super Bowl. NASA HQ would have to cut us some slack. The classified nature of our mission would prevent us from showing the public any of our payload activities. Our game films would be all that we could show.

We spent the rest of the day immersed in our Earth-observation experiment, taking photos for geologists, meteorologists, and oceanographers. For each of us, though, there was one very special Earth feature to photograph that wasn't on any of the scientists' lists . . . our hometowns. Even the other veterans on the flight, Jerry Ross and Hoot Gibson, had never seen their childhood homes from space. The orbits of our earlier missions had been too close to the equator. But *Atlantis* was crossing over all of America.

Albuquerque was an easy target to locate. The dark, winter-dormant flora of the Rio Grande River Valley contrasted well with the adjoining deserts, and Albuquerque's western border was formed by that river. I needed only to spot a few other landmarks to know I was approaching the city. There were the snowcapped peaks of the Sandia Mountains to the east and solitary Mount Taylor to the west. As it came into view, the city itself was a gray patch filling the terrain between the river and the mountains. It was impossible to see individual houses or even neighborhoods, but I could approximate the location of my childhood home. No longer was it on the edge of the city but rather deep in suburbia. Like other Sun belt cities, Albuquerque had grown up. But my mom still lived in the same house and I could imagine the thrill she would have felt if she could have looked up to see *Atlantis* passing overhead. There was no chance of that, though. The sun was too high.

I snapped a few photos and then Velcroed the camera to the wall. This was another sacred moment in my life and I didn't want to be distracted with setting an f-stop. I was looking into the cradle of my astronaut dream. There was no other place on the planet that held more memories for me. Two hundred and forty miles below were the deserts from which I had launched my rockets. Here was the Rocky Mountain West that had excited my imagination with its infinite horizons. Here was the sky I had navigated in a Cessna while making plans to be a test pilot and astronaut. Here was the place God had steered

Donna and me together. And, now, I was speeding over all of it in a spaceship.

Later we gathered around the window to watch the evening lights of Houston pass under us. The last rays of the setting sun were on *Atlantis,* so she would be visible as a bright star to anybody in the city who cared to look up. I wondered if someone had bothered to call our wives to tell them to watch for us. I would later learn that family escort Dave Leestma had. At the very moment I was staring downward, Donna was standing in an open field near our home and looking upward at our streaking star. After my return, she would tell me how the sight had overwhelmed her. "Mike, do you have any idea how amazing that was? *You* were in that point of light. I had to pinch myself to make certain I wasn't dreaming." I could appreciate her wonder. Every moment of orbit flight seemed like a dream to me, too.

Swine Flight went to bed without a care in the world . . . or off the world, for that matter. The dangers of ascent were behind us. We had already scored our mission success. Our only problem was a slow leak in *Atlantis*'s left inboard tire and that wasn't a big deal. MCC had noted it in their data and had directed us to program the autopilot to keep *Atlantis*'s belly pointed at the Sun. The heat was keeping the tire warm and its pressure up. We hoped the higher pressure would reseal the leak point. But, even if the tire went flat, we were scheduled to land on the Edwards AFB dry lakebed and would have its infinite runway to handle any type of steering problems after touchdown.

I fell asleep secure in the machine that surrounded me. This would be the last time on the mission any of us would feel safe.

CHAPTER 34

"No reason to die all tensed up"

The call from MCC was disturbing. During a review of launch video, engineers at KSC had seen something break off the nose of the right-side SRB and strike *Atlantis*. The concern was whether the object had damaged our heat shield, a mosaic of thousands of silica tile, a design feature that earned the shuttle its nickname, "Glass Rocket." The CAPCOM asked if anybody had seen any strikes during ascent or had noted any damage looking out the windows. "No" was our collective answer, but we did have a tool that would extend our vision to the shuttle's belly —the camera at the tip of the robot arm. Within several hours MCC validated a heat-shield survey procedure in the Houston sims and teleprintered it to me. I was going to get some unplanned arm time.

My heart was back in overdrive. Not only was I concerned about the possibility of heat-shield damage, I was also worried about the arm maneuver I was about to perform. It would put the RMS in very close proximity to the inboard portions of *Atlantis*'s right wing and fuselage, and I wouldn't win any friends if I caused damage while determining there had been none to begin with.

I swung the upper boom of the arm across the forward cargo bay and then tilted the lower boom over the right forward side of *Atlantis*'s nose. I swept the camera in a survey of that area, listening to Hoot's cursing as I did so. The exterior cameras on the shuttle had long been a source of frustration with astronauts. They easily bloomed and washed out while imaging areas in full sunlight. The glare from the black heat tiles was particularly troublesome to the camera optics, and Hoot fought with the aperture controls trying to get a decent view. He was finally successful and our TV revealed a checkerboard of black tiles. It was exactly what a pristine heat shield should look like. But as I moved the arm lower the camera picked up streaks of white. There was no mistaking what they were. The surface of every belly tile was jet black in color. Any white would be an indication of damage, an indication that the surface had been ripped away by a kinetic impact. As I continued to drop the arm lower we could see that at least one tile had been completely blasted from the fuselage. The white streaking grew thicker and faded aft beyond the view of the camera. It appeared that hundreds of tiles had been damaged and the scars extended outboard toward the carbon-composite panels on the leading edge of the wing. Had one of those been penetrated? If so, we were dead men floating. Damaged black tiles might still protect the vehicle. Even a missing tile should be survivable. But a hole in the leading edge of the wing would positively be fatal and we had no way to survey the entire wing edge. The arm wasn't long enough. (It was impact damage to a carbon panel on the left wing that would doom *Columbia* and her crew on February 1, 2003.)

In the cockpit a shocked silence gave way to shocked exclamations. I called MCC. "Houston, we're seeing a lot of damage. It looks as if one tile is completely missing."

The CAPCOM acknowledged my call. "We'll get back to you." MCC had our TV downlink so they were seeing what we were seeing, or so we assumed. I wondered if, at that very moment, there was a "Failure is not an option" speech being delivered as the flight director rallied the MCC team to deal with our situation. From our perspective it looked bad.

After a few minutes the CAPCOM came back with the MCC's analysis. "We've looked at the images and mechanical says it's not a problem. The damage isn't that severe."

Say what?! We couldn't believe what we were hearing. MCC was blowing us off. There was no discussion of having ground telescopes take some photos of *Atlantis* to possibly get a better view of the damage. There was no discussion of having us power-down the vehicle to give us the maximum orbit time to deal with the problem. There was no indication whatsoever that MCC thought we had a serious problem.

Hoot was immediately on the microphone. "Houston, Mike is right. We're seeing a lot of damage."

But after a short delay the CAPCOM came back with the original ho-hum assessment of "It's not a problem."

We all looked at one another in disbelief. Are they blind? Did they think the white streaks were seagull shit? It was obvious to us that we had taken a very bad hit. Maybe the image that was arriving on the MCC monitors was of poorer quality than what we were seeing. I tried to sharpen the image but was no more successful than Hoot had been.

Hoot tried again to convey the seriousness of what we were seeing and, again, MCC casually dismissed his concerns. We were in the Twilight Zone. *Who is this speaking to us and what have you done with the real MCC?* Maybe, I thought, MCC did know we had suffered major damage and were hiding that fact from us. There was precedence for that. On John Glenn's Mercury mission the MCC had an instrument indication that his heat shield had come loose from his capsule, but they kept the information from him. Since there was absolutely nothing he could do about a loose heat shield, and, if it was loose, he was going to die, they reasoned there was no reason to tell him the truth. Without giving him an explanation, they had instructed Glenn not to jettison his retrorocket pack in hopes its retention would help keep a loose heat shield in place. It turned out there had been nothing wrong with the heat-shield attachment and Glenn was angry with MCC for withholding information from him. Could MCC be hiding a deadly situation from us? I couldn't believe that. The MCC never gave up. If they suspected we had a serious problem, they would be pulling out every stop to get us home. It would be *Apollo 13* all over again. Their dismissive attitude must mean they believed we were okay. But I was going to be really suspicious if tomorrow's wake-up music was "Nearer, My God, to Thee" and there was a teleprinter message saying we didn't have to eat our broccoli.

Hoot set the mic aside, obviously frustrated and angry. No matter what MCC was seeing on their TVs, he felt they weren't seriously considering what *we* were seeing. At the moment I was glad our mission was classified. The public and press had not heard any of our discussion. Unless one of the family escorts told the wives of our problem they would not know about it. I prayed that was the case. There was nothing they could do and they didn't need the extra anxiety.

I couldn't sleep and floated upstairs to watch the sights. The autopilot was holding the shuttle's belly to the Sun, courtesy of our tire leak, so I had to move from window to window to get the best view. I was annoyed at the inconvenience . . . until our unique attitude served up a space sight I had never seen before. I was looking through the overhead windows in a direction that was precisely "down"-Sun when the upward-pointing attitude control thrusters fired. As their effluent of billions of ice crystals blossomed above the orbiter, a perfect shadow of *Atlantis* appeared and was carried into infinity at hundreds of miles per hour. The strikingly beautiful sight reminded me of Captain Kirk's *Starship Enterprise* dashing into warp speed. I kept staring, hoping the display would repeat but it did not. The Sun dipped below the horizon and *Atlantis* was plunged into preternatural darkness. But there was one certainty about orbit flight: the passing of one incredible space sight only meant the arrival of another. My eyes were drawn to the lime green curtain of an aurora borealis waving over the Arctic Ocean far to the north. It brightened and dimmed as the rain of magnetic particles from the Sun varied in intensity. I was still staring at this phenomenon when the long streak of a shooting star brought my mind back to our heat-shield problem. In just a few hours that would be *Atlantis* . . . a shooting star blazing across the Pacific with a tail of ionized gas a thousand miles long. We were locked in an aluminum machine that would melt at 1,000 degrees. On reentry the belly tiles would be subjected to 2,000 degrees. The nose and leading edge of the wings would see even hotter temperatures. Just a couple inches of silica and carbon fiber were all that protected us from immolation, and our camera survey had shown some of those inches had been ripped away. The heat would definitely be getting closer to that aluminum. And what about the missing tile? Could the wind use the cavity it created to grab the edge of adjacent tiles and strip even more off, just like roof shingles being sequentially stripped in a hurricane? Engineers had always assured us that was not possible, but then I was certain the SRB nose cone engineers would have assured us their work could not fail either.

Of one thing I was certain. If *Atlantis*'s wounds were mortal, our fortress cockpit would protect us long enough to watch death's approach. Certainly it would last long enough for us to see multiple warning messages as various systems were affected by the heat. We would probably live to experience the out-of-control tumble and breakup of the vehicle. Even after our fortress was penetrated by the incandescent heat, death would not be immediate. Our pressure suits would protect us from the loss of cockpit air. Only when the fire penetrated the fabric of our LESes would we die. If we were lucky, unconsciousness would come before the heat began to consume our flesh.

I kept returning to MCC's assessment for comfort. It was hard not to yield to their conclusion that we were going to be fine. I had never been associated with any teams as good as those that manned the MCC. But if *Challenger* had proven anything, it was that great teams do fail. A lot of very smart people had mishandled the O-ring issue that killed the *Challenger* crew. Were they now mishandling our heat-shield damage? Would an *Atlantis* presidential commission report end up containing the statement, "The crew radioed that the damage to their heat tiles looked serious, but in Houston their concerns were dismissed"?

The anxiety was exhausting and I finally gave in to Hoot's solution. The day before, as he floated to the windows to do some sightseeing, he said, "No reason to die all tensed up." I would do my best to relax and enjoy the sights.

CHAPTER 35

Riding a Meteor

"Fifty seconds." Hoot gave the time remaining until the OMS deorbit burn. I floated behind Jerry Ross and watched the countdown on the computer displays. As the burn execute time neared, I tightened my grip on Jerry's seat. The one-quarter G of the thrusting OMS engines was trivial but it was enough to put anything unrestrained on the back wall, me included.

Astronauts have great faith in the OMS engines. They are the essence of simplicity. They have no spinning turbo-pumps to worry us, not even an igniter to fail and jeopardize our lives. The fuel and oxidizer are pushed into the combustion chamber by helium pressure, and their chemical composition causes them to ignite on contact. No spark needed. Getting stuck in orbit would ruin your whole day, so having a deorbit engine that was as foolproof as possible eased one of our perennial worries.

At time-zero the cockpit shuddered under the hammer of the two engines. Small bits of food crumbs, which had escaped our cleanup, drifted to the back wall. Hoot and Guy watched the computer displays of helium pressures, temperatures, and other indications of engine performance. They were all nominal.

The burn, and its acceleration on our bodies, ended. We were back in weightlessness. Hoot checked the residuals, which indicated the error of the maneuver. They were negligible. Whatever fate awaited us, we were now irrevocably committed. The OMS deorbit burn had clipped 200 miles per hour from our speed and changed our orbit so its low point was into the Earth's atmosphere. There was no way we could climb back up to the temporary sanctuary of orbit.

According to the checklist I should have been strapped into the mid-deck seat, but there was nothing to do or see down there, so I had asked Hoot if I could hang out on the flight deck and shoot some video of the early part of reentry. I would get into my seat before the Gs got too high.

During the next twenty-five minutes we fell across the Indian Ocean, across the darkened continent of Australia, and shot into the night sky of the great Pacific basin. In our dive toward perigee, *Atlantis* gained back the velocity lost in the deorbit burn and added more. Shuttles achieved their peak speed on reentry, not ascent. The ride was as smooth and silent as oil on glass. The machine was on autopilot, with only her rear thruster jets active. Those were holding *Atlantis*'s nose 40 degrees high and presenting her wounded belly to the approaching atmosphere. Her aerodynamic control surfaces, the elevons on the wings and rudder on the tail, wouldn't be able to hold her in attitude until we were much deeper into the atmosphere. If the rear thrusters failed, we would slowly drift out of attitude, tumble, and die. But an RCS system failure was far down on our lists of worries. The thruster jets were just smaller versions of the OMS engines, using a simple blowdown helium pressure feed system with propellants that burned on contact.

Hoot called the descent. "Mach 25.1 . . . 340,000 feet . . . Guidance looks great." We were slightly faster than 25 times the speed of sound and at an altitude of 64 miles. Our little green bug was tracking perfectly down the center energy line. We were on course for Edwards AFB, still 5,000 miles away.

The atmosphere had thickened enough to become an obstacle to our hypersonic sled. Compression against *Atlantis*'s belly heated the air to a white-hot glow now visible from the front windows. I wondered what was happening underneath us. I had visions of molten aluminum being smeared backward like rain on a windshield. None of our instruments or computer displays showed *Atlantis*'s skin temperature. Only Houston had that data. I wondered if they would call us if they saw it rising. I hoped not. Such a call would definitely have us all tensed up as we died. Even Hoot wouldn't be able to laugh away that MCC call. I looked at our displays and meters, not sure which would be the first to reveal a heat-shield problem. They were all in the green. Hoot would later tell me his eyes were never long from the elevon position indicators, certain a left-right split in those instruments would be the first hint the right wing was melting.

With the nominal displays I was happy to consider that our worries might have been misplaced. Maybe we had been alarmists. Maybe the damage was minor, as MCC had indicated. I still couldn't believe that, but I prayed it was the case. I would have no problem offering the flight director, the CAPCOM, and everybody else an apology for having questioned their judgment.

"Three hundred ten thousand feet . . . still holding Mach 25 . . . some Gs starting to build."

Hoot didn't have to tell me about the Gs. Dressed in my LES I had added more than sixty pounds to my body. My zero-G-acclimated muscles were finding it difficult to bear that weight against even a small G-load.

I looked upward through the top windows. As I had seen on STS-41D a snake of plasma flickered over us and slithered into the black. Periodically it would pulsate in incandescent-bright flashes that filled the cockpit like camera flashes. I wished I had paid more attention to the reentry light show during my *Discovery* mission. Wasn't this plasma ribbon brighter? Weren't those flashes more frequent? Was it *Atlantis*'s vaporized skin enhancing the show? There were no calls of distress from Hoot or Guy and the radios were mute. If the heat was dissolving *Atlantis*'s belly, the damage had yet to reach a system sensor.

At 240,000 feet and Mach 24.9 the guidance system commanded *Atlantis* into a 75-degree right bank. She was high on energy and the autopilot was pulling her off course to increase the distance to landing. In that extra distance we would have more time to descend. We trusted the autopilot to turn us back toward the Edwards AFB runway at the appropriate time. It was impossible for a shuttle pilot to look out the window at a featureless ocean from 45 miles high while still 3,000 miles from the runway and manually modulate the orbiter's energy state. We had to trust the wizardry of accelerometers in *Atlantis*'s inertial measuring units to do that.

The plasma vortex intensified in brightness. I wondered how many earthlings were watching the celestial spectacle. The trail of superheated air would glow for many minutes after our passage. We were painting a white-hot arc from horizon to horizon to take away the breath of anybody watching, be they jungle island shaman or the crew of a supertanker.

The horizon came into view from Guy's window. The Earth's limb was tinted with the indigo of an impending sunrise.

"Mach 22 at 220,000 feet, a half-G."

I wasn't going to be able to stand much longer. My leg muscles were quivering. I had to get down the ladder to my seat. Before I left the flight deck, though, I craned my neck to see the Earth. The Sun had risen and painted a broken layer of clouds in flamingo pink. Just when I thought I had seen every "wow" scene in space, I was treated with a new feast for the eye. We were still traveling at a couple miles per second but had dropped to within forty miles of the cloud tops. The illusion was that we were accelerating, not slowing down. The clouds appeared to skim by at science-fiction speeds. The sight was a narcotic and I watched it until my zero-G-weakened legs couldn't take my weight any longer and I collapsed to the floor. It was beyond time to get to my seat. I crawled to the mid-deck ladder like a wounded infantryman and felt with my feet for the ladder rungs. In slow, deliberate movements I worked my way downward and into my seat. By the time I was strapped in, I felt as if I had just descended the Hillary Step with a Sherpa on my back. I was exhausted from working in a G-force that was one half of the Earth's pull.

I hated being downstairs. I was staring at a wall of lockers. There were no windows, no instruments. I felt claustrophobic. I could not imagine anything more terrifying than being in this room and hearing the death throes of a disintegrating shuttle while simultaneously having the lights and intercom go off, as had Ron McNair, Greg Jarvis, and Christa McAuliffe on *Challenger.* If the lights went dark and my intercom failed, I resolved to unstrap from my seat and try to climb upward so at least I could die looking out a window. There would be no reason to go to the side hatch and try our new bailout system. If I jumped out during aerodynamic heating, my body would be vaporized as quickly as a mosquito in a bug lamp.

I could hear rumbling through *Atlantis*'s structure. Again, I wished I had paid more attention on STS-41D. Was the rumbling normal wind noise?

We had glided 11,000 miles in the past 50 minutes but most of our deceleration was still ahead of us. Hoot's dialogue came quicker. "Mach 20; 210,000 feet; 1,190 miles to go . . . Mach 18; 200,000 feet . . . one-G . . . 185,000 feet; Mach 15.5 . . . 1.4 Gs."

We had passed through the hottest part of reentry and *Atlantis* was flying just fine. Without a hiccup she was completing her transformation from spaceship to airship. My doubts about our assessment of the tile damage were growing and, again, I was happy to be alive to have those doubts.

Now there was significant wind noise accompanied with heavy vibrations.

"Speed brakes coming out." We were ten minutes from landing.

At Mach 5 Guy activated the switches to deploy the air-data probes on the sides of the nose. The information they provided on airspeed and altitude would further refine *Atlantis*'s guidance.

"Mach 3.7 . . . 100,000 feet . . . I've got the lakebed in sight." *Atlantis*'s computers had done their job. After a glide spanning half the Earth, they had put a runway within our reach.

The wind noise outside my little room had become a freight train roar.

"Mach 1.0 . . . 49,000 feet." As *Atlantis*'s velocity dropped below the speed of sound our shock waves raced ahead of us, lashing the vehicle with a buzzing vibration. No doubt they were also sonic-booming the high desert and heralding our arrival to the wives.

We had finally entered the useable envelope of our crude bailout system. We were at subsonic velocity and below fifty thousand feet. If an escape became necessary now, I would pull the emergency cabin depressurization handle, followed by the hatch jettison handle. Then, I would unstrap from my seat, deploy the ceiling-mounted slide pole, clip my harness to a ring, and roll out. Of course all of this presupposed Hoot or the autopilot would be able to keep *Atlantis* flying in a straight-ahead, controlled glide. If the vehicle was in a tumble, the G-loads would pin us in the cockpit like bugs on a display board.

Hoot took control from the autopilot and banked *Atlantis* into a left turn toward final approach. Guy's calls of airspeed and altitude came like an auctioneer's. "Two hundred ninety-five knots, 800 feet . . . 290 . . . 500 feet . . . 400 feet . . . 290 . . . gear's coming." I heard and felt the nose landing gear being lowered. "Gear's down . . . 50 feet . . . 250 knots . . . 40 . . . 240 . . . 30 feet . . . 20 . . . 10 . . . 5 . . . touchdown at 205 knots." We were safely home. Our heat shield had held. I was anxious to look at it and see how much crow we'd have to eat.

After the standard postlanding cockpit visit by the flight surgeon, we changed into our blue coveralls and walked down the steps from the side hatch. NASA Administrator James Fletcher was present to greet us. We exchanged handshakes and then turned to inspect *Atlantis*. There was already a knot of engineers gathered at the right forward fuselage shaking their heads in disbelief. The damage was much worse than any of us had expected. Technicians would eventually count seven hundred damaged tiles extending along half of *Atlantis*'s length. It was by far the greatest heat-shield damage recorded to date. Some of the more severely damaged tile had been melted deeper than the initial kinetic penetrations. The area around the missing tile had been particularly brutalized and the underlying aluminum looked as if it had been affected by the heat. But there had been no "zipper" effect, as the engineers had promised. If any of them had been present, I would have flung my arms around them and kissed them.

We had taken seven hundred bullets and lived to talk about it. The damage had been sustained in the only place where it could exist and still be survivable. It started a few yards back from the carbon-composite nose cap and stopped just a few feet short of the right wing's leading-edge carbon panels. If any of those carbon-covered areas had been hit, we would have died as the *Columbia* crew would die fifteen years later, incinerated on reentry, except in our case the Pacific would have been our grave. Even the location of our missing tile proved fortuitous. By happenstance it covered an area where an antenna was mounted, and the underlying aluminum structure was thicker than in other locations. Had a different tile been blasted away, a skin burn-through might have occurred, allowing 2,000-degree plasma to run amuck inside *Atlantis*'s guts. God had watched after us. As Hoot turned from his inspection I heard him grumbling, "I'm going to be really interested in what MCC has to say about this."

A few days later we got to hear their story. The quality of the TV images in Mission Control had been very poor. The engineers had been convinced the damage was localized and minor. I wanted to say, "You should have listened to us," but they knew that. There was no reason to rub their noses in it. And then there was the nasty little fact that it didn't really matter. Even if MCC had determined the damage would positively result in vehicle destruction and our deaths, what could have been done? Nothing. The ticket home always entailed a flight through the blast furnace of reentry. We couldn't magically fly over, under, or around it. We couldn't have repaired any damage. There was no space station outpost for us to have escaped to. Our only hope would have been to have another shuttle come to our rescue and we wouldn't have had enough oxygen to wait for that. All we could have done was pitched the robot arm overboard and dumped every ounce of water and excess propellant to get our weight to a minimum to reduce the heat load, but even these efforts would have been in vain. Stripping the shuttle would have made the reentry just fractionally cooler. Just as it was with our contingency abort cue cards, any MCC recommendations would have just given us "something to read while we died."

MCC did have an explanation for the failure of the SRB nose cone. There had been a change in the manufacturing process, intended to improve the performance of the ablative material that protected the SRB from the aerodynamic heating it encountered during launch. At our debriefing of the incident at the Monday astronaut meeting, others wondered how many other slide rule jockeys were violating the prime directive of engineering: "*Better* is the enemy of *good enough*." We all wanted to shout from the rooftops, "If it's working, don't fix it!"

At this same debriefing, Hoot reacquainted the post-docs with the sick humor of the fighter pilot. During our mission, there had been a horrific earthquake in the Soviet Eastern Bloc state of Armenia, killing 25,000 people. The TV news was still showing images of masked workers pulling bodies from the rubble. "I know many of you have been very curious about our classified payload." Hoot paused until the room was hushed in expectation. "While I can't go into its design features, I can say Armenia was our first target." The military astronauts laughed. A handful of the post-docs cringed in disgust. Hoot tormented them further by adding, "And we only had the weapon set on *stun*!" The comment elicited more laughter and a few female darts of "Don't you guys ever grow up?"

CHAPTER 36

Christie and Annette

It was a few days after landing that Rhea Seddon bestowed another handle on Swine Flight. We became "The Grissom Crew." It was a play on a scene from the movie *The Right Stuff*. After Alan Shepard returned from his history-making flight as the first American in space, he and his wife had been hosted at the White House by JFK and Jackie. The movie dramatized how Gus Grissom and his wife had been expecting similar treatment when he returned as the second American in space. But it didn't happen. Runner-ups never slept at the White House. Rhea's "Grissom Crew" label of STS-27 was poking fun at the fact there was no White House invitation awaiting us in our in-boxes, whereas President and Nancy Reagan had received the STS-26 crew and their spouses. In a wonderful parody of the movie scene in which Mrs. Grissom laments her lockout, Rhea exaggerated her already severe Tennessee accent and swooned, "You mean, I won't get to meet Nancy and Ronald?!"

While it appeared we would remain invisible to the civilian world, we did have a "black world" postflight tour around the country. We visited the classified control center for our payload. We showed films of the satellite release and thanked everyone who contributed to the mission. It was all very staid and professional until Hoot presented a Swine Flight autographed photo to the unit commander. It was of the free-flying payload bearing Shep's inscription, *Suck on this, you commie dogs!* The group crowded around to see the photo. They couldn't wait to get it on their wall. Shep had made them feel like the warriors they were.

In a visit to Washington, D.C., we were invited to the Pentagon to brief the Joint Chiefs of Staff on our mission. My heartrate was as high walking into that office as it had been walking across the shuttle cockpit access arm. There was a veritable constellation of stars in the room, including five four-star generals and admirals. When a cute female lieutenant walked in, Hoot delivered a sotto voce "snort" in my ear. *Jesus,* I thought, *does he ever NOT live on the edge?* Chairman of the Joint Chiefs of Staff, Admiral William J. Crowe, Jr., was taking a seat not more than fifteen feet away and Hoot was snorting an aide-de-camp! Depending on your perspective it was either a new low or new high for

Swine Flight. I discreetly elbowed him. If I got the giggles from his antics now, that urinal-scrubbing assignment in Thule would be back in play.

In a mark of his leadership style, Hoot asked each member of the crew to say a few words about our mission tasks. I knew there were plenty of other egocentric commanders who would have hogged the stage for themselves. Not Hoot. After we all made our remarks and had taken our seats, Admiral Crowe asked his staff to rise and said, "I think we owe this crew a round of applause for their outstanding work," and five flag officers heartily responded. I was numb. The joint chiefs of staff of the United States military, led by their chairman—a total of twenty stars—were standing to applaud Mike Mullane. I could not have been more shocked if Hoot had stood up and announced he and Shep were gay lovers.

The day only got more unusual. We were driven to a classified location for an awards ceremony. As we followed our escort through multiple layers of security, I whispered to Hoot, "Maybe we'll meet Pussy Galore." He replied with a snort.

We were finally led into a walk-in vault where we were greeted by a senior government official. He offered his thanks for our work, then pinned the National Intelligence Medal of Achievement on each of our chests. Inwardly I laughed at the title. It sounded like an award for the brainless scarecrow in *The Wizard of Oz.* But it was a pride-filled moment for me, even exceeding what I had experienced in Admiral Crowe's office. I felt directly connected to America's defense in a way I had never felt in Vietnam or in my NATO forces tour. On STS-27 with the RMS controls in my hand, I had been at the tip of the spear.

The citation accompanying the medal reads

> . . . in recognition of his superior performance of duty of high value to the United States as Mission Specialist, Space Shuttle *Atlantis,* from 2 December 1988 to 5 December 1988. During this period, Colonel Mullane performed in a superior manner in deploying a critical national satellite to space. Colonel Mullane expertly operated the Remote Maneuvering System (RMS) to lift the satellite out of its cradle and position it for deployment of subsystems. The release of the system was performed so expertly that the spacecraft was left in a remarkably precise and totally stable condition. This allowed the activation sequence to continue expeditiously. Colonel Mullane's superior operation of the Shuttle Remote Arm, as well as his initiative and devotion to duty, led to the safe unberthing and deployment of a critical new satellite system crucial to our national defense and treaty verification. The singularly

superior performance of Colonel Mullane reflects great credit upon himself, the United States Air Force, and the Intelligence Community.

As the meeting broke up, I was looking forward to telling Donna about the award. It was as much hers as it was mine. She had earned it on that LCC roof. But my anticipation ended at the vault door. We were asked to hand back the medal. "Sorry, but this award is classified. You can't wear it publicly or talk about it. It won't appear on your official records. But if you are ever in town and want to come over and wear it in this vault, be our guests." Amazing, I thought. We had received a medal we could only wear in a vault. James Bond might have been able to tell Dr. Goodhead *(snort)* about his daring adventures, but we couldn't tell anybody about ours, not even our wives. (The award was declassified several years after the mission.)

No call ever came from the White House, but Swine Flight did score one gem of a PR trip into the civilian world. Dan Brandenstein decided that the STS-26 crew had overstayed their welcome in the "Return to Flight" spotlight and redlined them from the Super Bowl event. Our crew would make the trip to Miami and represent NASA during the Super Bowl XXIII halftime show.

Accompanied by our wives, we flew to Miami the day before the big game. That evening we were the guests of NFL Commissioner Pete Rozelle at a party . . . along with three thousand of his closest friends. The event was held in a convention facility and around its entryway balcony sat a clutch of harp-playing women dressed as cowgirl saloon hookers. The choreographer of that act had to have been on LSD. Later I noticed the ladies were faking their "plucks." When they took a break, the harp music continued. And their music wasn't the only thing being faked. Squadrons of silicone-stretched zeppelin breasts, mounted on Pamela Anderson look-alikes, cruised the room. Every escort service in the state must have sold out and then called Vegas for backup.

Mountains of food, rivers of booze, and live bands in every corner occupied the crowd. Donna and I loaded up our plates and searched for seats. I had the great misfortune to find two at a table with a fat, lavishly bejeweled gasbag of a woman who saw our arrival as an excuse to recount how rich and well-traveled she was. There was nothing on my guest badge to identify me as an astronaut and her glances at my cheap suit and Donna's middle-class wardrobe must have given her the impression I was some NFL groupie she could lord over. Her husband looked up from his food and communicated a "sucker" look to me. He never said a word, just ate, maintaining a grin on his face the entire time as if to say, "Good, now you've got her and I don't have to listen to that runaway pie hole of hers." I finally wearied of her nonstop travelogue and interrupted to say, "I'm an astronaut, so I've seen a little of the world, too." With that Donna and I rose from the table and departed, leaving her speechless, no doubt for the first time in her life.

The next day, January 22, 1989, found us in a skybox at Joe Robbie Stadium preparing to watch the San Francisco 49ers battle the Cincinnati Bengals. There was just one significant distraction: Billy Joel, who was to sing the national anthem, brought along his wife, Christie Brinkley, and they were ensconced in an adjoining skybox. Just a glass wall separated us. This arrangement created the greatest dilemma in the history of maledom. In front of me was the sports spectacle of the year. Joe Montana and Boomer Esiason were calling the signals for their teams. Seventy-five thousand fans were screaming. Even the STS-26 crew didn't create a moment like this. But a few yards to our right was Christie Brinkley, arguably one of the most beautiful women on the planet. What was a man to do?

We did what every man would have done. We watched both. As soon as the ball was blown dead, we'd all stare at Christie like pound dogs hoping for adoption. Then our heads would snap back for the next play. Our heads oscillated back and forth as if we were watching a tennis match. The rhythm was only interrupted long enough to get a beer. Could it get any better? Yes, it could. Turned out Christie was a fan of the space program. She came to our skybox to meet us. We all rose as if she were royalty, which of course she was. She was royally, freakin' beautiful. Hoot, Shep, and I generated more snorts than a hog farm.

"Are you guys astronauts?" We were all wearing our blue flight suits with patches reading "NASA" and other patches with renditions of space shuttles. Then there were our gold navy and silver air force wings, our Mach 25 patches, and American flag patches. We were either astronauts or Epcot Disney characters. It must have been the blonde in her asking the question.

Under my breath I whispered to Hoot, "I'll be anything she wants me to be." Hoot, no doubt, was thinking the same lecherous things I was thinking, but he played the gentleman and answered, "Yes."

"Have you been to the moon?"

I whispered fiercely to Hoot, "Tell her *yes*!" But he stuck to the truth, damn him. Maybe she would have taken us home if he had lied.

She began to walk down our ranks, smiling and asking questions. At any moment I expected her to say, "You mean you aren't the famous

STS-26 crew? How disappointing." But she didn't. She seemed pleased to meet us runner-ups.

To my amazement I noticed the rest of the crew were shaking her hand! Even Hoot. He must have stroked out when she walked into the room. There's no other way to explain his restraint. *What a bunch of weak dicks,* I thought. You shake Billy Joel's hand. You shake Commissioner Rozelle's hand. But you don't shake Christie Brinkley's hand.

When she came to me, I embraced her. Her arms came around my back and echoed the hug. Afterward she didn't even signal her bodyguard to stand between us (and no restraining order arrived later in the mail). I might have a medal I couldn't talk about, but I sure as hell was going to tell every male in the astronaut office what it was like to hug Christie Brinkley. The others could tell them what her handshake was like.

She remained to ask more questions while the game raged on behind us. Periodically we would hear waves of screams coming from the audience to signify some spectacular play, but with Christie in our company, who gave a shit? Finally, with a breezy wave and a promise, "See you at halftime," she left the box. Donna smiled at me. "I guess you don't want me to ever wash that flight suit again." I laughed and kissed her. Billy Joel had a dream of a woman, but so did I.

A short time later, while I was stretching my legs in the corridor behind the skybox, Billy stepped out for a cigarette break. We fell into conversation. I was struck by how nervous he seemed in this one-on-one situation. His eyes never held mine. It was as if I were talking with John Young or George Abbey. I asked him questions about composing music and he asked me questions about flying in space. I'm sure I was his inspiration for "We Didn't Start the Fire," but he had to go and use Sally Ride's name in the lyrics instead of Mike Mullane. I knew I should have written it down for him. I never asked him the one question I was burning to ask: "What's it like to sleep with Christie Brinkley?" I would bet he's been asked before.

As the game approached halftime, our crew was escorted down to the field to await our cue. While idling we met Annette Funicello and Frankie Avalon, who were on deck for their performance, a celebration of the 1950s teenybopper Florida beach movies. I marveled at Annette. At forty-six, the mother of several children, she had a shape a woman twenty years younger would have envied. Later, when she was performing the beach blanket dance routines that made her famous, she moved like she was back in bobby socks. I was impressed.

Christy Brinkley held true to her promise and approached us, this time to pose for photos. If she had known what degenerates Hoot,

Shep, and I were she probably would have maced us. All three of us were praying she would experience the first Super Bowl halftime show wardrobe malfunction. But, alas, that didn't happen. I made certain to stand next to her for our group photo. She put her arm around my waist, her hand accidentally slipping across my ass as she did so. I'm certain my hard body and Mel Gibson good looks broke up her marriage to Billy, but somehow the paparazzi never picked up on that.

We were finally signaled to step onto a platform that had been towed onto the field. A narrator briefly traced the history of the space program while various space scenes played on the Diamond Vision. Our part in the program concluded with our introduction as the most recent space shuttle crew. We waved to the audience and we were done. Then, to my amazement, nearby fans leaned over the railing to hand us paper for our autographs and gave us business cards to send photos. Others shook our hands, flashed thumbs-up, and shouted, "Go NASA!" These weren't space geeks asking for autographs. And it wasn't the white collar crowd, either. This was the proletariat cheering us on. It gladdened me to see such enthusiasm among average Joes and Janes. Apparently *Challenger* hadn't diminished the public's support of the space program, as many of us had feared it would.

That evening, the NFL hosted another open-seating buffet supper for its multitude of guests. Having been burned once, I decided to be very selective in finding a tablemate. Luck was with me. I noticed Annette Funicello and her husband were sitting at an otherwise empty table and steered for their company. I introduced Donna and we sat down for a very pleasant dinner. Annette was delightful. I pried stories out of her about her Mouseketeer days, including how she had been forbidden by Disney to wear a two-piece swimsuit in the movies and how she had received thousands of engagement rings through the mail from "love-struck teenage boys." I wanted to tell her none of those boys had been love-struck. Like me, they had all been *struck* by the topography of her sweater. But I knew if I offered that opinion I would have been struck by Donna.

Later, as Donna and I walked back to our hotel, she asked me which of the two celebrity women I had met that day, Christie Brinkley or Annette Funicello, was most captivating. Without hesitation I replied, "Annette."

Donna was surprised. "I thought for sure you would say Christie Brinkley. She's so much younger and so beautiful."

"Yeah, that she is. But at dinner tonight I kept thinking of all those times as a teenage boy I had watched Annette on the Mouseketeer TV program while fantasizing what was under the A and the E letters on her chest. And there she was, thirty years later, sitting right across the table from me and I was still fantasizing."

Donna laughed and offered up the familiar refrain, "Will you ever grow up?"

CHAPTER 37

Widows

After returning from Miami I fell into the standard postflight depression every astronaut experiences. As Bob Overmyer once said, "Being an astronaut is a ride on the world's biggest emotional roller-coaster. One day you're in orbit and talking to presidents, the next day you don't even have a reserved parking place." I was way down in the don't-have-a-parking-place dip. I felt certain it would be several years before I would have any shot at another flight. I was at the end of a long line. There were a hundred astronauts in the office. While I waited I would be assigned to one of the supporting roles of the business: working as a CAPCOM in MCC, evaluating software in SAIL, supporting payload development, or otherwise involved in another support function. While I appreciated the importance of the work, it was mundane compared with the exciting world in which mission-assigned crews lived. I had twice been a resident of that world and I knew. I didn't want to fly a desk. The word *retirement* was frequently on my mind. I knew Donna would welcome a decision to leave Houston. Every trip to the LCC roof was killing her. But, like the good Catholic wife she was, she would never put her feelings first. It would have to be my decision. Other TFNGs had already made theirs. Jim van Hoften, Pinky Nelson, Sally Ride, Dale Gardner, and Rick Hauck had all announced their intention to leave NASA, or had already done so. Unlike some of them, I didn't have the offer of a high-paying civilian job setting a deadline. I had done no job search. I had no plans for a next life whenever it started, other than it would start in Albuquerque. Donna's dad had passed away and left her the home in which she had grown up. That home and the Rocky Mountain West were magnets drawing us back.

Though there was no job to force a retirement decision, there were other factors pointing me to the NASA exit sign. NASA management was still an issue. After Abbey's departure, Don Puddy had been assigned to fill the FCOD position, an announcement that had been greeted in the astronaut office with stunned silence. Like Abbey, Don had gotten his start at NASA as an MCC controller in the Apollo program and ultimately served as a flight director for *Apollo 16*. He also served as a flight director for all three Skylab missions and the Apollo-Soyuz mission, and he was the reentry flight director for STS-1. Don was an exceptionally talented engineer and manager. But he wasn't a pilot. No matter how disliked Abbey may have been, we had all appreciated his four thousand hours of cockpit time. He had a personal appreciation of the issues of high-performance flight. We worried Puddy's lack of flying experience would seriously hobble him in his ability to handle astronaut concerns about shuttle nose-wheel steering, brakes, runway barriers, drag chutes, auto-land, and the many other pilot-specific issues surrounding shuttle operations. If there was ever a position that needed to be filled by an astronaut, it was chief of flight crew operations. Don Puddy's appointment made us suspicious that NASA's senior leadership still didn't want to deal directly with astronauts. Dick Covey told us a story that supported those suspicions. He had been in a meeting with Aaron Cohen, the JSC director, in which Cohen had polled his senior staff on whether or not the FCOD position should be filled by an astronaut, and the unanimous answer had been yes. The fact that an astronaut had not been assigned was proof to us that HQ had overruled Cohen. Our discontent with Puddy's assignment was so widely known that a week after the announcement, Aaron Cohen took the unprecedented step of coming to the astronaut office to curtly tell us the decision had been entirely his, without any HQ input. Nobody believed it. He then went on to list Puddy's attributes as a great manager, something that nobody was questioning. Puddy was exceptional. He just wasn't the best man for the job. An astronaut was.

I was tired of the us versus them. *Can't we all just get along?* The frustrations were a constant topic at the coffeepot and over '38 intercoms. When a news article appeared on the B-board about NASA Administrator James Fletcher's planned departure at the end of President Reagan's second term, an astronaut graffitied it with the comment *Two years too late.*

While Puddy's selection was a disappointment to astronauts, he instituted one critically important change—empowerment of the position of chief of astronauts. Shortly into Puddy's tenure, Dan Brandenstein unequivocally informed us that all flight assignments would originate with him and would then be successively approved or vetoed by Puddy, the JSC Director, Dick Truly, and the NASA administrator. Amazing! Someone in an astronaut leadership position was (gasp) . . . communicating with the astronaut corps! I was certain that the moon was in the Seventh House, Jupiter was aligned with Mars, and winged-swine were in the JSC treetops. If Abbey and Young had been present, they might have fainted. It had taken eleven freakin' years to hear this career-essential information—something John and George should have given us on day one.

Regardless of Dan's welcome leadership, there were other concerns. I could invest a couple years of my life working toward a third flight, only to have the rug pulled out from under me when a schedule change canceled my mission or another shuttle disaster interrupted the program or my health became an issue. In my last physical exam the cardiologist had seen an anomalous blip on the EKG traces. "It's no big deal, Mike. I'm not going to require you to fly on a waiver." But what if the blip became something serious enough to ground me?

I also considered what type of mission I might draw. I craved a spacewalk, but so did every other MS. Instead of an EVA, I might find myself on a Spacelab mission butchering mice and cleaning shit from monkey cages. The thought of waiting around a couple years only to end up as a space zoo janitor wasn't appealing.

And then there was the toll my career was taking on Donna. Though she wasn't about to use it against me, I was enough of a husband to at least think about it. And I was still learning about that burden. At a party I overheard a TFNG wife comment about the unaccompanied status of the *Challenger* widows. "There aren't a lot of men who would feel comfortable stepping into a dead astronaut's shoes." The observation hit me like a fist. I had never considered just how different the burdens of an *astronaut* widow were. I knew men. There weren't a lot of us capable of stepping into the shadow of a national hero—sure to wilt more than just an ego. Years later, at an astronaut reunion party, a TFNG widow told me, "I've dated, but nothing ever really develops." Men just couldn't deal with her deceased husband's astronaut title, she explained.

The unknowns, the fear, the burden on the family . . . they were all pointing to the NASA JSC front gate—time to drive out of it forever. I was within days of telling Brandenstein of my decision when the phone rang. It was Don Puddy. I was being assigned to another DOD mission, STS-36, only a year away. My sentence as an unassigned astronaut had lasted a month.

Puddy's call put me on a 6-G pullout from the bottom loop of the astronaut roller coaster. I now soared skyward. Every doubt, every fear about staying at NASA was gone. Within a year, I would once again be Prime Crew. When I told Donna the news, she smiled, but her eyes said volumes more. My announcement gut-shot her. I knew she wanted out. But she took it like a loyal soldier. She would be there for me no matter what.

Only a single aspect of STS-36 would ultimately be declassified, our orbit inclination. *Atlantis* would carry us into an orbit tilted 62 degrees to the equator, the highest inclined orbit ever flown by humans (it still remains the record). This wasn't the polar orbit planned for STS-62A (that would have been nearly a 90-degree inclination), but it meant we would get a view of more of the Earth's surface than any astronauts in history. Our orbit would almost reach the Arctic and Antarctic circles.

I would be one of two TFNGs on the mission. The other, J. O. Creighton (owner of the *Sin Ship* ski boat and Corvette my children found so alluring) would be the mission CDR. Like Hoot, J.O. was an exceptional pilot and leader. John Casper (class of 1984) would be the PLT. The other two MSes on the mission would be Dave Hilmers (class of 1980) and Pierre Thuot (class of 1985). In spite of his wimpy-sounding French name, Pierre was all-American. His astronaut nickname was Pepe.

Because our crew was at the back of the training line, Dave Hilmers and I had time to serve as the family escorts for STS-30. For the first time in my career, I was now the potential "escort into widowhood." The assignment allowed me another glimpse into the world of the astronaut spouse. At one of the beach house gatherings, Kirby Thagard broke into tears as she told the others that June Scobee had called to wish them all good luck. Just the name of a *Challenger* widow was enough to sheen the eyes of many of the women . . . at least when their own husbands were hours from launch.

Our days in Florida with the families were pleasant. Unlike some spouses who had earned reputations for treating the family escorts as personal butlers, the STS-30 wives were easy to work with. Because the mission CDR, Dave Walker, and one of the MSes, Mary Cleave, were single, there were only three spouses in our care: Kirby Thagard, Mary Jo Grabe, and Dee Lee. They adopted Dave and me into their extended families and we were guests at their numerous parties and other functions.

The crew was also a pleasure to work with. Mary Cleave was great fun. Now forty-two years old, she had a name tag on her flight suit reading, *Mary Cleave—PMS Princess*. She was another woman who wore the feminist mantle very lightly. When NASA HQ received a complaint from the California Democratic Party Women's Caucus about photos of Mary preparing a shuttle meal during her first flight (while the men were photographed doing technical work), Mary had laughed it off, saying, "I do shuttle windows and toilets, too." When the STS-30 mission slipped six days and the crew grew bored with TV, she requested a list of movies from a Cocoa Beach video-rental business and asked an astronaut to pick up her selections. As gags for the men on the crew, her choices included *Hollywood Chain Saw Hookers* and *All Star Topless Arm Wrestling*.

On the morning of STS-30's second launch attempt, May 4, 1989, Dave and I met with astronauts Bryan O'Connor and Greg Harbaugh. Bryan and Greg were at KSC supporting the mission and would also be on the LCC roof during the launch and available to help with the families if *Atlantis* was lost. Over breakfast the four of us reviewed NASA's contingency procedures: who would stand next to which family on the LCC roof, where the families would be temporarily gathered in the event of disaster, who would drive them to the KSC landing strip for their plane rides back to Houston. The wives had already been told to have their bags packed and ready to go before they were picked up for the trip to the LCC. This arrangement ensured they would not have to return to their condos and into a press feeding frenzy if disaster struck. A NASA official could retrieve the packed bags and bring them to the families at the KSC landing strip. But the procedure also meant the wives would have to unpack if the launch was scrubbed, and many wives had gone through the pack/unpack cycle multiple times for that reason. It was a pain but everybody understood the need.

Our breakfast table talk was devoid of emotion. Though we were planning our response to the death of friends and the widowing of their wives, our conversation was clinical and detached. We could have been talking about the logistics of a fishing trip. A review of the family escort procedures was just one more thing among thousands that had to be done as part of a shuttle launch. As I sipped my coffee, I thought of Donna's pre–STS-41D observation. It is, indeed, a strange business that plans so thoroughly for helping a woman into widowhood.

As *Atlantis* came out of the T-9 minute hold, Dave and I escorted the families from an LCC office to the roof. In the hallway we passed drawings done by the children of astronauts from prior missions. To keep the youngest kids of the families entertained during the interminable wait of a countdown, the LCC team provided poster board on which the children were encouraged to draw. After each mission that poster board was framed and hung in the hallway. The drawings served as another "widows and orphans" message for the team, a reminder of what was at stake.

On the walk to the roof the wives were talkative and it would have been easy to believe they were relaxed, but a glance into their eyes revealed otherwise. They were too large and darted too quickly. I had no doubt that the STS-41D and STS-27 escorts had seen the same look in Donna's eyes.

Steel folding chairs were set out on the roof but everybody was too nervous to sit. Portable speakers had also been deployed so the countdown could be monitored. Behind us, the 500-foot-high Vertical Assembly Building rose like a white cliff. In front of us was the route used by the 8-million-pound tracked crawlers to carry the stacked shuttles to their launch sites. The gravel road stretched eastward, its tan color bisecting the otherwise uniform green of the Florida lowlands. Three miles away the soaring lightning rod of Pad 39B, designed to protect space shuttles from lightning strikes, provided a sight line to *Atlantis*.

A broken layer of rainless clouds threatened a delay. While they posed no problem for an ascending shuttle, if an RTLS abort became necessary they would hide the runway and make the CDR's landing task more difficult. The countdown would be held at T-5 minutes until the weather improved. As I had done in the cockpit so many times, I now prayed fiercely for the count to resume. There was only one way to get the terror behind us . . . launch.

Dave, Greg, Bryan, and I circulated among the families translating the technobabble on the speakers. As word came that the weather was a go and the count was resuming, we faded to the rear of the group. This was a sacred family moment. The wives and the others needed to be alone with their thoughts and prayers, not feeling obligated to talk to us.

T-4 minutes. The two mothers, Kirby Thagard and Mary Jo Grabe, squeezed their children to their sides.

T-3 minutes. Several of the family members bowed their heads and closed their eyes, their interlaced fingers drawn tightly to their mouths. I was certain they were in prayer . . . as was I.

T-2 minutes. I could imagine the scene in the cockpit—the crew closing their helmet visors, cinching harnesses, exchanging good luck handshakes.

T-1 minute. The families were mute. One of the wives was shivering.

T-30 seconds. I looked at the kids and wondered how they would react to a disaster. *God, keep the crew safe!* It wasn't so much my prayer, as my demand.

The NASA voice took up the famous cadence. "T-minus 9 . . . 8 . . . 7 . . . go for main engine start . . . 6 . . . main engine start . . . 5 . . ."

A bright flash signaled SSME start. It was a sight that instantly brought excited shouts. Somebody clapped. The tension of the countdown had been broken and everybody felt a momentary, if premature, relief.

At SRB ignition *Atlantis* rose on promethean pillars of fire. The scene had a dreamlike quality to it. A 4½-million-pound machine was being borne upward on twin flames 1,000 feet long, and yet, there was no sound. That was being delayed fifteen seconds by the distance. The first noise to roll over us was the animal-like shriek of the SSMEs, which generated a new round of exclamations from the families. Six seconds later the SRB-generated noise came, a sound that made every listener wonder if the air itself was being tortured. It began as a rolling thunder, then quickly increased in decibel to a violent, ragged crackle. Birds jerked in midflight confusion. The noise echoed off the VAB wall and came back to shudder the LCC roof. From the parking lot below came the sound of car alarms, activated by the vibrations.

Atlantis entered the clouds and those gave momentary form to the shock waves of the SRB exhaust. They raced outward like the sonic waves of explosions. At booster burnout and separation the families cheered loudly. *Challenger* had forever stigmatized the SRBs and everybody was glad to see them, and the threat they represented, tumbling away.

With the twin rockets gone, the blue-white trinity of the SSMEs was all that marked the streaking machine. That fire slowly faded and within three minutes there was no sight or sound of *Atlantis*. Only the SRB smoke remained as a sign of her launch. That effluent had seeded the air so thoroughly with particulate that a cloud grew from it and showered the launchpad with an acid rain.

Now everybody was talking. The wives wiped tears and hugged one another. Through the excited chatter I kept an ear tuned to the speakers. I wouldn't be totally relieved until I heard the MECO call. At eight and a half minutes it came and I closed my eyes in prayer and thanked God there were no widows on that roof.

Back in the LCC, I called Donna to tell her of the successful launch and was shocked to find her sobbing in near hysteria. "Mike, I can't do it! I just can't do it again." She had watched the launch on TV and it had served as a terrifying reminder of what awaited her. She would have to make that T-9 minute walk again, probably multiple times, given my luck. For the first time in my life I was hearing my wife put herself first. But I wasn't going to step away from STS-36. That would never happen. I calmed her. "It'll be okay, Donna. Just this one more time and then it'll be over." I was confident she would rally. There were nine months until STS-36 would fly, which I hoped would give her enough time to shore up her emotional reserves.

But only six weeks later those reserves took another hit. We awoke on Father's Day to news that TFNG Dave Griggs had died the day before in the crash of a WWII aircraft while practicing for an air show. While his death was unrelated to shuttle operations, it was another grim reminder of the business of flying. Dave had left a wife, Karen, and two teenage daughters. One more time Donna and I drove to the home of a woman widowed in the prime of her life. One more time I watched Donna enter a sobbing clinch with a grieving friend.

After a service at St. Paul's Catholic Church, a group of us rendezvoused at the Outpost to continue Dave's wake. Kathy Thornton, a fellow crewmember on what would have been Dave's second shuttle mission, STS-33, brought one of the flower wreaths from the church and dropped it on the bar. With tears wetting her cheeks she sipped beer and remembered Dave with stories of their mission training. As I watched her, I wondered how many more tears would be shed for dead astronauts in this smoky dump of a bar. The sky was our Siren and no matter how we answered her call, in a plane or a rocket ship, she was always ready to kill us.

Though my assignment to STS-36 had buried all thoughts of immediate retirement, I continued to debate the course of my life after the mission was complete. Time and again I would resolve to tell Mike Coats (now the acting chief while Brandenstein was in mission training) that I would be resigning after STS-36, only to walk into something that would shatter that resolve. On one occasion it was an astronaut party at a local bar. Word spread through the crowd that the Long Duration Exposure Facility (LDEF), a large satellite launched before *Challenger*—which was to be retrieved and returned to Earth on STS-32—was seconds from being visible in the overhead twilight. We made a frantic dash for the doors. The bouncers and some redneck patrons interpreted this as the sign of a big fight moving into the parking lot. Imagine their surprise when they rushed outside to find thirty astronauts standing with their necks craned and fingers pointing at a bright point of light sailing overhead. I looked to my left at Bonnie Dunbar. In a month she would be the mission specialist at *Columbia's* robot arm controls, grabbing LDEF from space. *God, what an incredible business,* I thought. *How can I ever walk away from it?*

The same question arose at another social function, this one for a new group of astronaut interviewees. In their presence, I traveled through a time portal. A dozen years earlier I had been one of these eager young men, my eyes bright with the hope that I would miraculously make the cut, that I would be named an astronaut. They came to me, as I had gone to the vets in 1977, to ask what it was like to ride a rocket, what did the Earth look like from two hundred miles altitude? I could see it in their faces and hear it in their voices. They had been imprinted with a passion for spaceflight, just as I had been. How could I ever quiet that passion? How could I ever walk away from NASA?

I began the journey on December 18, 1989. After most astronauts had left for the day, I walked into Mike Coats's office and told him I was going to retire from NASA and the air force after STS-36. It was the most difficult decision of my life. There wasn't a eureka moment that had finally pushed me into it. Rather, it was a culmination of twelve years of "moments." Donna's telephone breakdown still weighed on me. My tour on the LCC roof with the STS-30 wives gave me a much better sense of what my launches were doing to her. And it wasn't just Donna's fear. My own fear had become a wearisome burden. How many times could I make the trip and survive?

The fear, the unknowns of the business, my doubts about NASA's management . . . all of it had conspired to propel me down the hall to Mike. But, even as I stood in front of him, I knew my decision was perilously balanced. I was like the circus acrobat tottering in a chair on top of a pole on top of a ball. The weight of a dust mote falling on my shoulder would be enough to send me toppling. If Mike had questioned my decision in the slightest manner—had he just said, "Are you sure?"—I suspect I would have immediately retracted my statement and walked away. But he didn't. As he continually flipped his pen, he confided to me that he had already made the same decision. He had told Puddy he would be leaving after his next flight. He probably heard me exhale. His decision endorsed my own. He didn't ask for an explanation, but I provided one. When I admitted that fear had a lot to do with it, he replied that it had been a huge factor for him, too—his own fear as well as Diane's. "That LCC roof wait is a torture." *There should be a monument to astronaut wives,* I thought. *It should feature the LCC with the countdown clock at T-9 minutes.*

I left Mike's office and called home. "I did it, Donna. I just told Coats we would be leaving after the mission." For a moment, Donna was silent. I had told her that morning of my retirement intentions, but I knew she didn't believe me. I had changed my mind too many times before. She understood the agony I was going through. She had seen the videos of me as a young teen running toward a parachuting coffee-can capsule. She knew the significance of my finger-worn *Conquest of Space* book. She had boxes of space memorabilia from my youth. She finally spoke. "Mike, I'm so happy. Thank you. I know you'll second-guess this decision to death, but it'll be okay. It'll all work out for the best. God has His plan." As I hung up the phone, I knew exactly what she was doing . . . lighting a candle of thanksgiving at her home shrine.

CHAPTER 38

"I have no plans past MECO"

I made my last flight to Kennedy Space Center as a member of a Prime Crew on February 19, 1990. With our T-38 afterburners tagging us with twin streaks of blue fire, we roared down the Ellington Field runway and shot into the dark. At our cruise altitude of 41,000 feet, we skimmed across the tops of massive thunderstorms associated with a cold front pushing through Dixie. Lightning illuminated the nimbus heads in colors of white and gray and blue. Fantastic shapes of electricity jumped from cloud to cloud. The charged atmosphere produced a St. Elmo's fire that painted the leading edge of our wings in a blue haze. As if that wasn't enough, I looked upward into a star-misted deep black. *God, how I'll miss this—the beauty and purity that is flight.*

At the KSC crew quarters, Olan Bertrand outlined the next three days of our lives. We were reminded to stay in health quarantine at all times, to not leave the quarters without telling Olan, to eat meals only cooked by the dieticians, to claim all meals on our travel voucher as "Government Furnished Meals." As employees of Uncle Sam we traveled on official orders, the itinerary of which read, *FROM: Houston, TX. TO: Earth Orbit.* All of our meals, transportation, and lodging would be provided by NASA, so we would only receive a standard per diem of a few dollars a day. A typical spaceflight earned an astronaut an extra $30 to $50 total.

We were also reminded not to carry anything personal on board the orbiter. When we climbed into *Atlantis,* everything on our persons, from the condoms on our penises to the LES helmets on our heads, was the property of the U.S. taxpayer. Gone were the days of astronauts stuffing their pockets with rolls of coins, corn beef sandwiches, and golf balls before their trips into space. In the early days of the space program these colorful items of contraband added a human interest touch to missions. Now they were forbidden, due in large part to an Apollo-era incident in which astronauts carried philatelic items they later sold. NASA felt it was inappropriate for crews to profit from items transported aboard a taxpayer-funded vehicle and imposed tight control over shuttle-flown material. Shuttle astronauts were restricted to twenty items in a Personal Preference Kit (PPK), which, together, could not exceed 1.5 pounds in weight. Weeks before launch, the PPK items were submitted to NASA HQ for approval and packed in a shuttle locker. In the PPK for my STS-36 mission, I had included the crucifix from my dad's casket, the wedding bands of my daughter Amy and her husband, Steve, a gold medallion that would be imprinted with the mission patch after landing, and some other small items of personal significance for the rest of the family.

Olan also reminded us to pack our clothes and wallets and mark the bag EOM (End of Mission) before leaving for launch. The items would be delivered to us at Edwards AFB after landing. "Also, include some civilian clothes in case of an abort." This was a standard request, but I knew some astronauts refused to pack their civvies out of a superstitious dread that to do so would *cause* an abort. I was one. If an abort occurred, I would just have to walk around Zaragoza, Spain, in my underwear.

As the VITT briefing ended, the family escorts arrived with our wives. They would be joining us for our midnight lunch. Unlike STS-41D and STS-27, both of which had banker's hours for launch windows, STS-36's window was from midnight to 4 A.M. This necessitated a killer sleep/awake schedule. We were going to bed at 11 A.M. and waking at 7 P.M. Breakfast was at 8 P.M., lunch at midnight, and supper at 6 A.M. A vampire kept better hours.

The wives were exhausted. Besides being Prime Crew spouses and wanting to see their men in the few opportunities remaining, they were also family entertainers for the week. Relatives and friends who were on a normal sleep schedule sought them out for KSC tour information, weather forecasts, and information on launch-day bus schedules. Cheryl Thuot and Chris Casper also had young children to deal with. If our wives were getting three hours of sleep a night, I would have been surprised.

They certainly didn't get much sleep this night. We met them again at the beach house for an L-2 barbecue dinner . . . at 8 A.M. Each of us was allowed an additional four health-screened guests, so this was a real crowd. In fact the gathering was too much. Introductions consumed a significant part of our time. The guests were also frantic to use this once-in-a-lifetime opportunity to get photos in every imaginable permutation: the STS-36 crew alone, the crew with wives, the crew with parents. All of this was being done on the outside deck so a billion no-see-um bugs also wanted to be in the picture. Photos were delayed while we slapped and danced and scratched them away. Inevitably we would pose for one shot and be into the next permutation when some old lady would scream, "I forgot to take off my lens cap. We'll have to pose that one again." Or, "I need to change film, so hold on." Or, "I left Aunt Betty's camera inside. Wait while I get it." Meanwhile I was thinking, *This is bullshit!* I had no patience for the extended families of the others and, no doubt, the other crewmembers were thinking the same thing as my mom jumped into firing position and started fiddling with her camera. I just wanted to be alone with my mom and children and Donna. I didn't want to share this time making small talk with people I had never met before.

During our first break from the Kodak moments I pulled my family to the beach. I would see Donna again, so I devoted my time to my mom and children. The kids were now old enough to make the beach house visit after being cleared by the flight surgeon. As I had been in previous launch good-byes, I was honest with them now about the risks. I didn't talk up the danger, but neither did I paint a Potemkin village for them. Since *Challenger*'s loss I was even more determined to keep them informed. I had heard that one of the older children watching *Challenger*'s destruction from the LCC roof had screamed, "Daddy, you said this could never happen!" I wanted my children and mom to know it could happen and to be prepared as much as possible.

I was filled with a father's pride as I watched my children. Pat was in the final weeks of his senior year at Notre Dame. He had attended the school on an ROTC scholarship and upon graduation was to be commissioned as a second lieutenant in the air force. He had matured into a real leader. He had also developed a wonderful wit. Friends who knew Pat and me would frequently joke, "The nut doesn't fall far from the tree." They were right. In many ways Pat was my clone—the most notable exception being his very good looks. But we did share the same crappy eyesight. As it had for me, his less than 20/20 vision was keeping him from air force pilot training. I had made calls to some general officer friends hoping they might know of a way for Pat to gain a medical waiver, but the Berlin Wall had come down, the Cold War was over, peace was going to reign forever, the air force had too many pilots, blah, blah, blah. All I heard were excuses. But I hadn't left it there. The *Challenger* disaster had shown me that dead astronauts had more cachet than live ones. Astronaut widows were consoled by presidents who told them to call if they needed anything. If I died on this mission I intended to posthumously use my celebrity status to get Pat into pilot training. Before leaving Houston, I had written a letter to fellow TFNG and USAF colonel Dick Covey on that topic. " . . . I'd like to thank you for taking care of Donna and the kids. I've always hated paperwork and I imagine dying generates more of it than anything else. At least I don't have to worry about it . . . I've told Donna to tell the president himself of Pat's desire to serve his country as a pilot. You might tell General Welch he's going to be getting the order from above, so he might just want to get ahead of the game by approving Pat's application right now . . ." Donna was holding the letter with my instructions to give it to Covey in the event of my death. (Visual acuity defects that are correctable have no flight safety impact—many military pilots wear glasses—and waivers have been given in the past for officers with glasses to enter pilot training.) There was also one other favor I had already asked of Covey: "If I die on this mission and there's anything left of my body, I don't want any of the female docs in the office doing an autopsy on me. I worry they'll get back at my sexist bullshit by telling everybody in the office I had a little dick." (A damnable lie!) Covey had a great laugh at that, but he promised to make my wishes known.

Amy, Pat's twin, was now married and living in Huntsville, Alabama. I didn't need to write any letters on her behalf. She was completely fulfilled as a wife and looking forward to the day she would be a stay-at-home mom. I had only recently come to accept her as she was. As obsessive-compulsive-West Point-engineer-astronaut fathers are apt to do, I attempted to fashion her into my own image and was frustrated by my repeated failures. I wanted her to graduate from college, but she dropped out after just one semester. I wanted her to have skills that would make her financially independent, if ever that was required, but she acquired none. Donna set me straight. "She's a sweet, good-hearted young woman. She doesn't want what you want. You just have to accept that." I finally had and was happy for her.

Laura was now nineteen and a freshman at DePaul University in Chicago majoring in the single degree field most guaranteed to drive an obsessive-compulsive-West Point-engineer-astronaut father into madness . . . theater. Laura wanted to be an actress. At least I had the experience of my older daughter to learn from—I accepted Laura's dream and enthusiastically supported her. I knew most theater degrees ended up being degrees in waiting tables, but it was her dream, and—as a man who had made a similar journey toward a long-odds prize—I wasn't about to discourage it.

As I walked the beach with my children beside me I was struck by the swiftness of life. The *Fiddler on the Roof* song "Sunrise, Sunset" came to mind. "Is this the little girl I carried? Is this the little boy at play?" It seemed like just yesterday I was running alongside bikes as the kids made their first no-training-wheels rides. Now, I walked with adults who would be doing the same thing with their children in the not-too-distant future. How swiftly fly the years.

I had missed a lot in those years, all in the name of career. How many times had I left for work with the kids asleep and returned after they were in bed for the night? How many evenings had there been no time to read to them or play with them because I had graduate school homework? How many birthdays had I missed? I didn't want to count. I had been a good father but not a great one. I had missed a lot that was irretrievable. The thought bolstered my confidence in the decision to retire from NASA. I didn't have many years left to build memories with my children and future grandchildren. I was forty-five years old—I was entering middle age. One glance in the mirror told me that. A proto-spare tire was faintly visible at my waist, and either my sink was going through puberty or my hair was thinning. I suspected the latter. I was only twenty-one years from the age of my dad's death. I had only thirty years left in an average life span. Of course, with a shuttle launch pending, all of those years were hypothetical. L-2 days to launch might mean two days until my death. So I gathered in this memory of my children as young adults . . . strong, healthy, attractive, their eyes and hearts set on the distant horizons of their lives.

Finally, our time was up. I hugged and kissed the kids. As I embraced my mom, she handed me a note. It was Psalm 91: *We live within the shadow of the Almighty, sheltered by the God who is above all gods. . . . Now you don't need to be afraid of the dark any more, nor fear the dangers of the day; nor dread the plagues of darkness, nor disasters in the morning.* I wasn't afraid of the dark or plagues. It was the "disasters in the morning" I feared.

Mom also handed me a card she had prepared and reproduced and was passing out to guests who had come to watch the launch. It was a prayer to Our Lady of Space. *Blessed Mary, Mother of our Savior and of mankind, watch over our men and women in space, and through your intercession, obtain for them protection from all harm and evil, grace to do God's will, and the courage and strength to fulfill their mission for the honor and glory of God. Amen. Watch over the crew of STS-36.*

That was my mom. She had the faith of ten people. Between her and Donna's prayers I should have been bulletproof. I just hoped Mom, the papist, didn't end up proselytizing to some Baptists in the family viewing area and find herself in a fistfight.

As they were climbing into the van for the ride back to the condo, Donna and Amy were wiping away tears. My mom, Pat, and Laura had their Pettigrew shields up. Their faces were pictures of worry, but they were dry-eyed.

Back at the crew quarters I went to the conference room to find that it had become a real bachelor pad. J.O., John, and Pepe had tossed money in the cash box, grabbed some beers, and were reviewing checklists while watching the Playboy Channel. I wondered how this was going to square with Our Lady of Space. Pepe's EKG recorder was on the table. As part of a life-science experiment, which would continue in weightlessness, he had been wired for heart data for the past week. I was glad I didn't have one. I could imagine what the docs would say when they saw my heartrate during one of my Prime Crew night terrors . . . *Holy shit!* They would never let me on the rocket. And I was *really* happy I didn't have to wear a bowel sound monitor. In their efforts to solve the mystery of space nausea, experimenters had requested some astronauts wear a microphone taped to their gut and a recorder on their hip to catch all the wheezes, pops, growls, and gurgles in their intestines. When the monitors were carried on secret DOD missions, the tapes could not be released to NASA until a USAF security official listened to them to be certain they hadn't captured a classified orbit discussion. That would be one heck of a job title to have on a résumé—bowel sound screener.

Pepe showed us his heart-activities log. The doctors had required him to record the time of every heart-affecting moment: bowel movements, meals, and each time he had intercourse. I noticed the "intercourse" column was blank. "Cheryl's not giving you any, huh?"

Dave Hilmers entered our company just as the TV was showing a topless Iowa coed doing naked handstands on a boat dock in a feature titled *Farmer's Daughter.* He was not happy. "Come on, guys . . . You shouldn't be watching this." I was surprised by Dave's complaint. Before I had been excommunicated from the office Bible study, I had learned he was one of the more religiously conservative astronauts. He wasn't the type of guy who would watch the Playboy Channel himself, but I had never heard him be openly critical of others who didn't share his belief system. Maybe it was the looming prospect of meeting Saint Peter and having to explain away the naked nineteen-year-old on the TV. I knew Dave was as scared as I was. While jogging with him, he had quoted a Scripture passage that he found helpful in calming his fears. At one of our crew parties, when asked about some postflight activity, he had dismissively replied, "I have no plans past MECO." Those six words spoke volumes. The threat to our lives was so real and immediate during the eight-and-a-half-minute ride to MECO, there was no reason to waste time with post-MECO plans. A post-MECO life was hypothetical.

Dave's criticism of our TV channel selection hung in the air. Our choice was either to appease him or watch the coed do naked handsprings off the dock and into a lake. It was a no-brainer. Pepe dismissed him. "Dave, a woman's body is a beautiful thing."

Hilmers shot back, "But not to be laughed at."

I defended Pepe. "We're not laughing. . . . We're lusting."

Dave just shook his head in resignation that our sorry souls were beyond saving. He picked up a checklist and began his review.

On L-1 we drove to the beach house for a midnight lunch with our wives and some NASA support personnel. There were no other family guests this time, thank God. On the drive I noticed J.O.'s cough was getting worse. It had been bothering him for the past several days. I wondered if the flight surgeon was aware of it. Knowing aviators, I doubted it. He probably wouldn't go to the doc until he coughed up a lung.

After the meal, the NASA guests departed and we were left with our spouses for the final good-bye. Donna and I bundled up and walked to the beach. Earlier at the crew quarters I had jogged under a crescent new moon, but it had set and the sky now offered a planetarium view of the winter constellations. There was enough starlight to illuminate the foam of the surf but the night air was chilly so we steered clear of the water. On the walk our eyes were continually drawn to the northern horizon. The xenon lights of Pad 39A sketched the salt air in shafts of white. *Atlantis* was being readied.

By now Donna and I were wizened veterans of shuttle launch good-byes. That's not to say this one was easier. It was actually harder. We had already confessed to each other that we were more terrified of this mission than either of the others because now there was an end in sight. It was hard not to recall all those Hollywood movies where the hero died on his last mission, and our last mission loomed. For me there was just one more time to surrender myself to the terrors of ascent. For Donna there was just one more T-9 minute climb to the LCC roof (or so she prayed). There was just one more adrenaline-draining launch. And then it would be over. We would finally close the NASA chapter of our lives and begin to write the post-MECO chapter. It would be a chapter filled with the fun and beauty of southwest living, of getting back to the deserts and mountains of our youth. It would be a chapter that would record the joys of watching Patrick and Laura follow Amy's lead into the bans of matrimony. And it would include the most wonderful part of life . . . grandchildren. Those were still glimmers in the eyes of our children, but we knew they would come. Yes, a wonderful post-MECO life awaited us. But to get there required one more "Go for main engine start."

With only the stars as witnesses, Donna and I kissed in a long and passionate embrace that we both knew could be our last.

By noon on February 21, 1990, I should have been an hour into sleep, but instead I was staring into the dark of my room. Even a sleeping pill hadn't done the trick. I could hear doors opening and closing, murmurs of conversation, J.O.'s worsening cough. I couldn't imagine he was going to be able to hide it from the flight surgeon and prayed it wouldn't generate a delay.

But it wasn't the faint noises keeping me awake. I was suffering a full-blown case of "last mission syndrome." I had experienced it in Vietnam. Then, the last mission had been identical to the other 133, but because it was my last, my fears had been magnified ten-fold. As I navigated the pilot to our reconnaissance target, I imagined every Viet Cong gunner in South Vietnam had my helmet in their sights. Now I envisioned every bug, glitch, and gremlin in *Atlantis*'s hardware and software was conspiring to end my life on STS-36. There was an SSME failure awaiting me, an APU fire, a turbo-pump ready to fly into a million pieces. My heart pounded in deep, thudding explosions. The adrenaline in my veins was eating the narcotic of the sleeping pill, like Pac Man munching through dots in his maze.

I got up, went to my desk, and switched on the light. I wrote to Donna and the kids:

21 Feb 1990, 12:22 P.M.
Dearest Donna, Pat, Amy, and Laura,

. . . I have no premonitions about this flight. I expect I will come home and we will pursue our other dreams in New Mexico. But there is no denying the danger. It is extreme, far beyond any other flying I've done. In spite of that I feel it is what I was born

to do. . . . If it is God's intention to call me, then I want you to know how much I have loved you before I leave. Darling, I have loved you. I haven't always shown it but I love you with all my heart and soul. You are the first and greatest gift God has given me. The children are the second. There are many others—my health, your health, the kids' health, my dreams, and the wings to reach those dreams. I want you and the children to remember that. I have been blessed beyond all human reason. Nobody should ever think that God was cruel to take me at this moment. Most people never come close to the self-fulfillment I've been blessed with.

I then wrote a separate letter to my mom:

21 Feb 1990, 12:47 P.M.
Dear Mom,

Just wanted to tell you for a last time how much I have loved you and Dad. You are the source of my mortal life and have filled it with such abundant love.

Please don't mourn my death. I've had the dreams of a thousand lifetimes fulfilled and died doing what I loved—to fly!

Your Psalm 91 was beautiful and I thank you for giving it to me and for praying so hard for me. Though in your grief you may feel God did not fulfill the promises of that Psalm, I think He did so wonderfully. He fulfilled every word of it.

Please continue to pray for me, as I will for you. I hope God allows me to get back to Dad. There's so much I would like to tell him. Also, tell Tim, Pat, Chipper, Kathy, and Mark that I loved each of them and wish them all the blessings I have known. I love you, Mom. Mike

I folded the letters and slipped them inside a pocket of my T-38 flight coveralls. Those were packed in my EOM bag and would be held for me at Edwards AFB. If I didn't make it back to claim the letters, I knew they would find their way to Donna and my mom. I climbed back into bed and somewhere in midafternoon a shallow sleep finally took me.

I awoke to the depressing news that the mission had been postponed for forty-eight hours. J.O. had been diagnosed with an upper respiratory infection. It marked the first time since *Apollo 13* that a mission had been delayed for crew health reasons. That J.O. had gotten sick didn't surprise me—we were all exhausted. The week of JSC quarantine to progressively shift our sleep cycle had been completely inadequate. And the sleeping pills were useless at inducing good periods of refreshing sleep. I was certain all of our immune systems were in chaos.

To prevent J.O. from infecting the rest of us, he was moved out of the crew quarters and into an unused room next door. He was being quarantined from the quarantine. At our 6 A.M. supper the rest of us put on germ masks and took him his meal. His room was an abandoned Apollo-era spacesuit facility painted brilliant white and illuminated with a full ceiling of intense fluorescent lights. Except for a small table, chair, and bed, the room was deserted. J.O. was seated at the table and, in the supernova lighting, he appeared remarkably like the old astronaut at the end of *2001: A Space Odyssey.* We wished him a quick recovery but none of us wanted to go anywhere near him. We placed his food tray on the floor and used a long-handled push broom to shove it close to his table and then immediately retreated from the room. He croaked a laugh at that.

J.O.'s sickness was just the beginning. John Casper began to feel poorly and went on medication. Dave Hilmers quickly followed. Then one of the mission support astronauts rushed from a briefing to vomit. The crew quarters had suddenly become a biohazard. Health-and-safety technicians entered in full-body moonsuits to swab the quarters for viruses. The flight surgeons also ordered urine and stool samples from all of us. I tried to ignore the turd request but I couldn't escape. I returned to my room from a briefing to find doctors Phil Stapaniak and Brad Beck had left me a not-so-subtle reminder—a Baby Ruth candy bar inside a collection container. My next bowel movement was into a Cool Whip dish. While I was forking a sample of the mess into a smaller container I screamed at the bathroom door, "This wasn't in the brochure!"

I sealed the smeared collection container in a plastic bag and dropped it in the garbage can. That contained more medical waste than a New Jersey beach: snotty tissues, other fecal collection bowls, packaging for antibiotics and decongestants, and an empty Pepto-Bismol bottle. STS-36 was being crewed by the walking wounded.

Pepe and I were the only healthy ones remaining. I prayed it would stay that way. If the bug slowly worked its way through all of us, there was the potential for a significant delay. If I was the last to be infected, I could get the giant screw. I could envision being pulled from the flight and a substitute MS taking my place. My old paranoia was back with me . . . that I would get to within hours of launch only to have the mission snatched away. In spite of the fact it was my third mission, that it wasn't going to make my astronaut pin any more "golden," and that

I was scared out of my wits, I still *needed* this mission as desperately as a heroin addict needs his fix. The flight surgeons were going to have to pry my jaws open to get a tongue depressor in my mouth.

CHAPTER 39

Holding at Nine and Hurting

Due to a bad weather forecast, the February 24 launch attempt was canceled before the gas tank was even filled. It was just as well since J.O. was still deathly ill. February 25 looked as if it would be the day. The weather forecast was good for the midnight opening of our launch window. J.O. looked and sounded like a consumptive, but he somehow managed to convince the flight surgeons he was okay. We headed for the suit room. For this flight I was wearing the diaper instead of the condom UCD. I was tired of worrying if the latex had slipped off. Maybe in my next life God would give me a penis more suited for spaceflight but for now I had to make do with what was there. The diaper was a more dependable choice.

We wore thermal underwear as an extra layer of insulation. Now that we had a bailout system, our pressure suits doubled as antiexposure suits and the long johns were intended to increase our survival time if we landed in the ocean. On the topic of surviving in the Arctic Sea during the dead of winter, we all had an Alfred E. Neuman, *What, me worry?* attitude. As Hoot Gibson had said on STS-27 (borrowing a line from *Butch Cassidy and the Sundance Kid*), "Hell, the fall is gonna kill you anyway."

Coverall-clad technicians awaited us in the suit room. Our bright orange LES pressure suits were draped on five La-Z-Boy–type recliners. Name tags on each recliner indicated which suit belonged to which astronaut, and every name but Pepe's was misspelled. He had snuck into the suit room earlier and made the bogus name placards in retaliation for our wearing of a prototype mission patch that bore his unusual French-Canadian-rooted name (Thuot) misspelled as Thout. We all laughed. Anything to ease the tension was welcome. I took a seat in the recliner marked "Mollane."

I removed my wedding band and put it in my glasses case. The ring could snag on something in an emergency escape. I would put it back on when I reached orbit. I had done the same thing for STS-41D and STS-27 and had come back alive. The act was now a ritual as much as a ballplayer's crossing himself in the batter's box. I had to survive to put the ring back on. God knew that and would protect me. Or, so I hoped.

The technicians helped me wiggle my feet into the bottom of the rear-entry suit. I then stood and zipped the integrated anti-G bladder around my gut. Next, I lowered my body and simultaneously inserted my arms into the suit's armholes while pushing my head through its neck "dam." The last time I had squeezed my head through an opening as tight as the LES had been at birth. The ring was intentionally made small so the rubber dam would clamp tight around the flesh of the neck and prevent the suit from filling with seawater.

After zipping the back of the suit closed, the technicians continued the dress-out. They laced my boots and buttoned my Snoopy-cap communication carrier on my head. Next came the pressure suit gloves. Finally the helmet was locked onto the neck ring. I was now fully dressed for the two things I never wanted to experience . . . a cockpit depressurization and/or a bailout from a wounded shuttle.

The techs connected their pressure test equipment and gave me my instructions. "Close your visor, turn on your O_2, take a deep breath and hold it." In the silence of the suit, my nervousness was obvious. I could hear my heart. It was squishing in my ears.

Next came a fully pressurized test. I gritted my teeth for this. As the suit filled with oxygen it stiffened to the consistency of steel. It felt as if a cinch was being run across each shoulder and through my crotch and then tightened until my prostate was in danger of being cleaved in half. Fortunately the test only lasted thirty seconds. I wasn't worried about dealing with the pain aboard the shuttle. If the suit ever pressurized in flight it would be because we had lost our cockpit atmosphere. In an emergency that dire, any suit pain would be insignificant.

The suit check was nominal and the technicians removed my gloves and helmet. They handed me a tray of items to stow in my pockets. I loaded a gas-pressurized space pen in the left-sleeve pocket and checked that my parachute knife was in my "pecker pocket," a sleeve on the inside of the left thigh. If I became entangled in my parachute and was being pulled to my death at sea, I could use the knife to hack at the shroud lines and hopefully save myself.

In my right thigh pocket I placed my spare glasses. In addition to my wedding band, I had secreted my mom's Psalm 91 in the case. Technically, the latter item was contraband, but it would only be discovered if

Jim Bagian and Sonny Carter were cutting the suit from my dead body. In that case I wouldn't care. In the same pocket I also loaded a radiation dosimeter, some aspirin for the zero-G backache that awaited me, and a Ziploc bag for stowing my used diaper once I was in orbit. I also included a barf bag. While I had yet to feel the slightest nausea in space, carrying the bag had become another ritual, as if that act alone prevented space sickness.

In my right ankle pocket I placed a mirror. The immobility of the suit made it impossible to look at the hose connections, so a mirror was needed. I also stowed my right pressure suit glove and tethered my bailout survival radio.

In my left ankle pocket I placed my left glove and more survival equipment: flares, strobe light, and flashlight. *Who am I kidding? The fall is gonna kill me anyway.*

As I was finishing with my stowage, some NASA photographers entered the room and snapped a few pictures. We all assumed casual, not-a-care-in-the-world, lying smiles.

I looked at my watch. Unfortunately we were fifteen minutes ahead of schedule. In this business you never wanted time on your hands. Mike Coats, who had been observing the suit-up, pulled out a deck of cards. He had been where we were and knew of the need to have every moment occupied. We gathered at a table for a couple hands of Possum Fargo, a dumb fighter-pilot card game that required no strategy—the worst hand won. As we played, Pepe observed that the flight surgeons seemed to have hung around him more than anybody else. "It was like they were two padres waiting to escort me to the gallows. I wonder if they thought I was going to panic or pass out or something."

Casper noted, "Pepe, you're always in a panic. How would they know?" He was right. Pepe was a twenty-four-volt guy in a twelve-volt world. He reminded me of a hummingbird in the way he darted at whatever he was doing, whether he was turning the page of a checklist, punching in a phone number, or flipping cockpit switches.

I added, "If they were watching for panic, they should have been at my seat." I told them of my pounding heart during the pressure test. Everybody nodded knowingly. We were all the same. Anybody who wasn't terrified getting ready to fly a space shuttle must have chased a couple Valiums with a fifth of vodka.

At my suggestion, we started playing draw poker. The game required more thought and was therefore a greater distraction. To the astonishment of all, I pulled two straights in just six hands. Incredible luck. As we were called for the bus, I hoped I hadn't used it all up.

In the elevator I noticed J.O. and Casper had net bags filled with flight surgeon–prescribed Afrin, throat lozenges, antibiotics, and other treatments. Casper held up his medicine bag and suggested a STS-36 motto: "Just say *maybe* to drugs." I wondered if this was a first in the space program . . . a commander and pilot carrying a small pharmacy as they headed for their machine. I wasn't worried. The autopilot would be flying us to orbit. If it wasn't, and J.O. was hand-flying us because of some malfunction, I knew he would do just fine. The threat of death has a way of focusing anybody, even a sick CDR.

It was 9:45 P.M. on a Sunday night when we stepped outside and headed for the astro-van. There were only a handful of NASA workers there to greet us, a fact that proved fortuitous for Pepe, as there were fewer people to laugh at his near fall. We had been instructed to stay close together on our exit so the photographers could get a *Magnificent Seven*–style group photo. Pepe was closest to the building door, and, as he passed it, his oxygen hose caught on the handle and he nearly walked right out from under his feet. It was great entertainment for the small crowd.

At the LCC the driver stopped to let Mike Coats off. He would get a ride to the Shuttle Landing Facility and serve as the weather pilot in the STA jet. Before stepping down, Mike had us all join our hands in collective prayer. We bowed our heads as he led, "God help you if you screw this up." It was a prayer that Dan Brandenstein had first composed and made famous among TFNGs. We all laughed, but knew that Mike spoke the truth. God better help us if we screwed up anything because management would shish-kebab our testicles if we did.

The van continued toward the pad and I sucked in every detail of the journey. The memories would have to last me the rest of my life, be that another few hours or decades. We sat facing one another, our suit-cooling units in the aisleway. In the high-humidity air the units produced a vapor that swirled around our feet. Several of us had one hand at our throat pulling back the rubber dam to allow an escape pathway for the cooling air being blown into the sides of our suits. Others solved the problem by inserting their space pens between flesh and rubber to create a flow path.

I watched the crew. Hilmers was quiet. I knew he was praying and that was more than fine with me. If God protected him, He would be protecting the rest of us Playboy Channel–watching children of Gomorrah. J.O. and Casper, still struggling with the effects of their illness, were subdued. Pepe and I were the motormouths, trying to hide behind our joking.

I observed, "With this swirling vapor at our feet, it's like being part of the STS-26 crew. Let's start singing 'I'm Proud to Be an American.' Come on, Dave, you know the tune." Dave Hilmers had been a "Return to Flight" crewmember.

Pepe quipped, "I forgot my badge. We're going to have to go back." We had left our NASA badges in our EOM bags—standard prelaunch protocol. With NASA security cars leading and trailing our van, the roadblock guards weren't about to stop us and ask for badges. It would be like the Vatican Swiss Guard stopping the pope mobile to check the badge of the guy in the funny hat.

Atlantis appeared above the darkened palmetto as an incandescent white obelisk. I couldn't imagine the gates of heaven appearing more brilliant or more beckoning. Everybody twisted in their seats to look and have their breath taken away. The scene instantly brought to mind Chesley Bonestell's painting *Zero Hour Minus Five* from my childhood book *Conquest of Space*. His winged rocket had been made of stainless steel but otherwise he had nailed the image. He had foreseen the soul-tugging drama of an illuminated spaceship standing ready against a star-filled sky.

As we drew nearer, the finer details of the pad appeared. The flame from the hydrogen gas waste tower streamed away in the wind. The same breeze snatched a vapor of oxygen from the tip of the gas tank. The white spherical supply tanks of liquid hydrogen and oxygen squatted on steel legs on either side of the pad like alien spaceships. The soaring finger of the lightning suppression mast seemed like an artistic touch, added merely to draw the eye skyward. The burnt orange of the ET and the American flag on *Atlantis*'s left wing provided the only color in what could have otherwise been an Ansel Adams photograph.

As we stepped from the crew van, the pad sights and sounds closed around me: the screeching hiss of the engine purge, the shadows playing on the vapors, workers marked with yellow light sticks hurrying to the booming call of the countdown, a light fall of snow from the maze of frosted cryogenic propellant lines. I crammed it all into my brain.

I stood at the edge of the gantry awaiting my turn for cockpit entry. I could feel Judy's presence. At this exact spot she and I had waited for our entry into *Discovery* . . . four times. My STS-27 flight had launched from Pad 39B, so this return to the 39A gantry was sort of a homecoming for me. I could see Judy's smile, her wind-whipped hair. I could hear her voice, "See you in space, Tarzan." I missed her. I missed them all.

Pepe came to my side. "Sure hope it all works."

I appended his comment. "I sure hope it all works *today.* I've got a sorry record for launching on a first attempt. This will be my seventh strap-in for a third ride into space."

As I walked toward the cockpit access arm I ran into John Casper, who was exiting the toilet. He was pulling his LES crotch zipper closed. I teased him, "We're going to have to call you Long Dong Casper. Nobody else has a lizard long enough to reach around the UCD, past the long johns, and out of the LES." He laughed. I was happy to help someone else relax, if only for a moment. I just wished someone would do it for me.

I got my wish in the White Room. One of the Astronaut Support Personnel (ASP) had placed a sign on the wall reading "Cut her loose!" This was the punch line of a particularly offensive joke—circulating among the Planet AD contingent—involving a naked woman bound to a bed. I chuckled. For five seconds I was able to forget what was about to happen.

Jeannie Alexander went to work securing me to the seat. Then she quizzed me on the components I would have to find in the event of a ground or bailout emergency. "Ship's O_2 connection?" It was on my left thigh. "Parachute disconnect?" My hands reached for my shoulders and I touched them. "Rip cord?" Another touch. "Barometric actuator lollipop? . . . Life raft actuator? . . . Overhead ground escape carabineer?" I found them all. She broke a light stick to activate its glow and Velcroed it to my shoulder. Now the fire-and-rescue people could find my body. She leaned over me and gave me a peck on the lips. "Have a great flight." She was another person I would remember for the rest of my life. As she left the cockpit she placed another light stick over the side hatch to show the exit in the event of a lights-out emergency. I heard the hatch close and soon my ears were popping as the cockpit pressure test started.

The wait began. My heart was in afterburner. The seat was a torture. My bladder distended toward rupture even as I was mad with thirst from my prelaunch dehydration efforts. And this time I faced a new experience certain to tax my reserves even more—I was launching in the downstairs cockpit. My dislike of the position had not changed since I last sat here on STS-27's reentry. I hated the tomblike isolation. I hated not having a window to look from or instruments to monitor. And I could never entirely expel from my mind the image of what it must have been like for Christa, Greg, and Ron in *Challenger*'s lower cockpit. There are some things best forgotten, but locked in the same box, I couldn't help but remember them.

God, let us go was my prayer. The thought of having to do this tomorrow made me nauseous. Maybe the cards had been a sign. Maybe seven would be my lucky number. At least I had Pepe's rookie complaints to entertain me. They came over the intercom in a nonstop flood: "Oh God, my back is killing me . . . My bladder is ready to explode . . . My stomach is being pushed out of my mouth . . . I got a cramp in my calf . . . I'm dying of thirst . . . I'm going to puke before I even get into space." J.O. jokingly asked how he would throw up in the LES while strapped to his seat. Ever the engineer, Pepe gave the question a moment of serious thought and replied, "I'll just roll my head around and puke in the back of my helmet."

Maybe I was just punch-drunk with exhaustion and fear but I found Pepe's constant dialogue hilarious. I was going to puke from laughing. J.O. warned, "Don't make me laugh, Pepe. I'll fall into another coughing fit." J.O. could barely talk without inducing a phlegmy hack.

After several waves of complaints, Pepe prefaced his next chapter with "Guys, I know I'm not a wimp, but . . ." and then continued the litany. John Casper picked up on the preamble and began to use it every time he had a complaint. "Guys, I know I'm not a wimp, but . . ." Soon the entire upstairs cockpit was doing it. Pepe heard my laughing cackle and continued his comedy routine by mimicking it . . . an explosion of rapid and high-pitched *hee, hee, hee*s. That got me giggling even more. If the launch director was listening he probably thought we had all gone insane.

Pepe's complaints faded and we looked for other ways to occupy our time and take our minds off our misery. We resorted to the old standby—roasting the flight surgeon. He was a captive audience, required to monitor our intercom but forbidden to speak to us directly unless we requested a conversation, and none of us were about to do that.

"I hear the doc's wife is having an affair with a chiropractor."

"And his daughter is sleeping with a malpractice lawyer."

"And his son is studying to be a malpractice lawyer."

There was a fake "Shhhhh . . . He might hear us."

"He's not listening. He's going over his stock portfolio."

"He's on the phone with his Hong Kong broker getting the fix on gold and the yen."

"He's probably phoning for a tee time."

"Hell, it's Sunday. He's not even there. Docs don't work on Sunday."

"Well, they don't work *sober* on Sundays."

Then we began to enumerate the perks the flight surgeons enjoyed. "They get hired by NASA as GS Infinities," a reference to the higher government service pay grade they were given.

"They get reserved parking places."

"They get preferential tee times."

"They just have to ask and women take off their clothes for them."

The banter finally ended and the intercom fell silent. Even Pepe got quiet. We all retreated into our own little chaotic worlds of pain and fear and prayer. Around T-45 minutes the range safety officer threw in the first wrench of what had been a smooth countdown. "RSO is no-go for blast." The blast to which he referred was the space shuttle being blown up. The RSO's computers had determined atmospheric conditions would amplify the power of the shuttle's destruction and jeopardize the safety of those around the LCC. His no-go call elicited groans and profanities in the cockpit. We'd reached the point of, *I'll kill anybody who gets in the way of our launch.* The RSO must have sensed the universal outrage at his no-go call and quickly reran the calculations to come up with acceptable numbers. "The RSO is go for blast." We all cheered . . . and laughed at the irony. We were cheering because a detonating shuttle would now kill only us and that was *good* because it meant the countdown could continue.

At T-5 minutes Casper started the APUs and the flight control system checkout followed. Everything was nominal and I was beginning to actually believe I had carried my luck from the card table to the cockpit.

"*Atlantis* . . . close helmet visors." Before complying with LCC's call I heard J.O. and John snort Afrin for a last time. I would be flying with a CDR and PLT on drugs.

I rechecked my harness. Other than look at a wall of lockers, it was all I could do. God, how I wished I was upstairs and had the distraction of the instruments. I had nothing whatsoever to do but dwell on my fear. I was the gas chamber victim waiting for the tablets to fall.

And then . . . "RSO is no-go for backup computer."

The intercom was immediately alive with our colorful assessments of the RSO: *bastard, asshole, sonofabitch!* We were now at the point of *I'll kill anybody AND THEIR WIFE AND CHILDREN AND MOTHER who gets in the way of our launch.*

The launch director ordered the countdown held at T-31 seconds in the hopes the RSO could clear his problem and the count could resume. But we couldn't hold for long with the APUs burning their fuel. A minute ticked away. *Come on . . . come on . . . fix your freakin' computer and give us a go for launch!* But as we waited, the liquid oxygen inlets on all of the SSMEs got too cold. The mission was scrubbed. I just melted into a formless blob. The suit technicians would have to look for me in the bottom of the LES.

Upon our return to the crew quarters we were offered the opportunity to go to the beach house and visit the wives. I called Donna and we both agreed we didn't want another beach good-bye. I could sense her complete exhaustion . . . mental and physical. I called my mom, the iron woman who had birthed six children and raised them with an invalid husband, and she was similarly incapacitated. The only silver lining to the scrub was that it reinforced my retirement decision. If stress was the killer the docs were saying it was, I was killing Donna, the kids, my mom, and myself with these launch attempts.

When the crew returned from the beach house, they found me in the conference room watching a movie. Pepe tilted his chair onto its back on the floor and lay in it to watch TV. "What the hell are you doing?" I was certain he had lost his mind.

"I've got to acclimate myself to lying in the orbiter. I was ready to die out there."

"Pepe, you're crazy. That's like practicing getting kicked in the balls. You'll never acclimate yourself."

But Pepe was not dissuaded. He remained in the reclined position throughout *Lawrence of Arabia*. I don't know how he did it. Only a gun to my head would have made me practice for tomorrow. I barely had the strength to lift a beer to my lips.

The next morning we relived it: Olan's Cajun face at my door, faking a smile for the photographers, having my nuts squeezed in the LES pressure test, confronting my fears on the drive to the pad, getting a kiss and a glowing light stick from Jeannie, laughing at Pepe's complaints, worrying about death, praying for life, and finally hearing, "*Atlantis*, the RTLS weather is no-go. We're going to have to pull you out." I didn't even have the strength to swear. This time the launch director decided to slip the mission by forty-eight hours to give everybody time to rest. Our next try would be on February 28.

Back in the crew quarters the techs stripped me out of the LES. After grabbing two beers from the kitchen, I walked to the bathroom, shed my long johns (reeking of sweat and faintly of urine), unfastened my diaper, and stood at the mirror. The craters under my eyes could have hidden a moon buggy. I wondered what a decent night of chemical-free REM sleep would feel like. It had been so long I couldn't imagine the experience. My neck was ringed red from the chafing of the LES neck dam. There were other suit tattoos: ruptured capillaries on the insides of my arms and bruises on my biceps from trying to move while the LES was pressurized. There were still multiple shaved and sandpaper-roughened hickeys on my chest from the EKG attachments applied during a prequarantine medical test. My thighs and calves had similar shaved and roughened patches of skin marking the attachment locations of sensors for a muscle-response test. The end of my penis was cherry red with what I could only hope was temporary diaper rash. Whatever it was, I wasn't about to bring it to the attention of the flight surgeons. If I had a urinary tract infection, it would come along for the ride. I had invested far too much in this mission to be pulled from it now. I entered the shower, stood under the cascading hot water, and drank my beer.

By the time we completed our debriefings the sun had risen and J.O. suggested we meet our wives at the beach house. I called Donna and this time we agreed it would be fun to get together.

The five of us entered the beach house living area to find it strewn with clothing: shirts, shoes, socks, panty hose, bras. There was even a bra swinging from the end of a ceiling fan. It was obvious we had entered a joke in progress. Sure enough, when we walked into the bedroom we found the family escorts, Hoot Gibson and Mario Runco, lying shirtless on the bed. Crowded next to them were all the wives, clothed but for their underthings, pretending to be *shocked* at our appearance. Everyone laughed, something we all needed as much as a good night's sleep.

Hoot teased us with the obvious point of the joke. "You guys are taking so long to get this mission going, your wives are developing some real *need* issues."

I threw it back in his face. "I'm not worried. You and Mario are navy officers. You have to be heterosexual to know what a woman needs. I'm surprised you guys aren't in a bedroom by yourselves."

Hoot and I had a well-deserved reputation for a disgusting synergy. Our exchanges devolved into more offensive comebacks and counter-comebacks until Donna finally hollered, "Enough! Will you guys ever grow up?" I had now heard that outburst from so many women so many times in my life, I thought it should be in Latin on the official shield of Planet Arrested Development—*umquam grow idiotum.*

The rest of the visit was relaxing. We had all been cured of the need to deliver a Bergman-Bogey good-bye at the water's edge, so we just sat around, drank beer, and traded stories. Pepe told us of his agony during the wait on the pad. Dave Hilmers shot him a hypothetical question: "Pepe, if NASA needs someone to replace an MS on the next flight, would you volunteer?" Pepe instantly replied, "Absolutely." His eagerness embodied the astronaut conundrum. Even as we waited on the pad, scared shitless and physically tortured, none of us could imagine not taking every offered mission.

When we returned to the crew quarters we were greeted by the local news showing a large, unmanned, French-built Ariane rocket blowing up shortly after liftoff from its South American pad. The story wouldn't have been covered anywhere else in America, but, on Florida's space coast, the competing French space program was news. The stations played the video again and again. There was no way Donna and the rest of the families could possibly miss it and I was certain the images of the flaming rocket falling into the sea would add to their anxiety. And that wasn't the end of it. That evening one of the networks was airing a docudrama on the *Challenger* disaster. The advertisements for that show were in all the newspapers and magazines, and the network was constantly hyping it. The wives were going to have to be sedated to get them to the LCC roof. With J.O.'s illness, the two scrubs, the Ariane blowing up, and the *Challenger* movie, it was a good thing I didn't believe in omens.

The evening of February 26 our crew flew to Houston for a refresher simulation. It had been so long since J.O. and John had practiced ascent emergencies, the mission trainers thought it would be a good idea to get them back in the JSC sim. I made the trip even though I had no duties associated with ascent. I just couldn't face the thought of sitting around the crew quarters all night with nothing to do. I had already watched more movies in the past thirteen days of quarantine than I had watched in the past thirteen years. I couldn't watch another. After landing at Ellington Field, I left the crew to their sim, drove home, watered the houseplants, and went running.

On the flight back to Florida I was stabbed with regret at my decision to leave NASA. The pain and fear that, yesterday, had provided validation for my retirement plans had been temporarily forgotten. Cocooned in the warm cockpit with the stars as a blanket, I wondered if I would ever find fulfillment outside of this business. There was an unknown scarier than space and I was fast approaching it . . . my post-MECO future.

This time I asked Jeannie to put light sticks over the single cue card Velcroed on the locker in front of me. The downstairs lighting was poor and I wanted the extra illumination to read the card. It outlined the procedures for a launchpad escape, for bailing out, and for a crash landing escape. I had every step committed to memory and didn't need the card but it gave me something to read during the wait. I also asked her to put a light stick next to the altimeter in front of me. In a bailout scenario, after pulling the emergency cockpit depressurization handle, I would watch the altimeter until it indicated we were below fifty thousand feet. Then, I would blow the side hatch and deploy the bailout slide boom. I would be the first out . . . into the ink black of a North Atlantic winter night and all the perils that it embodied.

Jeannie's face was beaded in sweat as she crawled over me to make my connections. Kevin Chilton, one of the ASPs, was the last to leave the cockpit. He pulled the pin that locked a safety cover over the cockpit depressurization and hatch jettison handles. Assuming we made orbit, I would reinsert the pin. He handed it to me. "Good luck, Mike."

"Thanks, Chilly. See you at Edwards."

I heard the hatch close, the mechanical *thunking* noise carrying a note of finality. A few minutes later J.O. watched from his port-side window as the last pad workers hurried across the access arm and entered the elevator. "The close-out crew just left. We're alone." J.O.'s observation reminded us that we sat at ground zero. Everybody else was racing to get away from the shuttle kill zone.

For the ninth time in my life I waited for launch. I was certain there would be a tenth time, tomorrow. The KSC weather was bad. I could feel the vehicle shaking in the wind and J.O. and John reported heavy rains lashing their windows from passing squalls. And it wasn't just the Florida weather that was a problem. Our two transatlantic abort sights—Zaragoza, Spain, and Morón, Spain (pronounced MORE-OWN)—also had weather issues. At T-9 the launch director held the count. God might have been punishing us for ignoring Dave's request to turn off the Playboy Channel.

Pepe's practice countdown in the briefing room chair had been useless in preparing him for another pad wait. It didn't take more than thirty minutes before he was once again entertaining us with his complaints. He ended one session with "My organs are shoving my diaphragm into my throat."

I replied, "You're wearing a diaphragm?" Everyone laughed so hard the engineers in MCC probably saw *Atlantis*'s vibrations in their accelerometer data. J.O. fell into a gagging wet cough. He was still not well, a fact that had made the *Houston Chronicle*. An unnamed source was quoted in that newspaper suggesting that J.O. was actually suffering from viral influenza. It wouldn't have surprised me if that was the case, but I was glad he was soldiering on. The longer our delay, the greater the chances I would become infected. (I would fall ill a day after landing.) I seriously doubted NASA HQ would hold the launch for my recovery, or the recovery of any infected MS, for that matter. As CDR and PLT, J.O. and John Casper were relatively irreplaceable. But with three MSes trained on the payload, any one of us was expendable. Given HQ's attention to the flight rate, I suspected management had already instructed JSC to have some MS substitutes standing by just in case. I prayed for a weather miracle.

Pepe outlasted me as the cockpit clown. He joked and complained without pause. Now he was reviewing all the movies we had watched during the past two weeks: "*Lawrence of Arabia, The Great Escape, How the West Was Won, The Terminator, Predator, Alien, Top Gun* . . ." We had seen more blood and guts than a meat packer. I resolved that the next movie I watched would be *Heidi*.

The T-9 minute hold dragged on . . . thirty minutes . . . an hour. Pepe gave us another item to consider. "I was just calculating . . . Since we started this never-go mission we've logged more than thirteen hours of on-the-back time. J.O. has even more time because he's first in and last out. In fact, J.O., you've been on your back for five hours just on this countdown."

"Thanks for cheering me up, Pepe."

I didn't have a body part that wasn't complaining. To ease the pain in my back, I loosened my harness and arched my hips upward. The restored circulation was heaven-sent but I couldn't hold the position for more than a moment. As my butt collapsed into the seat, a tide of cold urine squeezed from the diaper, climbed up my ass crack, and washed over my testicles. This was particularly disgusting knowing that, if we ever launched, I wouldn't see a shower for five days. If I did this tomorrow, which seemed certain, I was going to take my chances with the condom UCD.

As the groans and moans ricocheted among us, Dave Hilmers broke into song, "When the dog bites, when the bee stings, when I'm feeling sad, I simply remember my favorite things, and then I don't feel so bad." That threw me into the punch-drunk giggles again.

Pepe suggested a new song for Max-q (the astronaut band): "Holding at Nine and Hurting." It would have been a hit. At one time or another most astronauts have been there.

Rain continued to fall at KSC and the TAL weather looked grim, but observers at both places predicted a brief moment of acceptable launch conditions. The chances those moments would coincide were slim, but, with our launch window nearing a close, the launch director decided to give it a shot. He released the clock and we counted to T-5 minutes and held there. The wives were on the LCC roof. No doubt the rain made it even more miserable for them.

The wait extended. Even Pepe couldn't find anything more to say.

The only sounds were the steady breath of *Atlantis*'s cooling system and the irritating high-pitched whine of our pressure suit fans. The latter gave me a headache on top of my other pains. I refused to look at my watch, certain the digits were changing in quarter time. If there had been a glimmer of hope we would actually launch, the wait wouldn't have been so interminable. But we were all certain our investment in pain and adrenaline was going to be for naught. We would hold for the weather until the close of the launch window and then scrub. We would have to do it all again tomorrow.

I listened to the urgent voices of the launch controllers. Like us, they were exhausted and wanted to put the flight behind them and escape the inhuman sleep-work cycle. We were all gripped with a dangerous "launch fever," a headlong rush to get *Atlantis* flying. The sane one among us was our launch director, Bob Sieck. Nobody was going to stampede him into a wrongheaded decision. As he did a final poll of his LCC team he was calm, deliberate. Mr. Rogers singing "It's a beautiful day in the neighborhood" sounded manic compared with Bob's measured voice. Everybody listening wanted to jump in and finish his sentences. He was the perfect man for one of the most stressful jobs within NASA . . . and another person I would remember forever.

He polled the STA weather pilot and we heard Mike Coats reply, "Go." Next he polled the TAL weather pilot in Zaragoza, Spain, and got another go. There had been a blessed nexus of satisfactory weather conditions on both sides of the Atlantic. We were cleared to fly.

"*Atlantis,* we'll be coming out of the count in a few moments. It's been a real pleasure working with you guys. Good luck and godspeed."

I was shocked. For hours I had been convinced we would scrub. Now Casper was going through the APU start procedures. The clock was running. God had smiled on us. It had to have been Dave Hilmers's work. The rest of us reprobates didn't warrant any breaks from the Almighty.

I cinched my harness. My fear, which had ebbed with my certain belief the launch would be canceled, now roared over me like an avalanche. My mouth was metallic with it. My heart ran away with it. My hands shook with it. The palsy was a first for me. It had to be the combined effects of being downstairs and suffering from last-mission syndrome. Now, I was glad to be out of sight. Everybody knew everybody else was terrified, but nobody wanted to *see* their neighbor's fear, and trembling hands were a sure sign of it.

At T-2 minutes I closed my visor and turned on my oxygen. Again, I could hear J.O. and Casper snorting Afrin before they dropped their faceplates.

J.O. gave me a count. "One minute, Mike."

I squeaked out a "Roger."

"Thirty seconds, go for auto-sequence start."

"Fifteen seconds."

"Ten seconds . . . go for main engine start."

There was a heavy rumble followed by a 2-G slap. We were off. The rest of my life was just 510 seconds away.

CHAPTER 40

Last Orbits

At MECO I silently celebrated life. For the first time in what seemed an age, it occurred to me that I might live long enough to die a natural death.

We went to work on our mission activities, most of which I'm forbidden to describe. But the classified nature of both my DOD missions produced a mighty temptation for me. Riches and fame beyond anything any astronaut has ever achieved could be mine if I just told the world the *truth* . . . that on these hush-hush missions we actually rendezvoused with aliens. Given the vast population of conspiracy theorists, my claims would not be questioned. "Of course the government is hiding contact with aliens under the guise of military space shuttle operations," they would shout. I would be their hero for revealing what they have long suspected. Book and movie deals would net me millions. I would just need a convincing sperm-extraction and anal-probe story for my Barbara Walters interview . . . and to be able to look pained and violated as I told it.

On one occasion since leaving NASA, I did publicly make the "alien rendezvous" claim. I did it at Pepe's retirement ceremony. "Yes, we linked up with aliens," I told that audience, "and then had sex with them. It wasn't too bad after we got by the tentacles. Of course, Pepe, being a navy guy, picked the ugliest one."

One unclassified experiment aboard *Atlantis* proved immensely entertaining—a human skull loaded with radiation dosimeters. After returning to Earth those dosimeters would yield an exact measure of how much radiation was penetrating the brains of astronauts.

To reduce the creepiness factor of the experiment, the investigators had used a plastic filling to give the head an approximation of a face. The result was far more menacing than plain bone would have been. The face was narrow, cadaverous, with two bolts at the back of the skull looking like horns. Satan himself was riding with us. During a break in our payload work, I floated into a sleep restraint and extended my arms through the armholes, then ducked my head into the bag. Pepe and Dave taped the skull on top of the restraint so it appeared our friend had a body. (Your tax dollars at work.) They silently floated the bag to the flight deck and maneuvered me behind John Casper, who was engaged at an instrument panel. When he turned to find the creature in his face with arms waving, it scared the bejesus out of him. Later, we clamped Satan on the toilet. No doubt my desecration of the poor anonymous soul who had volunteered his body (and skull) to science has earned me a few more millennia in hell's fires.

With STS-36 I dodged the SAS bullet for the third time. Maybe, I thought, God had given me a free pass in space because I had vomited enough for ten men in the backseat of the F-4 during my early flying career. Whatever the reason, I was happy to stow my unused barf bag. John Casper looked as if he might need it, but maybe not for SAS. It could have been his meal of eggplant and tomatoes. Gag. The NASA dietician included it because John's other meal choices (heavy on butter cookies, M&Ms, and chocolate pudding) would leave him short of magnesium. I would have rather chewed on a magnesium flare. John hadn't eaten the entrails-looking dish, but just rehydrating it would have made me sick. Regardless of the cause, John was feeling poorly and called on Dave Hilmers to inject him with the antinausea drug Phenergan. NASA had all but given up on patches and pills to treat SAS and had converted to industrial-strength injectable drugs. I reminded John of the warning the potion carried, "Do not operate an automobile under the influence of this drug." He replied, "Lucky for me it doesn't say anything about operating a space shuttle."

Dave Hilmers was no doctor. He just played one in space. His premission training with needles had consisted of jabbing one into a piece of fruit. I would have to have been near death before I would have let a marine come near me with a needle. I expected to see blood and I wasn't disappointed. Dave accidentally moved the needle while it was embedded in John's ass and blood followed. A line of ruby planets shot from the wound like soap bubbles being blown out of a ring (giving new meaning to the term "airborne pathogen"). We chased the spheres with tissues. (Dave Hilmers must have been inspired by his needle work. After retirement from NASA he completed medical school and is now a pediatrician with a practice in Houston.)

While I had no nausea, I did experience the same painful backache from spine lengthening that I had encountered on STS-41D and 27. I also noticed the same Viagra effect. Every morning I would find myself painfully afflicted with a diamond-cutter erection, just like the geezers in the movie *Cocoon*. And I wasn't the only one dealing with this problem. On one reveille, as we all floated in our sleep restraints, Pepe looked at me and said, "I must have had a great dream about Cheryl [his "snort" cute wife]. I've got a terrific boner."

I smiled and replied, "I must have had a great dream about Cheryl, too."

Pepe laughed. "Damn you, Mullane! Keep my wife out of that filthy brain of yours."

Someday the blood shift of weightless flight will make for some very happy space colonists.

During the last sleep period of the mission, I stayed awake in the upper cockpit to soak up the space sights that would have to last the rest of my terrestrial life. I wanted to listen to music as I did so and searched for my NASA-supplied Walkman. It took me a moment to find it. The inside of the cockpit was covered with Velcro pads, and everything we carried, from pencils to cameras to food containers to flashlights, had Velcro "hooks" glued to them so they could be anchored to a pad. The only problem was remembering where you anchored everything. On Earth, nobody ever had to look on a wall or ceiling for a misplaced item. In space you did.

I put on headphones and inserted one of my personal-mix music tapes in the player (NASA allowed us six), then switched off the cockpit lights. Floating horizontally, I rolled belly up and pulled forward until my head was nearly touching a forward cockpit window. It was a trick Hank Hartsfield had taught me on STS-41D. With *Atlantis* in a ceiling-to-Earth attitude, my orientation had me lying facedown toward Earth. Though this attitude caused my body to brush against the ceiling instrument panels, which contained some of the most critical shuttle switches, I wasn't worried about bumping one out of position. All the switches were set between two wire wickets so they could only be accessed by a thumb and forefinger inserted between those hoops.

The real joy of my new position was the illusion it created. I could put my head so far forward that the shuttle's structure disappeared behind me. My view of Earth was completely unobstructed. It brought back memories of snorkeling in the Aegean Sea and watching the undersea life through my face mask. As I had then, I now had a powerful sense of being part of the element in which I was immersed, not a foreign visitor. When I steadied myself with my fingertips and then pulled those away, I would momentarily float free of any contact with *Atlantis,* enhancing the sensation of being a creature of space, not an astronaut locked in a machine.

To the strings of Beethoven's "Ode to Joy" I watched my planet silently move under me. But this time I was seeing it as never before. Not only was our orbit steeply tilted to the equator, we were also in one of the lowest orbits ever flown by a space shuttle. We were scarcely 130 statute miles above the Earth, approximately the distance from New York City to the eastern tip of Long Island, or Los Angeles to San Deigo. At this altitude the planet was hugely close and there were new details of its earth, sea, and sky to thrill me.

I could see the patina of the Earth's oceans. The wind-rumpled water gave them the texture of an orange rind, but in colors that varied with the angle at which the Sun stuck them. At high sun, the open seas were Crayola blue. At grazing angles, they reflected tones of gray and silver and copper. In places of exceptional water clarity, like the Caribbean, the dunelike humps and valleys of the seafloor were clearly visible, their white sand diluting the ocean blue to yield a striking turquoise. In the sheen of the Sun I could see evidence of the dynamics of the sea. There were circular eddies similar to the low-pressure-cloud swirls in the atmosphere. Boundaries between currents appeared as dark lines. Currents past headlands would create noticeably different downstream wave patterns, exactly like the ones I could see in clouds downstream from mountain ranges. In Persian Gulf anchorages I could make out the "dots" of supertankers and occasionally, in the glint of the Sun, I would catch sight of the V-shaped wake of one of these monsters under way. Later, as *Atlantis* was on the descending portion of an orbit deep into the southern hemisphere, I watched the miles-long bluish-green ribbon of a bloom of plankton. We had been told to expect to see these in the fertile waters approaching Antarctica. Farther south, a flotilla of icebergs sailed on currents like so many ships of the line.

At the southern limit of her orbit, *Atlantis*'s nadir came within three hundred miles of the coast of the Antarctic continent, now in late summer. I pulled a pair of gyroscopically stabilized binoculars from their Velcro anchor and peered southward. The pole was nearly 1,800 miles distant, so I had no view of it. Instead, I focused on the rugged coastal mountain chains. The occasional black of a windswept cliff was the only color in an otherwise sheet-white topography.

Atlantis curved northward and began her 12,000-mile fall toward the opposite end of the Earth. It was a remarkable physics that kept me on this godly merry-go-round. We were literally falling. Just as a thrown ball falls in a curve, *Atlantis* was on a curving trajectory to impact Earth. But impact never came because the Earth's horizon was continually bending out of the way. *Atlantis*'s engines had thrown her onto a falling curve that matched the curvature of the Earth. In my upper-cockpit perch, I had no sense of that fall but in the windowless mid-deck I had experienced brief moments in which the sensation had been overwhelmingly powerful. The day before, I had been seized with an illusion that the mid-deck cockpit floor was steeply tilted and if I didn't grasp something I would slide down it. Try as I might I could not convince myself that I would not fall. I actually seized the canvas loop of a foot restraint to keep from sliding off my imaginary cliff. The sensation was so distracting I finally abandoned the mid-deck and floated upstairs. The view of the Earth's horizon immediately eradicated any sense of the fall.

The ocean under *Atlantis* was now the Pacific. The sun dropped and its terminator light painted a scattering of cumulus clouds in coral pink. In the darkness that followed I looked spaceward to the unfamiliar stars of the southern hemisphere. The Magellanic Clouds were visible as hazy smudges. A quarter moon rose. Seen through the thick part of the atmosphere, the orb was severely distorted, appearing boomerang in shape, an effect of the light-bending qualities of the air. The crescent tips were squeezed inward and the greater surface bulged outward. Only after rising above the atmosphere did the crescent appear normal. Then, it cast a spotlight of silver across the water. Except for its grand scale, the sight was identical to watching the moon rise over the sea from a Cape Canaveral beach.

Just twenty-two minutes after leaving Antarctica's seas, *Atlantis* passed over the equator and I was treated to the never-ending light show of the intertropical convergence zone. Here, the trade winds of the northern and southern hemispheres mixed in equatorial heat and humidity to produce perpetual thunderstorms. The nimbus clouds took on the appearance of sputtering fluorescent lightbulbs, so continuous was the lightning within them.

Atlantis crossed Central America in less than a minute and I looked ahead to America's East Coast. In a six-minute passage, the city lights of the entire seaboard passed by my window: Key West, Miami, Jacksonville, the cities of the mid-Atlantic, then Washington, D.C., Philadelphia, New York, Boston, and Portland. The lights sprawled over the darkened continent like so many yellow galaxies.

Twenty-two minutes north of the equator, *Atlantis* brushed the Arctic Circle. The deep night of winter in the northern hemisphere made it ideal for viewing the lights of the aurora borealis. I watched them grow and collapse in their ephemeral, spiritlike dance. Streamers of emerald green and fuchsia waved as if rippled by the wind. The lower end of one curtain took on an intense glow, like the head of a comet, its attached streamer trailing away like a sun-blown tail. The lights were so captivating I watched them until they were just a haze on the receding horizon, and I was happy to know I had tickets for the next show starting in ninety minutes.

I moved to the back cockpit to enjoy a different light show . . . the atomic oxygen glow engulfing *Atlantis*'s payload bay. The low-orbit space through which *Atlantis* plunged was not empty. We were in the outer reaches of the Earth's atmosphere, which contained billions of atoms of UV-altered molecular oxygen known as atomic oxygen. The wind they produced was vanishingly thin, but it was enough to react with the shuttle's windward surfaces to cause a Saint Elmo's–like fire. The glow was so intense it appeared we had flown into a hazy alien fog. Every affected surface was covered to a depth of several feet. If we had not been warned of the phenomenon, I would have worried we had passed into the Twilight Zone and our spaceship had been transformed into a ghost ship. We had been damned by the curse of the skull man.

Atlantis curved over northern Europe toward another forty-five minute day. If ever there was a music composition perfect for watching the beauty of an orbit sunrise, that composition would be Pachelbel's Canon. As the violin melody played on my Walkman, the rising sun painted the horizon in twenty shades of indigo, blue, orange, and red. God, how I wanted to stop and just hover.

I took off my headset and watched the Earth in silence. I also wanted to *hear* spaceflight and seal that memory in my mind. The cabin fans stirred the air with their constant soft whoosh. From downstairs I could hear the muted clatter of the teleprinter printing out checklist changes and weather reports for tomorrow's reentry. Someone coughed. The UHF radio captured the gibberish of a foreign pilot talking to his controller somewhere below.

I inhaled the smell of *Atlantis*. There was no evidence of the humanity that inhabited her, no odor of our bodies, our food, our waste, our emesis. The engineers had done a remarkable job of filtering the air. The only "smell" was that of unnatural sterility. I missed the scents of rain, desert, and sea . . . and I had only been away from the Earth for four days. I wondered if engineers would ever be able to package smells of our home planet so that Martian pioneers could remember their roots. For their sakes, I hoped so.

I took a moment to look around *Atlantis*'s cockpit and capture that memory, knowing that when I crawled from her side hatch tomorrow it would be for the last time. The windows and floor were the only surfaces not covered with switches, controls, circuit breakers, computer monitors, or TV screens. Cue cards dotted the panels. Bound checklists were similarly scattered on Velcro pads. Twelve years ago, I had been overwhelmed with the machine's complexity. Now, the cockpit was as familiar and comforting as my living room.

I turned and looked forward. The PLT's seat belt hovered like a charmed snake. The three computer screens were off. No reason to waste power during a sleep period. My eyes touched on the life-and-death switches I had so often feared might play a part on one of my missions: the abort selection switch, the SSME shutdown buttons, the BFS engage buttons. I would never need any of them and I thanked God for it.

I floated back to the forward windows. The orbits continued . . . 25,000 miles, 90 minutes, one sunrise, one sunset, a brush with the Arctic Circle, a brush with the Antarctic Circle. At each equatorial crossing *Atlantis* passed 1,500 miles west of its prior transit, an effect of the Earth's eastward spin underneath our orbit. In circuit after circuit, I was seeing a different sea, a different land, a different sky. I watched North African deserts stretch to the horizon in dunes as perfectly spaced as ripples in a pond. I passed over snowy Siberian forests as virgin as the Garden of Eden. I saw the green vein of the Nile and the white-tipped chaos of the Himalayas and the Andes. I saw perfect fans of alluvial debris debouching onto desert floors, each a signature of millions of years of mountain erosion. I thrilled to shooting stars and the stellar mist of space and twinkling satellites and the jewel that was Jupiter. I saw the Baikonur Cosmodrome, *Sputnik I*'s launch site, with the nearby Aral Sea appearing oil black against the winter white of the Kazakh Steppes. A few turns later the desert-lonely lights of Albuquerque came into view and I marveled at how those two places, so geographically distant from each other, had been inexorably linked in my life. I passed over every unimproved road my parents had ever dared, every mountain I had ever climbed, every sky I had ever flown. With the music of Vangelis and Bach and Albinoni as a sound track, I watched the movie of my life.

CHAPTER 41

The White House

Our first order of postlanding business was to review our mission film and edit two separate movies, one intended for security-cleared eyes only, the other for the public. Because of the secrecy surrounding our orbit activities, the latter had little in it. We wanted to include the fun video we had taken of our satanic crewmember in hilarious poses, but Dan Brandenstein squelched that. "If we keep showing on-orbit pranks, headquarters is going to assume control of editing our postflight movies. They're getting pissed the press only shows us screwing off in space." We thought it was bullshit, but understood Dan's position and honored it. The world would never see Beelzebub clamped on a shuttle toilet.

Our postflight travel was similar to that of STS-27. I journeyed to places I can't mention to be congratulated by people whose office titles are similarly unmentionable. I received another National Intelligence Medal of Achievement from another "black world" Wizard of Oz that I could only wear in a vault. This citation (declassified years later) reads:

. . . Colonel Mullane's superior performance led to the safe deployment and successful activation of a system vital to our national security. The singularly superior performance of Colonel Mullane reflects great credit upon himself, the United States Air Force, the National Aeronautics and Space Administration, and the Intelligence Community.

At one of our stops some spooks hosted us to a candlelit dinner in their black-world building. The office secretaries acted as servers since no caterers could enter. We showed our mission movie and, lubricated by wine, I added my own editorial comments. As space video of the Boston–Cape Cod area was shown, I injected, "Moscow doesn't have as many communists as are living in this picture." There was a peal of laughter. Hank Hartsfield would have been proud.

The highlight of our meager postflight PR tour was a visit to George Bush, Senior's White House. We were shocked by the invitation. STS-36 had been virtually ignored in the press. There were no women on the crew, no minorities, no firsts of any kind that might have turned out the press to cover a presidential handshake. Whatever the reason, the invitation was sincerely appreciated.

We met the president in the Oval Office, taking seats in sofas set around a coffee table. Mr. Bush sat in a nearby chair. The questions he asked indicated that he was well briefed on our mission. But it was hard to carry on a conversation. A steady stream of aides and secretaries were constantly coming to his side to get answers to questions and his signature on documents. I wondered if the man was ever alone, even on the toilet.

I knew my air force master sergeant dad was watching from heaven, his chest puffed up with button-busting pride. It was a proud moment for me, too. What my crewmates and I had done on STS-27 and STS-36 would probably remain classified for decades. We were the most invisible of astronauts. Nobody would sing "I'm Proud to Be an American" while we were raised on a platform before the cheering masses. Our names would never be in the lyrics of a Billy Joel song. But this was infinitely better. I was standing in the Oval Office of the White House while the president of the United States shook my hand and thanked me for my contribution to America's security.

Later, we gathered behind the president's desk to have a crew photo taken. The desktop was littered with documents bearing red-striped "Top Secret" covers. John Casper whispered, "Mike, look at his notepad." I did. On it was written "Gorb dinner?"—obviously the president's self-reminder about something associated with the upcoming visit to Washington by the Gorbachevs. I whispered back to John, "Maybe he's looking for a joke to loosen up things at a state dinner. Why don't you suggest a golfing joke with a cow's ass in the punch line?"

"No" was John's terse reply.

After we finished the classified discussions, Mrs. Bush ushered in our wives to meet her husband. We all posed for photos with the First Family. The president gave each of the crew a pair of cuff links embossed with the presidential seal and the wives received a stick pin with the same logo.

It was a beautiful May day and the doors to the Rose Garden were open. At one point during the photo session a bumblebee joined us and hovered around the president's brightly colored tie. An aide shooed it away, and it found another target . . . a secretary who obviously had a phobia of buzzing insects. She screamed, threw a sheaf of papers in the air, and began to run in circles, flailing at her hair and trying to escape the insect. This was hardly a scene I expected to witness in the presidential Oval Office. I whispered to Pepe, "I sure hope she doesn't fall on the button labeled 'DEFCON 1.'"

We left the president to his never-ending work and followed Barbara Bush on a tour of the White House. If I had not been aware she was the First Lady, I would have never guessed it from her behavior. She was talkative, witty, and completely devoid of any air of celebrity. She reminded me of my mother. I could easily picture her baiting a hook or hoisting a beer or throwing another log on the campfire.

We stepped into an ancient elevator for a trip to the upstairs living quarters. With five astronauts, five wives, Mrs. Bush, and an assistant, we were cheek to jowl in the small volume. Mrs. Bush was directly behind me and I did my best to resist being crushed into her front. Before the elevator door closed, Millie, the first dog, somehow managed to wiggle under our feet to make it an even tighter squeeze. As the box crept upward, the silence was total. In spite of Mrs. Bush's easy manner we were all very self-conscious of her company. To occupy the uncomfortable seconds we watched the elevator indicator panel with the same intensity as an astronaut watching a space rendezvous. Some of us moved slightly to accommodate the dog. Chris Casper, John's wife, finally cracked under the oppressing silence. She nervously offered an icebreaker—"Oh, I feel it between my legs." While it was obvious she was referring to Millie's wagging tail, the words hung over our sardined group like really bad flatulence. A reference to anything between a woman's legs was tough to comment on in polite company, much less in the company of the First Lady of the nation. Chris quickly realized her mistake and tried to recover by amending her words. She nervously added, "I mean I feel the dog between my . . . er . . . my legs."

It was just too much for me to keep my mouth shut. She had served up a ball just begging to be spiked. I couldn't resist. "Are you sure it's not John's hand?" I inquired. My comment elicited a few snickers and an elbow jab from Donna. As had frequently been the case in my life, I immediately wished the joker in me would have kept quiet. *What was Mrs. Bush thinking?* I wondered. Maybe this time I had gone too far.

I need not have worried. As regret shot through my brain, I felt Mrs. Bush's hand lightly pat me on a butt cheek as she said, "*That's* John's hand." Then she winked at Donna and said, "I've got him right where I want him." I was stunned. She was a Mike Mullane clone. She couldn't let a perfect setup fall to the sand—she had to nail it.

Upstairs her joking continued. She halted in front of a painting of some daughters of a forgotten nineteenth-century president. "What do you think about this portrait?"

We were all mute. The women in the painting had a striking resemblance to hogs wearing wigs and gowns. They were creatures right off of Dr. Moreau's island of horrors. As our collective silence was fast approaching embarrassment, Mrs. Bush took the heat off and answered her own question. "This is the ugliest painting I've ever seen. The women were part of the First Family, for God's sake. They could have requested some artistic license. What were they thinking? For my official portrait I intend to get an artist who will make me look good."

She led us to a room with a view of people waiting to begin their White House tour. The crowd screamed in delight and grabbed their cameras when they saw Mrs. Bush waving. She was a queen who deported herself in every way as a commoner.

She was also a proud mother and grandmother. On every table and mantel were framed photos of her family. I didn't see a single photo of her posed with any of the multitude of stars she had certainly met in her life. Clearly her VIPs were her children and grandchildren. She spoke of her philosophy of life: "In your old age you will never regret the contract never signed, the trip never taken, the money never earned, but you will definitely regret it if your children turn out poorly because of neglect." She used Ronald Reagan as an example. "He's a wonderful man but he has four children who won't speak to him." Maybe she was giving us the unsolicited advice because she could see in our eyes how driven we were. If there was ever a collection of men vulnerable to neglecting their families, it was astronauts.

We sat for tea and cookies and she told us stories about some of the people she had met and unusual places she had traveled. She volunteered her thoughts on a controversy in which she was embroiled and that was being given significant press coverage. She had been invited to give the commencement address at Wellesley College, but, after accepting, some of the students had organized a movement to disinvite her. These women considered her a poor role model since her only identity was through her husband. Apparently, for them, being a wife and mother were not qualifying credentials for a commencement speaker. Mrs. Bush was completely gracious and accepting of their dissent, but from the first moment Donna had seen the story in the newspaper she had been furious. Donna had spent her life as a wife and mother and didn't consider herself a second-class woman for having done so. I worried she was going to offer an opinion to Mrs. Bush along the lines that those Wellesley girls were just a bunch of small-minded, immature bitches, but she maintained her composure. Fortunately Donna didn't have my hair-trigger mouth.

After tea, Mrs. Bush led us downstairs to finish our tour, giving us a running commentary on the history of the rooms we passed. But she skipped over some recent history I was privy to. An astronaut who had made an earlier White House visit had told of entering a room in the company of Mrs. Bush and being brought to a sudden halt by the overpowering stench of fresh dog shit. Everybody had quickly fixated on the source . . . Millie's deposit. The astronaut witness had recounted how a silence as heavy as the odor had enveloped their group. Nobody wanted to acknowledge the obvious, that Millie had desecrated the carpet. But, without missing a beat, Barbara Bush turned to look at her astronaut visitors and jokingly warned, "If I read about this in the *Post* tomorrow, you're all dead meat!"

Mrs. Bush would have fit perfectly into our TFNG gang. I could see her at the Outpost and Pete's BBQ and on the LCC roof. There are some things the trappings of wealth and power and great political office can never dissolve. Among these are the bonds of the military family. As the wife of a WWII naval aviator, Barbara Bush had long ago experienced everything we had lived and were continuing to live . . . fear, the heartache of hearing "Taps" played over friends' graves, and consoling grieving widows and fatherless children.

As we walked away, I thought of those dissident Wellesley women. They had been right about one thing—Mrs. Bush shouldn't have been invited to speak at their commencement merely because she was the First Lady. Any woman could be one of those. Rather, she should have been invited because she was a member of the Greatest Generation, because she had kissed her man off to war and been left to wonder if she would ever see him again, because—as the loving and supportive wife of a WWII naval aviator—she had done her part to save the world. Those were commencement address qualifications for any college, even Wellesley.

CHAPTER 42

Journey's End

In May 1990, I retired from the USAF and NASA in an astronaut office ceremony attended by thirty or so of my peers. The gathering was held in the main conference room where, twelve years earlier, I had first heard John Young welcome our TFNG class. Dressed in my air force uniform, with my ribbons and the astronaut wings I had flown in space pinned to my chest, I accepted the Air Force Legion of Merit from USAF Major General Nate Lindsay. Nate had become a close friend over the course of my two DOD missions and I was honored he and his wife, Shirley, had taken the time to fly to Houston to attend the ceremony. Donna, my mom, and my son, Pat, were also in attendance. Even my Pettigrew genes couldn't completely subdue the emotions that stirred in my soul at the sight of Pat. I could feel my throat tightening and my eyes welling. I began my air force career at my commissioning on the Plain at West Point in 1967. At that time, Donna and my mom had each pinned on one of my second lieutenant butter bars. Now, my twenty-two-year-old son, dressed in his air force uniform and wearing the same virginal rank, was shaking my hand and hugging me.

I kept my comments brief knowing everybody had to get back to work. Somewhere there was a countdown clock urgently marking the weeks to the next launch. I thanked everybody for their years of support, making a special reference to Pat, Amy, Laura (the girls had been unable to attend), and my mom. I saved my greatest praise for Donna. Tears threatened to douse her cheeks. I then concluded with the observation that I was the third generation of my family to have seen combat. My maternal grandfather had served in France in WWI and my dad in WWII. I had done a tour in Vietnam. I offered the hope my children's generation would never see a war. At that time it seemed like a sure bet. The Soviet Union had ceased to exist. How could there ever be another threat to America as great?

That night, the astronaut office hosted a going-away party for Donna and me at a local restaurant. Beth Turner, one of the office secretaries, obtained a life-size cardboard rendition of a studly bodybuilder and placed it at stage center. She covered the face with my astronaut photo and the crotch with a sequined jockstrap stuffed with something flattering. Against this backdrop Hoot Gibson roasted me with stories of my botched T-38 landing in Brewster Shaw's backseat, my near death experiences while performing STS-1 chase duties with "Red Flash" Walker, and my intercom comment from STS-27— "The RSO's mother goes down like a Muslim at noon." He also recounted how a group of female DOD security secretaries, tasked with declassifying our STS-27 audiotapes, had been confused by my multiple references to the "Anaconda." They had assumed it might be a secret code word for our payload. I had to explain to them it was a Swine Flight euphemism for *penis.* As the crowd laughed at Hoot's stories (he was so damned good at *everything*), I thought of how all astronauts long to leave a memorable and heroic legacy. Hoot had defined mine . . . screwing up a T-38 backseat landing, slandering the mother of the man who was two switches away from killing us on Swine Flight, and introducing some sweet young innocents to the disgusting humor of Planet AD. Oh well, I guess it could have been worse.

Hoot finally ended the roast by embracing me in a cheek-to-cheek hug, an act of physical affection that surprised me. But I understood. Like warriors back from the battle, we were intimately bound by our own unique duels with death, by the incommunicable experience of spaceflight.

The audience applauded, the youngest astronauts being the most enthusiastic. It had been the same way back in my freshmen days. *Why don't these old farts just leave or die or something?* I was now the old fart and my departure was freeing up one more seat into space. For silver-pinned astronauts, that was something to applaud.

Back home Donna and I talked long into the night. I tried to convey to her my everlasting gratitude for the life she had given me, but how do you say thanks for a dream? I tried with "I'm glad you walked out of that party in 1965 to kiss me." I don't think I could have said it better than with those few words. But for that kiss, my life would have been different.

As sleep was approaching, I thought there was one other thing I had to do before I walked out of NASA. I needed to hitch a ride to KSC.

I stopped outside the launchpad perimeter fence, where the tourist buses parked, and stepped from the car. The visitors center was closed and tours had ended so I knew I wouldn't be disturbed. *Columbia* was being prepared for her tenth mission and was almost completely hidden by the rotating service structure. Only her right wing and nose and the tips of the SRBs were visible. I had wanted to drive to the pad and take the elevator to the cockpit level, but I knew that would have been a bureaucratic hassle. Even astronauts weren't free to move through security checkpoints. So my last view of *Columbia* would be as the tourists saw her: from a quarter mile away, wrapped in her steel cocoon, hardly looking like a spaceship at all.

The sun had recently set and the pad xenon lights were on. The wind brought muted loudspeaker calls to my ear and the techno-talk spun me back to the summer of 1984, which had been filled with so much fear, disappointment, and joy. But mostly it was the joy of August 30 that now sharpened in my mind's eye. My heart accelerated at the memory of engine start. I could feel the rattle of max-q and see the fade-to-black as *Discovery* raced toward her orbit. Hank's voice was as clear in my brain now as it had been six years earlier: "Congratulations, rookies. You're officially astronauts." I could hear the cheers of Judy, Mike, and Steve at the realization our silver pins had undergone the alchemy of fifty miles altitude and been transformed into gold.

I got back in the car and steered for the astronaut beach house. My last moments as an astronaut had to end on that beach. No other place conjured up more memories or more emotions than its sands. Its quiet solitude and proximity to the infinity of the sea and sky gave my soul a release unattainable anywhere else.

I pulled into the driveway, climbed the stairs, and opened the door. The house was deserted, as I knew it would be. Except for prelaunch picnics and spousal good-byes, few astronauts or NASA officials ever visited the facility.

Nothing had changed since my STS-36 visits. In fact, nothing had changed since my first beach house visit twelve years earlier. A framed abstract painting, which suggested a collision of multiple sailboats, hung on a wall. It had probably been selected by the same decorator who had chosen an exploding volcano for crew quarters wall art. (We were astronauts, for chrissake. What would be so wrong with some space and rocket photos?) The mantel of the fireplace was still crowded with various liquor and wine bottles. Some had probably been emptied by Alan Shepard, Neil Armstrong, Jim Lovell, and other legendary astronauts. On the windowsills and end tables were shells, sand dollars, and other flotsam collected by generations of astronauts and their spouses. I was sure their beachcombing, like Donna's and mine, had merely been a distraction from that impending final good-bye. The small den was still crowded with the same Ozzie and Harriet–era furniture: orange vinyl chairs, orange vinyl sofa, faux-wood coffee and lamp tables, and ceramic light fixtures decorated with splatters of, what else, orange paint. A small television, old enough to have captured the 1960s Gemini launches, sat on another imitation-wood piece.

I walked to the kitchen, ignored the desiccated carcass of a roach on the countertop, shoved a few dollars in the honor cash box, and liberated a Coors from the refrigerator. I sipped on that as I continued my tour in the back bedroom. It held another astronaut artifact, the convertible sofa bed Hoot and Mario had used for their high jinks with the STS-36 wives. I knew the bed had supported more than just that one prank—at a Houston party one tipsy TFNG wife had jokingly complained, "I hated doing it at the beach house on a bare mattress." I looked in the closet. There was still no linen. If Hoot and the wives had thought about the multidecade *special* use of that particular piece of furniture, I doubt they would have climbed onto it. Even dressed in an LES, I wouldn't have sat on it.

On the other side of the den was the dining/conference room and I entered it. A large table dominated the area. An easel holding a blackboard and chalk sat at one end. The board featured a hieroglyphics of engineering data from a premission briefing on a prior shuttle launch. It was easy for me to imagine the crew in the surrounding chairs hanging on every word of the VITT presenter, praying he wouldn't use the D word . . . delay. I would never miss those worries.

Finally, I walked to the door and paused for one last moment to allow the memories to congeal and be sealed in my brain. As it was with every step of this journey, I was seeing a part of my life I would never see again. In a few more weeks I would be a civilian outsider with no more ability to access this beach house than one of the tourists on a KSC bus tour. With a lump in my throat I switched off the light, closed the door, and headed down the crumbling concrete walkway to the beach.

The breeze was cool and I zipped my jacket and took a seat in the sand. As far as I could tell, I was the only living being on the planet. Even the gulls had retired for the night to their hidden nests. The only sound was the respiration of the surf.

I had no agenda. I just wanted some time with my thoughts, wherever they might take me. And they immediately took me to the land of doubts. For the millionth time I wondered if I was doing the right thing leaving NASA. Even at this late hour, I knew my decision was reversible. I could walk back into the beach house and call Brandenstein and tell him I'd changed my mind and would like to stay at JSC as a civilian mission specialist. I knew he would make it happen. After my retirement ceremony, I had run into him at the bathroom urinals and he had said, "Mike, you should stay. I'm running out of MSes." But I knew if I returned to Houston with the news I had changed my mind, it would kill Donna. My decision stood. Now was the time to leave. My astronaut career was over.

Joy was the next emotion to overcome me. I was a three-time astronaut. My pin was gold. Sputnik had set me on a life journey toward the prize of spaceflight, and I had gained that prize. It had not been easy. I started the journey without pilot wings, when only pilots were astronauts. I did it without the gift of genius. But God had blessed me through his earthly surrogates: my mom, my dad, and Donna. Every step of the way, they were at my side, physically and spiritually, giving me the things I needed to ultimately hear my name being read into history as an astronaut.

Mom and Dad gave me the gift of exploration. They tilted my head to the sky. They supported my childhood fascination with space and rockets. In dealing with my dad's polio, they were living examples of tenacity in the face of great adversity. On countless occasions I had needed that example to persevere in my journey. I needed it to survive the rigors of West Point, to survive airsickness in the backseat of the F-4, to survive graduate school and flight test engineer school.

Donna was the other great dream-maker in my life. She never wavered in her support . . . ever . . . even though the journey had been difficult and terrifying. She assumed the role of single parent to our three children to give me the focus I needed for the journey. She waited for me through a war. She buried friends and consoled their widows and children. She came to accept my limitations as a husband—my sometimes blind selfishness for the prize. She endured the terror of nine space shuttle countdowns, six beach house good-byes, six walks to the LCC roof, an engine start abort, and three launches. Throughout my journey she was my shadow . . . always there next to me.

I thought of the NASA team upon whose shoulders I had been lifted into space. While I had serious issues with some of NASA's management, I had only the greatest respect and admiration for the legions who formed the NASA/contractor/government team . . . the schedulers, trainers, MCC team members, the USAF and other government personnel associated with my two DOD missions, the Ellington Field flight ops personnel, the admin staff, the flight surgeons, the suit techs, the LCC teams, and thousands of others.

I considered how my NASA experience had changed me. I walked into JSC in 1978 as a cocky military aviator and combat veteran, secure in my superiority over the civilians. But watching Pinky Nelson steer his jet pack across the abyss of space toward the malfunctioning Solar Max satellite humbled me. Hearing Steve Hawley joke in the terrifying first moments of our STS-41D abort, "I thought we'd be higher when the engines quit," was another lesson. I learned that the post-docs and other civilians had skill and courage in spades, and I admired and respected them all.

By far, the greatest personal change my NASA experience had wrought was in my perception of women. I learned that they are real people with dreams and ambitions and only need the opportunity to prove themselves. And the TFNG women did. Watching a nine-month-pregnant Rhea Seddon fly the SAIL simulator to multiple landings was a lesson in their competence. Watching video downlink of her attempting an unplanned and dangerous robot arm operation to activate a malfunctioning satellite was a lesson. Watching Judy perform her STS-41D duties was a lesson. Knowing Judy might have been the one to turn on Mike Smith's PEAP in the hell that was *Challenger* was a lesson. Through their frequent displays of professionalism, skill, and bravery, the TFNG females took Mike Mullane back to school and changed him.

It was impossible to sit on this beach and not think about *Challenger.* The ocean that churned at my feet was more of a grave for that crew than anything in Arlington Cemetery. Why them and not me? As January 28, 1986, receded into the past, that question loomed larger and larger in my consciousness. There had been twenty TFNGs with the identical title—mission specialist. One in seven of us had died. It could have been any of us aboard *Challenger.* Why wasn't it the atoms of my body rolling in the beach house surf? It was the unanswerable question survivors everywhere asked . . . the soldier who sees the friend at his side take a bullet, the firefighter who watches the house collapse on his team, the passenger who missed her connection to the fatal flight. For some reason, known only to God, we had all been given a second life.

And where would I journey on the ticket of my second life? I still didn't know. I had yet to do a job search. I just didn't have a passion for anything in the civilian world. I was facing what every retiring astronaut faces—the reality we had reached the pinnacle of our lives. We groped above us searching for the next rung on the ladder of life and it just wasn't there. What does a person do for an encore after riding a rocket? Whatever it was, we would have to climb *down* that ladder to reach it. No matter how much money we made or what fame we acquired in our new lives, we would never again be Prime Crew. We would never again feel the rumble of engine start or the onset of Gs or watch the black of space race into our faces. We were forever earthlings now. It was a sobering thought, but I knew I would adjust. I would find a challenge somewhere. If there was one thing my mom and dad had taught me, there were plenty of horizons on the Earth I had yet to look over.

I swallowed the last of my beer and rose from the sand. As I turned, my eyes were seized by *Columbia's* xenon halo. Over the black silhouettes of the palmettos, the salt-laden air glowed white with it. She awaited her Prime Crew. I envied the hell out of them.

Epilogue

In my post-MECO life I found an unlikely horizon to explore—I became a professional speaker. Given my early adventures at the podiums of America, that might seem like a disaster waiting to happen but I've learned to corral my Planet AD tendencies and fake normalcy. With a microphone in my hand I am a model of political correctness. Hoot would never recognize me. I deliver inspiring, motivational, and humorous programs on the subjects of teamwork and leadership. I learned the good, the bad, and the ugly about those topics while at NASA.

This book has been another horizon I had to sail over. There has always been a secret chamber in my soul where the flame of literary creation has flickered. In high school I loved it when teachers assigned term papers, a fact I kept quiet, knowing my classmates would have beaten me to death had they known. Sometimes my prose would be seriously misplaced, as when I devoted a paragraph in my science fair report to the beautiful sunset that had been a backdrop to one of my homemade rocket launches. I was teased by my fellow junior scientists for that. Of course, ego played its part in my literary quest—I wanted to tell *my* story. But I had noble objectives, too. I wanted the world to understand the joy and terror that astronauts and our spouses experience. I know other astronaut authors have attempted to convey the same thing and, no doubt, many will try in the future. This has been my best shot at it. Finally, I wanted to tell the world a little about my mom and dad. Heroes like them are rare and they deserve a measure of immortality between the covers of a book.

My mom would not live to see herself in these pages. On Memorial Day 2004, she was diagnosed with advanced pancreatic cancer and died on July 4 at age seventy-nine. I was the one who told her of the doctor's prognosis—that she had just a few weeks to live. She took the news with her characteristic courage. She didn't utter a word of dismay or shed a single tear. She merely shrugged her shoulders, as if I had just told her she had nothing more serious than a stomach virus, and said, "Well, I've had a great life." This from a woman who endured the terror of her husband serving in WWII, who raised six children with that same man in a wheelchair, and who was further cheated when she was widowed at age sixty-four. It hardly sounded like a "great life." But my mom always saw the glass as half full and smiled and laughed her way through life until her last conscious moment. As one of my brothers said, "Mom set the bar damned high on living and dying." That she did. As I sat with her in the ebbing days of her life, random images from that life flashed in my brain. I saw her squatting next to a campfire, cooking pancakes and bacon. I saw her pouring my dad's urine from a milk carton into a motel toilet. I saw her handing over the stainless-steel extension tube of her vacuum cleaner so I could fashion it into a rocket. I saw her "mooning" the camera during her wait for the launch of STS-36. She had sewn the mission number on the rear of her "good luck" green briefs and, at the beach house, had bent over to show the unique cheerleading sign on the billboard of her sixty-four-year-old backside.

With me and two of my brothers holding her hands she died at home and was laid to rest in the same grave as my father at the Veterans Cemetery in Santa Fe, New Mexico. I placed another set of my shuttle mission decals on the new grave marker. They were Mom's missions, too.

As I write these words, only five TFNGs remain on active duty with NASA: Fred Gregory, Steve Hawley, Shannon Lucid, Anna Fisher, and Steve Nagel. All of them are in administrative positions and will probably never fly in space again. The space history books are closed on the TFNGs. But our class wrote some remarkable entries in those books:

First American woman in space: Sally Ride.

First African American in space: Guy Bluford.

First Asian American in space: El Onizuka.

First American woman to do a spacewalk: Kathy Sullivan.

Most space-experienced woman in the world: Shannon Lucid, with a total of 223 days in space, including a six-month tour on the Russian Mir space station.

While flying the MMU, Bob Stewart, Pinky Nelson, Dale Gardner, and Jim van Hoften became some of the only astronauts to orbit completely free of their spacecraft.

On STS-41C, TFNGs were part of a crew that completed the world's first retrieval, repair, and re-release into space of a malfunctioning satellite. On STS-51A, TFNGs played key roles in the first capture and return to Earth of a pair of crippled satellites.

Rick Hauck commanded the first post-*Challenger* mission. Hoot Gibson commanded the first shuttle–Mir space station docking mission. Norm Thagard became the first American to fly aboard a Russian rocket when he was launched to the Russian Mir space station. TFNG

Dick Covey commanded the first repair mission to Hubble Space Telescope (HST) to correct its flawed vision. Dan Brandenstein commanded STS-49, a mission to capture and repair a massive communication satellite stuck in a useless orbit. The mission involved an emergency three-person spacewalk, the only such spacewalk ever conducted, and was one of the most difficult shuttle missions in history.

TFNGs logged nearly a thousand man-days in space and sixteen spacewalks. Five became veterans of five space missions (Gibson, Hawley, Hoffman, Lucid, and Thagard). The first TFNGs entered space in 1983 aboard STS-7. Steve Hawley became the last TFNG in space sixteen years later, when he launched on his fifth mission, STS-93, in 1999. TFNGs were ultimately represented on the crews of fifty different shuttle missions and commanded twenty-eight of those. It is not an exaggeration to say TFNGs were the astronauts most responsible for taking NASA out of its post-Apollo hiatus and to the threshold of the International Space Station (ISS).

There are twenty-nine of the original thirty-five TFNGs still living. Besides the loss of the *Challenger* four and Dave Griggs's death, Dave "Red Flash" Walker, a veteran of four shuttle flights, succumbed to natural causes at the age of fifty-seven. Dave was the pilot who scared the holy bejesus out of me during the 1981 STS-1 chase plane practices. He had teased death so often, I had come to believe he was bulletproof. I had failed to consider cancer.

We almost had to bury Hoot Gibson in 1990 when he was involved in a midair collision while racing his home-built plane. The other pilot died but Hoot was able to land his severely damaged machine and walk away. If ever there has been a pilot who has worn out a squadron of guardian angels, that would be Hoot. He and Rhea Seddon now live in Tennessee with their three children. Hoot flies for Southwest Airlines and Rhea is the assistant chief medical officer at Vanderbilt Medical Group at that Nashville university.

In my retirement I have noted the deaths of other astronauts whose life paths intersected mine. Sonny Carter's death was particularly shocking. Sonny had been one of the STS-27 family escorts and Donna had relied on his calming presence during her LCC waits for that mission. He was never without a smile and a positive word. On April 5, 1991, while on the way to give a NASA speech, he died in the crash of a commercial airliner. The manner of his death was a gross violation of the natural order—it was expected that an astronaut dying in a plane would do so as a crewmember, not as a passenger. Sonny was twice cheated . . . in death at age forty-three and while belted into a passenger's seat.

Bob Overmyer died in his retirement while flight-testing a small plane. I was one of the CAPCOMs for his STS-51B flight. During that mission, a communication glitch allowed the crew's private Spacelab intercom to be momentarily broadcast to the world. It included a panicked call from Bob to his lab scientists: "There's monkey feces floating free in the cockpit!" I later teased him that he was probably the first marine in the history of the corps to ever use the word *feces.* He laughed at that. Bob was dead at age fifty-nine.

Astronaut-scientist Karl Henize, whom I had worked with on my very first astronaut support job—the dreaded Spacelab—died in his retirement at age sixty-six of respiratory failure while attempting to climb Mount Everest. He is buried on the side of that mountain at 22,000-foot elevation.

Besides these astronaut deaths, I noted other passings. Don Puddy, who replaced George Abbey as chief of FCOD and who approved me for my STS-36 mission, died of cancer in 2004 at age sixty-seven. Jon and Brenda McBride's son, Richard, died in a plane crash while undergoing navy flight training. Brewster and Kathy Shaw suffered a horrific loss, too. One of their college-age sons was murdered in a random carjacking. If it is possible for a soul to audibly scream, mine did at that news. No death, not even the ones sustained in the *Challenger* and *Columbia* tragedies, affected me as much. Every parent understands.

Gene Ross, the ever-present and ever-amicable owner of the Outpost Tavern, died in 1995. He didn't live to see his bar immortalized on the silver screen. Disney would use it as a backdrop for a scene in the 1997 movie *RocketMan* and a portion of the movie *Space Cowboys* would be filmed inside the cluttered, smoky cave.

Under its new management, the Outpost has seen a few minor changes. The shell-covered parking lot has been leveled. Gone are its bunker-buster craters. And a small red neon light proclaiming "The Outpost Tavern" now decorates a side of the building. But for that, the structure still appears abandoned and ready for demolition. The only improvement to the interior has been the addition of modern bathrooms. The old toilets—one-hole closets with tilting floors and rusted porcelain fixtures—had been intimidating enough to prompt Donna to once comment, "I would rather pee in the outside bushes than sit on an Outpost toilet seat." The interior of the bar remains a time capsule from the halcyon days of the TFNGs: Photos and posters of smiling astronauts and mission crews still cover the walls and ceiling. My NASA genesis photo is still there. In a display of Texas pride, Gene Ross put up the photos of all the Texas-born astronauts in the entryway next to the bikini-girl-silhouette saloon doors. Epochs of cigarette smoke and grease have put a yellow film over those photos but I'm still visible as the thin, dark-haired, thirty-two-year-old astronaut candidate I was in 1978. Whenever a trip takes me to Houston, I always make it a point to visit the Outpost. I will sit at the bar, order a beer, and listen to the TFNG ghosts whisper the stories of joy and heartbreak that have been written there.

John Young retired from NASA on December 31, 2004, after a forty-two-year career that included six space missions covering the Gemini, Apollo, and shuttle programs. He twice flew to the moon, landing on it on *Apollo 16*. In a NASA press release John was praised as an "astronaut without equal." You will never hear me say otherwise.

George Abbey was appointed director of the Johnson Space Center by NASA Administrator Daniel Goldin on January 23, 1996 (no doubt putting the fear of God in those who had celebrated, too enthusiastically, his JSC departure in 1987). Five years later he was "reassigned" by Goldin from that position to NASA HQ to serve as Goldin's senior assistant for international issues. The press noted that the announcement of Abbey's JSC termination came after close of business on a Friday and with little description of the responsibilities of his new title, signatures that the change was actually a firing. Some speculated that cost overruns on the ISS program had prompted Goldin to remove George. He retired from NASA on January 3, 2003, after a nearly forty-year career with the agency.

I was last face-to-face with John and George in 1998 at the twentieth anniversary of the TFNG class. We traded empty hellos and then separated. I was no longer their hostage and would not pretend friendship.

As an outsider I watched the shuttle program fully recover from *Challenger.* Though the STS never recaptured its Golden Age, it did achieve an average of seven missions per year throughout most of the 1990s. Among its more significant post-*Challenger* missions were the launch of Hubble Space Telescope, nine missions to the Russian Mir space station, and multiple missions in support of the assembly and resupply of the International Space Station. The latter was being constructed in partnership with the Russians. The godless commies had become our friends. Even Bill Shepherd, who had penned the *Suck on this, you commie dogs* inscription on a photo of our STS-27 payload, would morph into *Comrade* Shepherd and fly a five-month ISS mission with two Ruskies.

The shuttle continued to experience near misses with disaster, providing more evidence that it would never be truly operational. One of the closest calls occurred on STS-93. During the early part of ascent a small repair pin in the combustion chamber of one SSME came loose and impacted the inside of the engine nozzle, puncturing its cooling jacket. Just as a hole in the radiator of an automobile will cause a leak of engine coolant, *Columbia's* nozzle damage was doing the same thing. As she roared upward, she was bleeding coolant. But in *Columbia's* case the coolant was also the engine fuel. The shuttle's liquid hydrogen plumbing system circulates that supercold fluid around the engine nozzles before the hydrogen is burned. *Columbia* was headed into orbit in danger of running out of gas. Fortunately the damage and the resulting leak were small. The propellant loss resulted in an early engine shutdown, but *Columbia* still achieved a safe orbit only seven miles lower than planned.

The nozzle damage turned out to be just one of the near misses for the STS-93 crew. Five seconds into flight an electrical system short circuit resulted in the failure of several black boxes controlling two of the SSMEs. Backup engine controllers, powered by a different electrical system, took over the control of those engines and there was no impact to their performance. But for eight and a half minutes, two of *Columbia's* engines were just one failure away from shutting down and forcing the crew into an ascent abort. The source of the short circuit was later isolated to an exposed wire.

Another shuttle near miss occurred on STS-112 when a circuit failure resulted in only one set of the hold-down bolt initiators firing at liftoff. In the launch sequence the hold-down bolts are exploded apart just milliseconds prior to SRB ignition so the rocket is completely free of the ground when the boosters ignite. Had the redundant initiators in the hold-down bolts not fired, *Atlantis* would have been still anchored to the pad at SRB ignition. The machine would have destroyed herself trying to rip free of the bolts.

STS-93 and STS-112 were saved by system redundancy, but there was another recurring problem on shuttle launches for which there was no redundancy to provide protection. Insulation foam was shedding from the gas tank and striking the orbiter. The phenomenon was first noted on STS-1 and was subsequently documented by photo imagery on sixty-four other shuttle missions. Hank Hartsfield and Mike Coats had observed it on our Zoo Crew flight in 1984. This foam-shedding anomaly was a violation of a design requirement, just as the pre-*Challenger* SRB O-ring erosion had been a design violation. Nothing was supposed to hit our glass rocket, not even something as seemingly innocuous as the foam from the ET. But as hit shuttles kept returning to the Earth safely, engineers became ever more comfortable with accepting the design violation as nothing more than a maintenance issue—the foam strikes were requiring a handful of damaged tiles to be replaced between missions. The "normalization of deviance" phenomenon that had doomed *Challenger* in 1986 had returned to infect NASA and blind management to the seriousness of the foam loss problem. On January 16, 2003, eighty-two seconds into the flight of *Columbia,* a briefcase-size piece of foam, weighing approximately one and a half pounds, shed from the ET and struck the Achilles' heel of the shuttle heat shield, one of the wing leading-edge carbon panels. The impact blasted a hole of indeterminate size in that carbon. The damage had no effect on ascent and *Columbia* safely reached orbit. The site of the impact was not visible from the cockpit windows and the crew remained oblivious to the fact that their shuttle was mortally wounded. It could not survive reentry.

On the ground NASA engineers were aware of the foam strike—KSC cameras had recorded the incident. But these same engineers had no idea what, if any, damage had occurred and since *Columbia* was flying without a robot arm, they could not direct the crew to remotely survey the site (as we had been able to do on STS-27). A handful of engineers requested their management to ask the Department of Defense to use its photographic sources to acquire images of the impact site. Had these photos or a crew spacewalk determined *Columbia* could not survive reentry, there was a reasonable chance *Atlantis* could have been hurriedly readied for launch on a rescue mission. The *Columbia* crew would have then donned spacesuits and transferred to *Atlantis,* and *Columbia* would have been abandoned in orbit. But key managers dismissed the photo request and never ordered a spacewalk. On February 1, 2003, *Columbia* would burn up on reentry, killing her seven-person crew.

I was in northern New Mexico at the time of the disaster, visiting my daughter and her family. Had I known of the reentry trajectory, I could have stepped outside and watched *Columbia* pass nearly overhead. But I was not an eyewitness. I received the news from TV: "The space shuttle *Columbia* is overdue for landing at the Kennedy Space Center." Images of *Columbia*'s fiery destruction soon followed. As I watched them I couldn't help but visualize what the crew had experienced. I had no doubt their fortress cockpit had kept them alive during the out-of-control breakup of their machine. Just like the *Challenger* crew, they were trapped. Their backpack-parachute bailout system was useless at the extreme altitude and speed. And I couldn't help but visualize the families. They would have been waiting at the KSC Shuttle Landing Facility, giddy in anticipation of having their loved ones safely on the ground and in their arms. They would have been chatting happily about the parties and postflight trips that were planned. Then an escort into widowhood would have come to their side to tell them the news. Their husbands and wife, fathers and mother would not be coming home.

I wasn't affected by *Columbia*'s loss as deeply as *Challenger*'s. I had only a passing acquaintance with a few members of the crew. But I was still heartbroken. I stepped from my daughter's house, walked into the adjacent desert hills, and began my prayers. Even as I was saying them, atoms of *Columbia* and her crew were quietly and invisibly settling to Earth around me.

The final report of the *Columbia* Accident Investigation Board (CAIB) would read remarkably like the *Challenger* report issued seventeen years earlier. In fact, in some key paragraphs of their document, the CAIB could have plagiarized the Roger's Commission report nearly word for word. The only edits required would have been to substitute "External Tank" for "Solid Rocket Booster" and "foam-shedding" for "O-ring erosion." Workplace cultural issues, including overwhelming pressure to keep shuttle launches on schedule, had, again, resulted in NASA mishandling repeated evidence of a deadly design flaw.

I have been too long removed from NASA to make any firsthand comment on those cultural issues or the leadership failures they suggest. Nor can I predict whether the agency will be able to fix itself . . . though I see reason for hope. The shuttle team's meticulous response to the heat-shield damage sustained by *Discovery* on the first post-*Columbia* shuttle mission (STS-114) and the agency's intention to keep the shuttle grounded until the maddeningly persistent ET foam-shedding problem is fixed suggests NASA has made safety its top priority. The question is, "Can this reinvigorated safety consciousness persist through the remaining life of the space shuttle program?" It didn't last after *Challenger,* as *Columbia*'s loss attests. Perhaps new NASA administrator, Dr. Michael Griffin, is a leader who can keep the agency focused on safety. I pray so. There have been enough families devastated in this business, not to mention the disastrous impact on America's manned space program that another shuttle loss would precipitate.

In Senate testimony, Dr. Griffin has said he intends to retire the shuttle by 2010, arguing, "The shuttle is an inherently flawed system." He's right. It is an outrageously expensive vehicle and lacks a viable crew escape system. A well-led and adequately funded team might still have been able to safely operate even this "flawed system," but the old NASA lacked both leadership and money.

Griffin continued, "We all know that human perfection is unattainable. Sooner or later there will be another shuttle accident. I want to retire it before that can occur." His plan is to fly the shuttle a maximum of nineteen times—eighteen for ISS support and one for Hubble Space Telescope repair. My sympathies go out to the most junior astronauts who have been warned by NASA that they may never earn their gold pin on the shuttle because of the limited number of missions remaining. They are living what had been my greatest fear . . . that I would remain an astronaut in name only.

In all likelihood the craft that will replace the shuttle will be a capsule launched atop some type of booster rocket, possibly a reuseable shuttle SRB augmented with a liquid-fueled upper stage. It's back to the future. The capsule will probably accommodate a four-person crew and be more sophisticated than those of the Apollo program, but with the same type of tractor escape rocket design to pull astronauts to safety in the event of a booster failure. Future astronauts will return to Earth under parachutes.

If all goes according to Griffin's plan, on a day in late 2010, a reentering space shuttle will sonic-boom KSC for the last time. For the last time a pilot will take the stick of a winged spaceship and guide it to a runway landing. For the last time we will hear the call, "Houston, wheel stop." The space shuttle will be history, retired at age thirty. I suspect every TFNG will be watching . . . and remembering. I certainly will.

Political correctness finally neutered the astronaut corps . . . or, perhaps, males from Planet AD have gone extinct. Several veteran NASA secretaries confided in me that contemporary astronaut parties are "boring." I can believe it. When an astronaut applicant recently called me for insight into the interview process, I was shocked to hear her say that a resident astronaut had already warned her, "Drinking alcohol is frowned upon." (No telling what the corps would say about imbibing in helium.) While I have never been one to believe alcohol is necessary to have fun, the comment hints that there is a *new* astronaut on the block, as good with a stick and throttle as any before but less flamboyant and more mainstream than the TFNGs. It doesn't surprise me. The current civilian astronauts were born into an America that is politically correct in the extreme and the pilots now come from a military that is more sober and religious. So, besides the males from Planet AD, maybe the wild and wooly Right Stuff astronaut—that astronaut who lives life at the edge of the envelope, be it at happy hour or in a cockpit—has also gone the way of the dodo.

The last TFNG reunion occurred in 1998, our twentieth anniversary.

Most of the men and all of the surviving women were present. The women seemed least changed, though I'm sure makeup and Clairol had a lot to do with that. The men, me included, were showing our age with expanding waistlines, receding hairlines, and liver-spotted foreheads. A few men sported new wives, though none of those seemed to be of the "trophy" variety. They were mature and pleasant. The rest of the wives were aging gracefully but their days of giving us men a "six nipples under glass" show were, sadly, gone.

Before dinner, Rick Hauck led us in a moment of silence to remember our fallen friends, then gave a short presentation that included a recap of some of the significant history written by our group. We each received TFNG T-shirts bearing thirty-five small caricatures of our individual likenesses. The shirts also featured the past-tense headline "We Delivered." It was an update to the original 1979 TFNG T-shirt, which had displayed the same caricatures and the title "We Deliver." The TFNG class had, indeed, *delivered* for NASA and America.

Before scattering to our hotels we posed for a class photo. I sensed a renewed closeness in the assembly. It wasn't the Knights-of-the-Roundtable closeness we had once shared—that level of camaraderie had forever ended when the first Abbey flight assignments had winnowed us. But the white-hot fierceness of our competition had been cooled by the years. We were all gold-pinned astronauts; most of us gold-plated several times over. We were all bound by an experience singularly unique in the history of man . . . spaceflight. As we stood for our reunion photo, fewer than four hundred earthlings had ever flown into space. Even the fraternity of those who had summited Mount Everest was more than twice as large. The exclusivity of the astronaut experience would forever be a force that would pull TFNGs together.

I occasionally run into a TFNG in my travels. I once crossed paths with Hoot Gibson in his capacity as a Southwest Airlines pilot and had cause to regret it. In the late 1990s I was a passenger on a flight he was piloting. As the jet reached cruise altitude, he announced over the intercom that "world-famous astronaut Mike Mullane was aboard and would be happy to sign autographs." To ensure my distress, he added my seat number. A line formed and a few old ladies grabbed their cameras for photos. I wanted to leap from the plane to escape the severe embarrassment. *Better dead than look bad.*

The three TFNG *Challenger* widows have successfully moved on with their lives. As Lorna Onizuka shared with me, "We stubbed our emotional toes along the way, but I think we've all come through the tragedy as happy, content, and successful women and mothers." Lorna thinks it was the mothering instinct that got everybody through the worst days—each of them had to place their children first and didn't have time to be emotional cripples. "My children saved my life," was Lorna's assessment.

The "man repellent" factor of the astronaut-widow thankfully did not endure. June Scobee and Jane Smith remarried. Lorna and Cheryl McNair remain single but Lorna says they both have vibrant social lives. Lorna says, "I've shared my life with a special man for more than ten years." She laughs as she recounts some of the problems of reentering the dating scene as a mother of two. "When I wasn't home and a man would call, my oldest daughter would ask him if he was bald." For some reason that daughter had a "bald men need not apply" attitude. Lorna's youngest daughter would ask a male caller if he smoked cigars, which was her criterion for rejection. And both daughters would tell men they had to have Mom home by the ten o'clock news. If Lorna's happy, upbeat attitude is representative of the other *Challenger* survivors, as she believes it is, they are doing quite well.

Donna and I are approaching our sixtieth birthdays. We both weigh more, sag more, and forget more than we did in those euphoric, intoxicating early TFNG days. But life has been good . . . *grand,* really, because we have been blessed with six wonderful and healthy grandchildren. Pat and Wendy, Amy and Steve, and Laura and Dave have all given us two grandchildren each: Sean and Katie, Hanna and Meagan, and Noah and Gwyneth. While holding our first grandchild, I asked Donna, "Would you have ever thought we'd be telling our kids to have more sex?" As the saying goes . . . "If I had known grandkids were so much fun, I would have had them first."

Donna and I also just passed our fortieth anniversary . . . not wedding, but rather the anniversary of that fateful first kiss of January 3, 1965. We celebrated with a glass of wine and were asleep by 9:30 P.M. We each got married for the wrong reasons, but we somehow endured long enough to fall in love.

The astronaut beach house is still standing and I hope it is forever preserved for future generations of astronauts. It sits on sacred ground. The spouses of the *Challenger* and *Columbia* crews last held the love of their lives on its sands. No doubt some future crew spouses will hold dear the memory of their last beach house visit, too, for it will include a memory of the last time they embraced their lovers. It is the nature of spaceflight that more crews will perish. Even if NASA can fix its culture, the complexity of the machines and the unforgiving environment of space will claim more astronauts.

Another place sacred to astronauts was created after I retired. In 1991 the Astronaut Memorial, funded largely by the sale of Florida *Challenger* license plates, was dedicated at the KSC Visitors Center. Whenever I visit that center, I always make it a point to walk to the memorial. It consists of a large matrix of granite panels bearing the names of all astronauts who have died in the line of duty. Those names have been chiseled completely through the stone to allow mirrors set behind the panels to reflect the sunlight through the etchings. The entire panel assembly automatically rotates to follow the sun and continuously catch its light. There are now twenty-four names in the granite, the earliest being Theodore Freeman, killed in 1964 in the crash of his T-38 jet, and the latest being the *Columbia* Seven.

On my visits to the memorial I will take a seat on a bench and stare at the four TFNG names the panels bear . . . Francis "Dick" Scobee, Judith A. Resnik, Ellison S. Onizuka, and Ronald E. McNair . . . and remember the last moment I saw them.* They were walking to a sim wearing Prime Crew smiles. It is how I will always remember them . . . young, happy, *soaring* with the knowledge they were next up. I will remember each of them in my prayers. I will also include prayers for their spouses and Judy's family. The life those spouses and parents knew also ended on January 28, 1986, but nobody ever etched their names on a monument.

From the memorial I will walk to a nearby full-scale space shuttle mock-up. Metal platforms have been installed around the display so tourists can climb up and walk through the cockpit. I will anonymously join a group of families and watch them take photos and listen to them marvel at the complexity of the switch panels and the cramped volume. Invariably my attention will be drawn to a child among them. In his or her amazed young face I will be transported back to 1957. I am standing in my front lawn with the identical expression, watching *Sputnik I* twinkle through the terminator.

September 7, 2005
Albuquerque, New Mexico
www.mikemullane.com

*The panel only bears the names of astronauts who died in the line of duty. For that reason Dave Griggs and Dave Walker are not memorialized on the panel.

Glossary

AB—Afterburner. The throttle position that increases the thrust of a jet engine by burning additional fuel at the back of the engine.

AD—Arrested Development. The state of many military aviators, the author included.

ADI—Attitude Director Indicator. An instrument that shows aircraft or spacecraft attitude relative to the Earth's horizon.

AFB—Air Force Base.

AOA—Abort Once Around. A launch abort in which the shuttle makes one orbit of the Earth and lands in the United States.

AOS—Acquisition of Signal. A call to the crew that indicates the shuttle data stream is being received at Mission Control.

APU—Auxiliary Power Unit. A hydraulic pump on the space shuttle. There are three APUs powering three hydraulic systems on the orbiter. There is nothing "auxiliary" about the shuttle's APUs. They are the primary power source for the hydraulic systems. The "auxiliary" is a holdover aviation term. It refers to similar units that back up the primary engine-driven hydraulic pumps on jet aircraft.

ASP—Astronaut Support Personnel. Astronauts who help the mission crew strap into the space shuttle and who assume control of the shuttle cockpit from a just landed astronaut crew.

ATC—Air Traffic Control. Facilities on the ground that monitor aircraft in the air.

ATO—Abort to Orbit. A launch abort in which the shuttle flies into a safe orbit after an engine failure.

BFS—Backup Flight System. A backup computer that will take over control of a space shuttle. The BFS is engaged by the depression of a button on the top of the commander's or pilot's control sticks.

CAIB—*Columbia* Accident Investigation Board. The board appointed to investigate the loss of the space shuttle *Columbia*.

CAP—Crew Activity Plan. The checklist that specifies which crew activities are to be performed at what point in the mission.

CAPCOM—Capsule Communicator. The astronaut in Mission Control who talks to astronauts in space.

CDR—Commander. The astronaut who occupies the front left seat of a launching/landing space shuttle and who has overall responsibility for the mission.

CNO—Chief of Naval Operations. A four-star admiral who has overall responsibility for the United States Navy.

DEFCON—Defense Condition. The status of American military forces, from peacetime (DEFCON 5) to fully prepared for war (DEFCON 1).

DOD—Department of Defense.

DPS—Data Processing System. The computer heart of a space shuttle.

EMU—Extra-vehicular Mobility Unit, i.e., a spacesuit.

EOM—End of Mission. Used in reference to the end of a space shuttle mission.

ESA—European Space Agency. The European equivalent of NASA.

ET—External Tank. The orange fuel tank attached to the belly of a launching space shuttle. It carries 1.3 million pounds of liquid oxygen and 227,000 pounds of liquid hydrogen for the three liquid-fueled engines at the back of the orbiter.

EVA—Extra-Vehicular Activity. A spacewalk.

FCOD—Flight Crew Operations Directorate. The organization at Johnson Space Center having overall responsibility for crews involved in flight operations, including T-38, Vomit Comet, and shuttle flight operations. The astronaut office falls under the domain of FCOD.

FDO—Flight Dynamics Officer. The Mission Control position that oversees all aspects of the shuttle's trajectory and vehicle maneuvers from liftoff to landing.

GIB—Guy-in-Back. Military slang for the backseat occupant of a two-place fighter aircraft.

GLS—Ground Launch Sequencer. A computer in the Launch Control Center at Kennedy Space Center that controls a shuttle countdown until thirty-one seconds prior to liftoff, at which time the shuttle's own computers assume control of the countdown.

GPC—General Purpose Computer. One of five IBM computers that form the shuttle's electronic "heart." One of these is the BFS computer. *See BFS.*

GS—Government Servant. The title of civilians working for the government. A number system, e.g., GS-9, indicates the rank of the worker.

GWSA—George Washington Sherman Abbey.

HQ—Headquarters.

HST—Hubble Space Telescope.

ICOM—Intercom. The system used by astronauts to talk to one another when they are in the LES or separated between the upper and lower cockpits or between the shuttle cockpits and a Spacelab module.

IFR—Instrument Flight Rules. A term used in aviation to indicate a pilot is following the directions of an air traffic controller on the ground.

INCO—Instrumentation and Communication Officer. The MCC controller responsible for the command and data links between the MCC and the space shuttle.

ISS—International Space Station.

IUS—Inertial Upper Stage. A large Boeing-built booster rocket used to lift satellites into their final orbits and to accelerate space probes out of Earth orbit.

IVA—Intra-Vehicular Activity. Usually used as a crewmember title, i.e., IVA crewmember. A crewmember who helps spacewalkers prepare for a spacewalk and monitors them while they are outside the spacecraft.

JSC—Johnson Space Center in Houston, Texas.

KSC—Kennedy Space Center in Florida

LCC—Launch Control Center. The Kennedy Space Center team that directs the countdown and launch of a space shuttle.

LCG—Liquid Cooling Garment. A netlike long underwear worn under a spacesuit and that holds a maze of small tubes that circulate chilled water to prevent spacewalkers from overheating.

LDEF—Long Duration Exposure Facility. A bus-size satellite launched on shuttle mission STS-41C in 1984 and retrieved and returned to Earth by STS-32 in January 1990. LDEF carried several hundred passive experiments to understand the effects of space exposure on various materials.

LES—Launch/Entry Suit. The orange-colored spacesuits that astronauts wear for launch and reentry. These suits would automatically pressurize if there was a cabin pressure leak.

LOS—Loss of Signal. A call to the crew that the shuttle will soon be out of contact with Mission Control. Usually the call is given in a countdown format, as in, "*Atlantis*, you'll be LOS in two minutes."

LOX—Liquid Oxygen.

Mach—The engineering term for the speed of sound. Astronauts wear a Mach-25 patch indicating they have traveled twenty-five times the speed of sound.

max-q—An engineering term for the point in flight when an aircraft or spacecraft experiences the maximum aerodynamic pressure. Max-q (where the M is capitalized) is also the name of the astronaut band. Though there have been several generations of astronaut band members, the band name remains the same.

MCC—Mission Control Center. The Johnson Space Center team that directs a shuttle mission from "tower clear" (the moment the shuttle rises above the launchpad) until the "wheel stop" call at landing, at which time control is returned to Kennedy Space Center.

MDF—Manipulator Development Facility. A full-scale simulation of the Canadian robot arm and shuttle cargo bay in a building at Johnson Space Center.

MEC—Master Events Controller. A black box on the space shuttle that controls critical events like the commands to jettison the booster rockets and the empty gas tank.

MECO—Main Engine Cutoff. The moment in a shuttle launch when the three liquid-fueled engines shut down.

MLP—Mobile Launch Platform. The "launchpad" on which the space shuttle is stacked and that is carried to either Pad 39A or B by a massive tracked crawler.

MMU—Manned Maneuvering Unit. A space jet pack. An MMU has high-pressure gas thruster jets that allow an untethered astronaut to fly short distances from the space shuttle.

MS—Mission Specialist. Astronauts trained for mission payload activities, e.g., using the robot arm, doing a spacewalk, conducting experiments, etc.

MSE—Military Space Engineer. Department of Defense personnel flown on some DOD missions.

MSFC—Marshall Spaceflight Center in Huntsville, Alabama.

NASA—National Aeronautics and Space Administration.

O_2—Gaseous oxygen breathed by astronauts.

OFT—Orbital Flight Test. The first four space shuttle flights. After these were successfully concluded, the STS was proclaimed *operational.*

OMS—Orbital Maneuvering System. Two six-thousand-pound-thrust liquid-fueled engines at the tail of the orbiter. These are used for the final boost into orbit, the brake from orbit, and for large orbit changes.

PAM—Propulsion Assist Module. A solid-fueled rocket motor attached to the bottom of a communication satellite to lift it to a 22,300-mile-high equatorial orbit.

PAO—Public Affairs Officer. An MCC position filled by NASA's representative to the public.

PEAP—Personal Emergency Air Pack. A portable container of breathing air, which astronauts would use in a ground escape through toxic fumes.

PLBD—Payload Bay Doors. The clamshell doors that cover the space shuttle payload bay.

PLT—Pilot. The pilot astronaut who sits in the right front seat during a shuttle launch and landing. Like the mission commander, the PLT is trained to fly the shuttle.

PPK—Personal Preference Kit. The twenty items of personal significance that NASA permits astronauts to fly in space.

PR—Public Relations. Refers to all things associated with NASA's interface with the public.

PROP—Propulsion. An MCC controller who monitors the shuttle RCS and OMS propulsion systems.

PS—Payload Specialist. A "part-time" astronaut trained for a specific experiment. PSes are not career NASA astronauts and receive only safety and habitability training on the shuttle.

RCS—Reaction Control System. A system of forty-four small rocket motors on the tail and nose of the orbiter that control the vehicle's attitude and are also used in small orbit changes, e.g., during the final stages of a rendezvous or separation from a deployed satellite.

RHC—Rotational Hand Controller. The "stick" used to rotate the tip of the robot arm about a point. The CDR's and PLT's control sticks, used to maneuver the orbiter, are also referred to as RHCs.

RMS—Remote Manipulator System. The Canadian-built robot arm operated from the rear cockpit of the orbiter. It is used to capture and release satellites, maneuver spacewalking astronauts and cargo, and for vehicle inspections (through its end-mounted TV camera).

RSLS—Redundant Set Launch Sequencer. The software module in the shuttle's computers that controls the final thirty-one seconds of a shuttle countdown.

RSO—Range Safety Officer. A USAF officer who monitors a shuttle launch and is prepared to blow up the vehicle if it goes out of control and threatens a civilian population center.

RSS—Range Safety System. The explosives aboard the solid rocket boosters and the external gas tank and the supporting electronic equipment that would be used to blow up an out-of-control space shuttle.

RTLS—Return to Launch Site Abort. A launch abort in which the space shuttle returns to land at the Kennedy Space Center.

SAIL—Shuttle Avionics Integration Laboratory. An electronic lab in which shuttle software can be evaluated. SAIL has a replica of the shuttle cockpit.

SAS—Space Adaptation Syndrome. Space sickness.

SEAL—Sea, Air, Land. An acronym for an elite navy force that is trained for special covert operations against the enemy.

Sim Sup—Simulator Supervisor. The team leader who prepares scripts of malfunctions to train astronauts and MCC controllers. The Sim Sup's team inputs malfunctions and evaluates the response of astronauts and the MCC to simulated emergencies.

SLF—Shuttle Landing Facility. The 15,000-foot-long runway at Kennedy Space Center used by landing shuttles.

SMS—Shuttle Mission Simulator. The primary simulators at Johnson Space Center for training astronauts to operate the shuttle systems and respond to emergencies.

SRB—Solid Rocket Booster. Twin boosters attached to the sides of the external gas tank. The term "solid" in the title refers to the propellant, which has the consistency of hard rubber.

SSME—Space Shuttle Main Engine. A liquid-fueled engine at the back of the orbiter that burns the liquid oxygen and liquid hydrogen carried in the external gas tank. There are three SSMEs at the tail of the orbiter.

STA—Shuttle Training Aircraft. A Gulfstream business jet modified to have the landing characteristics of a shuttle. Pilot astronauts (CDRs and PLTs) train for shuttle landings in the STA.

STS—Space Transportation System. A fancy name for what the public would call the space shuttle. The STS is made up of the winged vehicle (the orbiter), the solid-fueled rocket boosters, and the external gas tank.

TAL—Trans-Atlantic Landing abort. A launch abort in which the shuttle makes an emergency landing at an airport in Europe or Africa.

TDRS—Tracking and Data Relay Satellite. A satellite used by NASA to relay commands, data, and astronaut voice communication between the orbiter and MCC.

TFNG—Thirty-Five New Guys. The nickname adopted by the astronaut class of 1978. The name is a play on an obscene military acronym FNG (F***ing New Guy), used to describe someone new to a military unit.

THC—Translational Hand Controller. A square-shaped controller that can be moved in or out, up or down, and left or right. These control inputs will produce the corresponding movement at the tip of the robot arm. The CDR and PLT also have THCs that will fire the orbiter's thrusters to move it in the direction commanded.

UCD—Urine Collection Device. A condom/nylon bladder arrangement or an adult diaper worn by astronauts on the three occasions when they cannot use the shuttle toilet: launch, spacewalks, and reentry/landing.

UHF—Ultra-High Frequency. A radio frequency.

USAF—United States Air Force.

USMC—United States Marine Corps.

USN—United States Navy.

VAB—Vertical Assembly Building. The 500-foot-high building originally used to prepare the *Saturn V* moon rockets. The shuttle stack is completed in the VAB before being transported to the launchpad.

VFR—Visual Flight Rules. An aviation term referring to flights where the pilot is responsible for his/her own clearance from other aircraft and objects.

VITT—Vehicle Integration Test Team. The team at Kennedy Space Center that supports the checkout of the orbiters as they are prepared for a mission.

WETF—Weightless Environment Training Facility. A large swimming pool used by astronauts to train for spacewalks.

WSO—Weapons Systems Operator. The air force crewmember (usually in two-place fighters like the F-4 or F-111) who is responsible for navigation, electronic warfare, and weapons status. WSO is used interchangeably with GIB (guy-in-back).